DMRC

Delhi Metro Rail Corporation

Junior Engineer

Electronics Engineering

DMRC
Delhi Metro Rail Corporation

Junior Engineer
Electronics Engineering

Previous Years Solved Papers
(10 Sets)

3 Actual Solved Papers & 7 Practice Papers

G K Publications (P) Ltd

CL MEDIA (P) LTD.

Edition : 2019

ISBN : **978-93-88426-19-0**

Typeset by : *CL Media DTP Unit*

Administrative and Production Offices

Published by : **CL Media (P) Ltd.**
A-45, Mohan Cooperative Industrial Area,
Near Mohan Estate Metro Station,
New Delhi - 110044

Marketed by : **G.K. Publications (P) Ltd.**
A-45, Mohan Cooperative Industrial Area,
Near Mohan Estate Metro Station,
New Delhi - 110044

For product information :
Visit ***www.gkpublications.com*** or email to ***gkp@gkpublications.com***

PREFACE

'Junior Engineer Recruitment Examination' is a combined two-stage examination conducted by DMRC for recruitment to Delhi Metro Rail Corporation. The sheer number of candidates, highly technical nature of exam and the question paper pattern make this examination as highly sought after by candidates aspiring for a prestigious government job.

The entire examination is Computer Based Test (CBT) comprising of two papers. Candidates qualifying both Paper – I and Paper – II with a high merit within the zone of consideration (in both papers separately) as decided by DMRC are called. The final list is based on the overall merit of CBTs of both Paper – I & Paper – II together. Based on this list, candidates are further shortlisted category-wise for Medical Examination/GD & Interview, at designated locations.

GK Publications has been the **"Publisher of Choice"** to students preparing for GATE, ESE and other technical test prep examinations across the country. GKP's **DMRC 2019 Exam Prep Series** provides a wide range of study material – classified into 7 guides and 7 previous years' solved papers books to simplify the entire preparation.

This book contains content from every trade related to the exam such as General Awareness, General Intelligence & Reasoning, Domain Knowledge and General English. These books have been thoroughly updated as per the latest pattern and syllabus (as prescribed by the exam conducting body) to provide everything you need to boost your scores.

GKP has also launched an android app to provide you an update on all upcoming vacancies in the technical segment. Moreover, the app also has a lot of added content to aid your preparation in the form of vocab builder, GK quiz and so on.

We hope this effort of ours will be helpful in achieving your dreams and securing a government job. If you have any suggestions on improvement of this book, you can write to us at gkp@gkpublications.com.

All the Best!

Team GKP

CONTENTS

SOLVED PAPERS

PRACTICE PAPERS

SOLVED PAPER 2014

ELECTRONICS ENGINEERING

Read the following instructions carefully before you begin to answer the questions. This Booklet contains questions in English.

INSTRUCTIONS

1. This Booklet contains 180 questions in total comprising of the following three parts:

Part – I:	Aptitude & General Awareness	[1 – 45 Questions]
Part – II:	Technical	[46 – 120 Questions]
Part – III:	General English	[121 – 180 Questions]

2. All questions are compulsory and carry 1 mark each.

3. Before you start to answer the questions you must check this Booklet to ensure that it contains all the pages and see that no page is missing or repeated.

4. For every wrong answer 1/3 marks will be deducted.

5. You are supplied with the OMR-Sheet separately.

6. Answers must be indicated by completely blackening one of the corresponding ovals on OMR-Sheet against the relevant question number, by black/blue ball pen only. Answers which are not indicated by black/blue ball pen will not be awarded any mark.

7. The OMR-Sheet must be handed over to the Invigilator before you leave the Examination Hall.

8. Failure to comply with any of the above instructions will render a candidate liable to such action/penalty as may be deemed fit.

SOLVED PAPER 2014

APTITUDE & GENERAL AWARENESS

Directions (Question No. 1 to 3) : Study the following information and answer the questions given below it.

A blacksmith has five iron articles M, N, O, P and Q each having a different weight.

- M weighs twice as much as N.
- N weighs four and a half times as much as O.
- O weighs half as much as P.
- P weighs half as much as Q.
- Q weighs less than M but more than O.

1. What of the following is the lightest in weight?

(*a*) M (*b*) N

(*c*) P (*d*) O

2. Q is lighter in weight than which of the other two articles:

(*a*) M, P (*b*) M, N

(*c*) O, M (*d*) None of these

3. Which of the following articles is the heaviest in weight?

(*a*) N (*b*) O

(*c*) M (*d*) Q

4. The difference between the simple and compound interest on a certain sum for 3 years at 10% p.a. is ₹ 46.5, then which of the following is the sum?

(*a*) ₹ 1600 (*b*) ₹ 1500

(*c*) ₹ 1700 (*d*) ₹ 1800

5. By selling 18 chocolates, a vendor loses the selling price of 2 chocolates. Find his loss percent:

(*a*) 8% (*b*) 12%

(*c*) 15% 4) 10%

6. The length, breadth and height of a room in the shape of a cuboid is increased by 10%, 20% and 50% respectively. Find the percentage change in the volume of the cuboid:

(*a*) 65%

(*b*) 75%

(*c*) 85%

(*d*) None of these

7. Keshav is travelling by car at the rate of 40 km/h. After 80 km he rests for 20 minutes. How long will he take to cover a distance of 240 km:

(*a*) 6 hours 50 minutes

(*b*) 6 hours 40 minutes

(*c*) 7 hours

(*d*) 7 hours 20 minutes

8. A job has to be completed by 12 boys in 15 days. If three boys are absent from the first day, then by what percentage should the remaining boys increase their rate of working to complete the job?

(*a*) $22\frac{1}{3}\%$ (*b*) $22\frac{1}{2}\%$

(*c*) $33\frac{1}{3}\%$ (*d*) None of these

9. The income of Anil, Firoz and Ketan are in the ratio 7:9:12 and their spending are in the ratio 8:9:15. If Anil saves 1/4 of his income, then the saving of Anil, Firoz and Ketan are in the ratio:

(*a*) 59:99:69 (*b*) 56:99:69

(*c*) 56:96:69 (*d*) None of these

10. Ramesh started a business with a capital of ₹ 2100. After 4 months he admitted another partner Sujeet. What amount should Sujeet put in, So that the profit may be divided equally at the end of the year?

(*a*) ₹ 3750 (*b*) ₹ 3350

(*c*) ₹ 3150 (*d*) None of these

11. Pointing to a photograph of a boy Lokesh said, "He is the son of the only son of my mother". How is Lokesh related to the boy?

(*a*) Uncle (*b*) Father

(*c*) Grand father (*d*) Brother

12. If 'dog' is 'cat', 'cat' is 'rat', 'rat' is 'mat', 'mat' is 'cow', then which one of the following is not an animal?

(*a*) Mat (*b*) Dog

(*c*) Rat (*d*) Cow

13. Complete the series by choosing the correct option.

ELFA, GLHA, ILJA, _____, MLNA

(*a*) KLMA (*b*) KLLA

(*c*) KMLA (*d*) LKLA

14. In some cars, there is the warning on external rear view mirrors "Objects are closer than they appear in mirror". What type of mirrors are used in such cases?

(*a*) Plane; (*b*) Convex;

(*c*) concave; (*d*) None of these.

15. Equinox is the position when sub solar point fall's on:

(*a*) Equator (*b*) Cancer

(*c*) Capricorn (*d*) Pole

16. The super computer of India Param Yuva–II has been developed by:

(*a*) Tata Consultancy Services

(*b*) Center for Development of Advanced Computing

(*c*) Reliance Info Comm.

(*d*) National Informatics Center

17. Which one is not an epic?

(*a*) Beowulf (*b*) Illiad

(*c*) Ramayana (*d*) Canterbury Tales

18. In history, the name of prince Khurram, came to be known as:

(*a*) Jahangir (*b*) Shahjahan

(*c*) Aurangzeb (*d*) Akbar

19. Public Accounts Committee is constituted .... year and it consists of not more than .... members consisting of .... members elected by Lok Sabha and not more than .... members from Rajya Sabha. .... was appointed the chairman of Public Accounts Committee recently:

(*a*) Every, 22, 15, 7, K.V. Thomas

(*b*) Every fifth, 18, 13, 5, M.M. Joshi

(*c*) Every second, 25, 18, 7, L.K. Advani

(*d*) Every, 20, 15, 5, Arun Jaitley

20. Identify among the following the Chinese investigating Journalist who was awarded the prestigious 'Ramon Magsaysay Awards for 2014:

(*a*) Saur Marlina Manurang

(*b*) Wang Canfa

(*c*) Hu Shuli

(*d*) Fu Qiping

21. This is a ground penetrating radar and will provide centimeter scale resolution of the geologic structure of the subsurface. This is among the seven selected payloads that will be sent to Mars 2020 rover mission:

(*a*) MEDA (*b*) RIMFAX

(*c*) SHERLOC (*d*) MOXIE

22. "Vistara" is the brand name of:

(*a*) Newly launched 4G mobile phone

(*b*) Newly launched Airlines

(*c*) Newly launched Housing Scheme by Union Govt. for BPL families

(*d*) Newly launched Health Awareness Scheme by Govt. of India

23. The film, 'Bhaag Milkha Bhaag', based on the life of Milkha Singh, an ace runner, was directed by the same person who directed

(*a*) The Bandit Queen (*b*) Pan Singh Tomar

(*c*) Rang De Basanti (*d*) Lagaan

24. Identify the wrong match:

(*a*) Haldia Port : West Bengal

(*b*) Tuticorin Port : Tamilnadu

(*c*) Kandla Port : Gujarat

(*d*) Paradip Port : Andhra Pradesh

25. Sanskrit Granth 'Hitopedesh' was written by:

(*a*) Chaitanya (*b*) Narayan Pandit

(*c*) Kalidas (*d*) None of these

26. 15 Children are standing in a row facing North. Ravi is to the immediate left of Prabha and is 8th from the left end. Arjun is second from the right end. Which of the following statements is not ture?

(*a*) Prabha is 7th from the right end

(*b*) There are four children between Prabha and Arjun

(*c*) There are five children between Ravi and Arjun

(*d*) Arjun is 13th from the left end

27. The proposals for the partition of India and Pakistan were contained in the:

(*a*) Cripps Mission Proposals

(*b*) Cabinet Mission Proposals

(*c*) Mountbatten Plan of 3rd June 1947

(*d*) None of the above

28. A card is drawn from a pack of well shuffled deck of 52 cards and its probability is 1/4 then which of the following is possible:

(*a*) Getting a "card of hearts"

(*b*) Getting a black card

(*c*) Getting a Jack

(*d*) None of these

29. A candidate for election to the Rajya Sabha must be not less than:

(*a*) 35 Years of age (*b*) 25 Years of age

(*c*) 30 Years of age (*d*) 40 Years of age

30. Curie point is a temperature at which:

(*a*) Matter becomes radio active

(*b*) A magnetic material undergo a sharp change in their magnetic properties

(*c*) A metal loses conductivity

(*d*) Transmutation of metal occurs

31. Gross National Product (GNP) of a country is sum of the market value of all goods and services produced in a year:

(*a*) Plus export Earnings

(*b*) Plus Market value of imports

(*c*) Minus foreign aid

(*d*) Plus net income from abroad

32. 60% of the students in a school are boys. If the number of girls in the school is 300, then the number of boys is:

(*a*) 500 (*b*) 300

(*c*) 450 (*d*) 750

33. It is required to get 40% marks to pass an exam. A candidate scored 200 marks and failed by 8 marks.

What were the maximum marks of that exam:

(*a*) 530 (*b*) 540

(*c*) 520 (*d*) None of these

34. In a division sum, a student took 63 as divisor instead of 36. His answer was 24. The correct answer is:

(*a*) 36 (*b*) 42

(*c*) 32 (*d*) None of these

35. The perimeter of an isosceles triangle is 42 cm and its base is $1\frac{1}{2}$ times each of the equal sides. Find the length of each side of triangle:

(*a*) 12 cm, 12 cm, 18 cm

(*b*) 6 cm, 6 cm, 9 cm

(*c*) 18 cm, 18 cm, 24 cm

(*d*) None of these

36. A quadrilateral with a pair of opposite sides parallel is called:

(*a*) Parallelogram (*b*) Trapezium

(*c*) Polygon (*d*) None of these

37. Find the total surface area of a cylinder having radius of the base 14 cm and height 30 cm.

(*a*) 3872 cm^2 (*b*) 4872 cm^2

(*c*) 3782 cm^2 (*d*) None of these

38. A sum of ₹ 8,500 is to be divided among 5 men, 6 women and 8 boys in the ratio of 10:7:1. The share of one woman will be:

(*a*) ₹ 85 (*b*) ₹ 385

(*c*) ₹ 850 (*d*) ₹ 595

39. The average weight of 8 persons is increased by 2.5 Kg. This happens when one of the eight who weighed 56 Kg is replaced by a new man. The new man weighs:

(*a*) 61 Kg. (*b*) 70 Kg.

(*c*) 76 Kg. (*d*) 80 Kg.

40. If A = 26, SUN = 27 then CAT = ?

(*a*) 24 (*b*) 57

(*c*) 58 (*d*) None of these

41. Anuraag is 40 m South-West of Vikram. Chiraag is 40 m South-East of Vikram, then Chiraag is in which direction of Anuraag?

(*a*) North–East (*b*) South

(*c*) East (*d*) South–East

42. In a certain code, BRAIN is written as . * % ÷ # × and TEAR is written as + $ ÷ % . How is NEAR written in that code?

(*a*) $ × # % (*b*) ÷ + $ %

(*c*) × $ ÷ # (*d*) × $ ÷ %

43. Complete the following series.

3, __, 17, 24, 31, __, 45, __

(*a*) 10, 36, 50 (*b*) 10, 37, 56

(*c*) 10, 38, 52 (*d*) None of these

44. How many times from 4 AM to 4 PM the hands of clock are at right angles?

(*a*) 24 (*b*) 20

(*c*) 22 (*d*) 18

45. Which of the following diagrams indicates the best relation between bakery, cake and biscuit?

(*a*)  (*b*) 

(*c*)  (*d*) 

TECHNICAL

46. Which one of the following statements is correct? Digital modulation techniques are used in satellite communication systems since:

(*a*) They are easier to handle

(*b*) Large bandwidth utilization is possible

(*c*) They have a higher spectral efficiency

(*d*) They are less prone to interference

47. How many bits are required to encode 32 level PCM:

(*a*) 6 (*b*) 5

(*c*) 4 (*d*) 3

48. The noise is more dominant in which of the following:

(*a*) PAM (*b*) PWM

(*c*) PPM (*d*) FSK

49. The phase velocity of wave propagating in a hollow metal wave guide is:

(*a*) Greater than the velocity of the light in free space

(*b*) Less than the velocity of the light in free space

(*c*) Equal to the velocity of the light in free space

(*d*) Equal to the group velocity

50. The magnitude of the open-circuit and short circuit input impedance of a transmission line are 100 ohm and 25 ohm respectively. The characteristic impedance of line is:

(*a*) 25 ohm (*b*) 50 ohm

(*c*) 75 ohm (*d*) 100 ohm

51. The line of sight communication requires transmit and receive antenna to face each other. If the transmit antenna is vertically polarized, for the best reception the receive antenna should be:

(*a*) Horizontally polarized

(*b*) Vertically polarized

(*c*) At 45° with respect to the horizontal polarization

(*d*) At 45° with respect to the vertical polarization

52. A network contains linear resistance and ideal voltage source. If the value of all the resistors are doubled, then the voltage across each resistor is:

(*a*) Halved

(*b*) Doubled

(*c*) Increased by four times

(*d*) Not changed

53. Twelve 1 ohm resistances are used as edge to form a cube. The resistance between two diagonally opposite corners of the cube is:

(*a*) 5/6 ohm (*b*) 1 ohm

(*c*) 6/5 ohm (*d*) 3/2 ohm

54. An eight bit digital data 10101100 is fed to an ADC. The reference voltage is +10V. The analog output voltage will be:

(*a*) 1.05V (*b*) 6.74V

(*c*) 10.10V (*d*) 5.15V

55. A series RLC circuit has a resonant frequency of 1 KHz and a quality factor Q = 100. If each of the R,L and C is doubled from its original value, the new Q of the circuit is:

(*a*) 25 (*b*) 50

(*c*) 100 (*d*) 200

56. The input to a coherent detector is DSB-SC signal plus noise. The noise at the detector output is:

(*a*) The in-phase component

(*b*) The quadrature-component

(*c*) Zero

(*d*) The envelope

57. Gunn diode is a:

(*a*) Negative resistance device

(*b*) Positive resistance device

(*c*) High noise device

(*d*) Low frequency device

58. The intrinsic carrier concentration of silicon sample at 300° K is $2.5\times10^{16}/m^3$. If after doping, the number of majority carriers is 5×10^{20} $/m^3$, the minority carrier density is:

(*a*) $1.25 \times 10^{12}/m^3$ (*b*) $0.125\times10^{12}/m^3$

(*c*) $2.5\times10^{20}/m^3$ (*d*) $0.5\times10^{4}/m^3$

59. The unit of $\nabla \times H$ is:

(*a*) Ampere (*b*) Ampere/meter

(*c*) Ampere/meter^2 (*d*) Ampere-meter

60. A PIN diode is:

(*a*) A metal semiconductor point-contact diode

(*b*) A microwave mixer diode

(*c*) Often used as a microwave detector

(*d*) Suitable for use as a microwave switch

61. 11001, 1001 and 111001 correspond to the 2's complement representation of which one of the following sets of number?

(*a*) 25, 9 and 57 respectively

(*b*) –6, –6 and –6 respectively

(*c*) –7, –7 and –7 respectively

(*d*) –25, –9 and –57 respectively

62. To couple a coaxial line to a parallel wire line, it is best to use a:

(*a*) A slotted line

(*b*) Balun

(*c*) Directional coupler

(*d*) Quarter wave line transformer

63. Which of the following is used extensively where lowest power consumption is necessary?

(*a*) CMOS (*b*) NMOS

(*c*) PMOS (*d*) Any of the above

64. Reflex klystron is a:

(*a*) Low power generator

(*b*) High power oscillator

(*c*) Low gain amplifier

(*d*) Not an oscillator

65. If the radiated power of AM transmitter is 10 KW, the power in the carrier for modulation index of 0.6 is nearly:

(*a*) 8.24 KW

(*b*) 9.26 KW

(*c*) 8.47 KW

(*d*) 9.6 KW

66. The current through 8 ohms branch is:

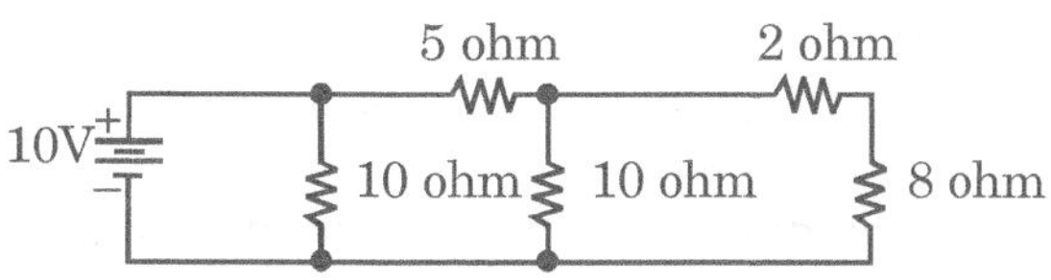

(*a*) 1A

(*b*) 0.5A

(*c*) 1.5A

(*d*) None of these

67. JFET in properly biased condition acts as a:

(*a*) Current controlled current source

(*b*) Voltage controlled voltage source

(*c*) Voltage controlled current source

(*d*) Impedance controlled current source

68. The circuit shown in figure is best described as a:

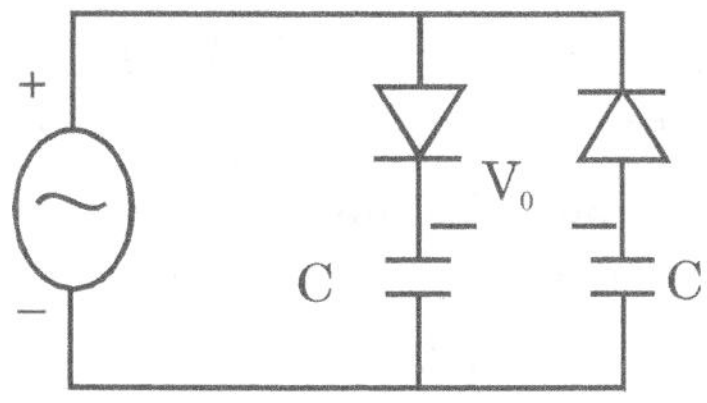

(*a*) Bridge rectifier

(*b*) Ring modulator

(*c*) Frequency discriminatory

(*d*) Voltage doubler

69. Which amplifier will be preferred for highest gain?

(*a*) Darlington pair

(*b*) Cascade amplifier

(*c*) Cascode amplifier

(*d*) Depends on the circuitry

70. The Boolean expression for the truth table shown is:

A	B	C	f
0	0	0	0
0	0	1	0
0	1	0	0
0	1	1	1
1	0	0	0
1	0	1	0
1	1	0	1
1	1	1	0

(*a*) $B(A+C)(\bar{A}+\bar{C})$ (*b*) $B(A+\bar{C})(\bar{A}+C)$

(*c*) $\bar{B}(A+\bar{C})(\bar{A}+C)$ (*d*) $\bar{B}(A+C)(\bar{A}+\bar{C})$

71. Without any additional circuitry, an 8:1 MUX can be used to obtain:

(*a*) Some but not all Boolean functions of 3 variables

(*b*) All function of 3 variables but none of 4 variables

(*c*) All functions of 3 variables and some but not all of 4 variables

(*d*) All functions of 4 variables

72. The resolution of a potentiometer should be:

(*a*) Zero (*b*) Low

(*c*) High (*d*) Infinite

73. Negative feedback in amplifier:

(*a*) Improves the SNR at input

(*b*) Improve SNR at output

(*c*) Reduces distortion

(*d*) Decreases bandwidth

74. Spectral density expresses:

(*a*) Average voltage

(*b*) Average current

(*c*) Average power in a waveform as a function of frequency

(*d*) None of these

75. Which of the following is used as a data selector?

(*a*) Encoder (*b*) Decoder

(*c*) Multiplexer (*d*) De-multiplexer

76. If a signal x(n)= x(−n), then it is called ____ signal:

(*a*) Odd (*b*) Energy

(*c*) Power (*d*) Even

77. The response of an LTI/LSI system is given by the ______ of input and impulse response:

(*a*) Convolution (*b*) Correlation

(*c*) Superposition (*d*) None

78. Companders are used in communication systems to:

(*a*) Compress bandwidth

(*b*) To improve frequency response

(*c*) To improve signal to noise ratio

(*d*) None of these

79. I_{CB0} in a transistor can be reduced by reducing:

(*a*) IB (*b*) VCC

(*c*) IE (*d*) Temperature

80. Light is confined within the core of a simple optical fiber by:

(*a*) Refraction.

(*b*) Total internal reflection at the outer edge of the cladding.

(*c*) Total internal reflection at the core cladding boundary.

(*d*) Reflection from the fiber's plastic coating.

81. The complement of AB + BC′ + CD′ is:

(*a*) A′CD + B′C′ + B′D

(*b*) A′C′ + BC + AB′D′

(*c*) AC + BC + ABD

(*d*) A′C′ + B′C′ + A′B′D′

82. The following expression when simplified will become XY(X′YZ + X′Y′Z′ + XY′Z)

(*a*) 0 (*b*) 1

(*c*) –1 (*d*) X

83. In a PCM system of telemetry, the quantization noise depends on:

(*a*) The sampling rate and quantization level

(*b*) The sampling rate only

(*c*) The number of quantization level

(*d*) Information provided is not sufficient

84. In PCM system, if we increase the quantization levels from 2 to 8, the relative bandwidth requirements will:

(*a*) Be doubled (*b*) Remain same

(*c*) Be tripled (*d*) Becomes eight times

85. The modulation system inherently most noise resistant is:

(*a*) Frequency modulation

(*b*) Pulse width modulation

(*c*) Pulse code modulation

(*d*) Phase modulation

86. Which of the following microphone does not require polarizing current?

(*a*) Crystal microphone

(*b*) Condenser microphone

(*c*) Carbon microphone

(*d*) All of the above

87. The large signal bandwidth of an opamp is limited by its:

(*a*) Loop gain (*b*) Slew rate

(*c*) Output impedance (*d*) Input frequency

88. In an unclocked R-S flip flop made of NOR gates, the forbidden input condition is:

(*a*) R = 0, S = 0 (*b*) R = 1, S = 0

(*c*) R = 0, S = 1 (*d*) R = 1, S = 1

89. A phase-locked loop (PLL) is a feedback circuit consisting of a:

(*a*) Phase detector. (*b*) Low-pass filter.

(*c*) VCO. (*d*) All of these

90. A 339 IC is an example of a fourteen-pin DIP that can be made to function as a ________:

(*a*) Comparator (*b*) 555 timer

(*c*) D to A converter (*d*) Ladder network

91. Filters with the ___ characteristic are useful when a rapid roll-off is required because it provides a roll-off rate greater than –20/dB/decade/pole:

(*a*) Butterworth (*b*) Chebyshev

(*c*) Bessel (*d*) Elliptical

92. Which of the following applications include a phase-locked loop (PLL)circuit?

(*a*) Modems (*b*) AM decoders

(*c*) Tracking filters (*d*) All of these

93. Rectification efficiency of a fullwave rectifier without filter is nearly equals to:

(*a*) 51% (*b*) 61%

(*c*) 71% (*d*) 81%

94. What is the voltage resolution of an 8-stage ladder network?

(*a*) $V_{ref}/128$ (*b*) $V_{ref}/256$

(*c*) $V_{ref}/512$ (*d*) $V_{ref}/1024$

95. In which region is the operating point stable in tunnel diodes?

(*a*) Negative-resistance

(*b*) Positive-resistance

(*c*) Both negative- and positive-resistance

(*d*) Neither negative- nor positive-resistance

96. Which of the following diodes is limited to the reverse-bias region in its region of operation?

(*a*) Schottky (*b*) Tunnel

(*c*) Photodiode (*d*) Rectifier

97. Which of the following semiconductor materials is (are)used for manufacturing solar cells?

(*a*) Gallium arsenide (*b*) Indium arsenide

(*c*) Cadmium sulfide (*d*) All of these

98. If $Y(S) = \frac{s^2+1}{s}$, the network has:

(*a*) 1 H inductor and 1 F capacitor in parallel

(*b*) 1 H inductor and 1 F capacitor in series

(*c*) 1 H inductor and 1 Ω resistor in series

(*d*) 1 H inductor and 1 Ω resistor in parallel

99. When the temperature of a doped semiconductor is increased, its conductivity

(*a*) Decreases

(*b*) Increases

(*c*) Does not change

(*d*) Increases or decreases depending on whether it is p-type or n-type

100. Schering bridge is used to measure

(*a*) Frequency (*b*) Inductance

(*c*) Capacitance (*d*) Mutual Inductance

101. A step voltage E is applied to a series R-L circuit. The rate of change of current is maximum at t = _:

(*a*) Zero (*b*) Infinity

(*c*) L/R (*d*) R/L

102. PLAs, CPLDs, and FPGAs are all which type of device?

(*a*) SLD (*b*) PLD

(*c*) EPROM (*d*) SRAM

103. Holding current for an SCR is best described as:

(*a*) The minimum current required for turn-off

(*b*) The current required before an SCR will turn on

(*c*) The amount of current required to maintain conduction

(*d*) The gate current required to maintain conduction

104. What type of temperature coefficient do thermistors have?

(*a*) Positive

(*b*) Negative

(*c*) Either positive or negative

(*d*) None of these

105. In a microprocessor:

(*a*) One machine cycle is equal to one clock cycle

(*b*) One clock cycle consists of several machine cycles

(*c*) One machine cycle consists of several clock cycles

(*d*) One machine cycle is always less than one clock cycle

106. In 8085 microprocessor with memory mapped I/O which of the following is true?

(*a*) I/O devices have 16 bit addresses

(*b*) I/O devices are accessed during IN and OUT instructions

(*c*) There can be a maximum of 256 input and 256 output devices

(*d*) Logic operations can not be performed

107. A blank EPROM has:

(*a*) All bits set to logical 0

(*b*) All bits set to logical 1

(*c*) Half the total number of bits set to 0 and remaining half to logical 1

(*d*) Either (*a*) or (*b*)

108. One application of a digital multiplexer is to facilitate:

(*a*) Code conversion

(*b*) Parity checking

(*c*) Parallel-to-serial data conversion

(*d*) Data generation

109. The coefficient of coupling between two coils is 0.45. The first coil has an inductance of 75 mH and the second coil has an inductance of 105 mH. What is the mutual inductance between the coils?

(*a*) 3.54 mH (*b*) 7.88 mH

(*c*) 39.9 mH (*d*) 189.3 mH

110. The fast carry or look-ahead carry circuits found in most 4-bit parallel-adder circuits:

(*a*) Increase ripple delay

(*b*) Add a 1 to complemented inputs

(*c*) Reduce propagation delay

(*d*) Determine sign and magnitude

111. To multiply a number by 8 in 8085 we have to use RAL instruction:

(*a*) Once (*b*) Twice

(*c*) Thrice (*d*) Four times

112. An energy signal has G(f)= 10. Its energy density spectrum is:

(*a*) 10 (*b*) 100

(*c*) 50 (*d*) 20

113. Feedback factor at the frequency of oscillation of a Wien bridge oscillator is:

(*a*) 3 (*b*) $\frac{1}{3}$

(*c*) $\frac{1}{29}$ (*d*) $\frac{3}{29}$

114. A 4-bit R/2R digital-to-analog (DAC) converter has a reference of 5 volts. What is the analog output for the input code 1010:

(*a*) 0.3125 V (*b*) 3.125 V

(*c*) 0.78125 V (*d*) –3.125 V

115. A quarter wave line open circuited at far end behaves as:

(*a*) Inductance (*b*) L and C in parallel

(*c*) Capacitance (*d*) L and C in series

116. In a directional coupler:

(*a*) Isolation (dB) equals coupling plus directivity

(*b*) Coupling (dB) equals isolation plus directivity

(*c*) Directivity (dB) equals isolation plus coupling

(*d*) Isolation (dB) equals (coupling) (directivity)

117. In a circular waveguide the dominant mode is:

(*a*) TE_{01} (*b*) TE_{11}

(*c*) TE_{20} (*d*) TE_{21}

118. In a klystron amplifier the input cavity is called:

(*a*) Buncher (*b*) Catcher

(*c*) Pierce gun (*d*) Collector

119. A loss less line of characteristic impedance Z_0 is terminated in pure reactance of $-jZ_0$ value. VSWR is

(*a*) 10 (*b*) 2

(*c*) 1 (*d*) Infinity

120. The velocity factor of a transmission line depends on:

(*a*) Temperature

(*b*) Skin effect

(*c*) Relative permittivity of dielectric

(*d*) None of these

GENERAL ENGLISH

Directions (Question nos. 121 to 128) : Read the following passage carefully to answer these questions given below it. Certain words / phrases are printed in bold to help you locate them while answering some of the questions.

Management is a set of processes that can keep a complicated system of people and technology running smoothly. The most important aspects of management include planning, budgeting, organizing, staffing, controlling and problem solving. Leadership is a set of processes that creates organizations in the first place or adapts them to significantly changing circumstances, Leadership defines what the future should look like, aligns people with that vision and inspires them to make it happen despite the obstacles. This distinction is absolutely crucial for our purpose here. Successful transformation is 70-90 percent leadership and only 10-30 management. Yet for historical reasons, many organizations today do not have much leadership. And almost everyone thinks about the problem here as one of managing change.

For most of this century, as we created thousands and thousands of large organizations for the first time in human history, we did not have enough good managers to keep all those bureaucracies functioning. So many companies and universities developed management programs and hundreds and thousands people were encouraged to learn management on the job. and they did. But people were taught little about leadership. To some degree, management was emphasized because it is easier to teach them leadership. But even more so management was the main item on the twentieth century agenda because that is what was needed. For every entrepreneur or business builder who was a leader, we needed hundreds of managers to run their ever-growing enterprises.

Unfortunately for us today, this emphasis on management has often been institutionalized in corporate cultures that discourage employees from learning how to lead. Ironically past success is usually the key ingredient in producing this outcome. The syndrome, as I have observed it on many occasions goes like this: success creates some degree of market dominance which in turn produces much growth. After a while keeping the ever larger organizations under control becomes primary challenge. So attention turns inward and managerial competences are nurtured. With a strong emphasis on management but not leadership, bureaucracy and an inward focus take over. But with the continued success the result mostly of market dominance, the problem often goes un-addressed and an unhealthy arrogance begins to evolve. All of these characteristics then make any transformation efforts much more difficult.

Arrogant managers can over evaluate their current performance and competitive position, listen poorly and learn slowly. Inwardly focused employees can have difficulty seeing the very forces that present threats and opportunities. Bureaucratic cultures can smother those who want to respond shifting conditions. And the lack of leadership leaves no force inside these organizations to break out of the morass.

121. Which of the following is Not the characteristic of Bureaucratic culture?

(a) Managers listens poorly and learn slowly

(b) Managerial competences are nurtured

(c) Employees clearly sees the forces that present threats and opportunities

(d) Prevalence of unhealthy arrogance

122. Which of the following is similar in meaning of the word "smother" as used in the passage?

(a) Suppress (b) Encourage
(c) Instigate (d) Criticize

123. Which of the following characteristics help organizations in their transformation efforts?

(a) Emphasis on leadership but not management
(b) A strong and dogmatic culture
(c) Bureaucratic and inward looking approach
(d) Failing to acknowledge the value of customers and shareholders

124. In the passage management is equated with:

(a) Organisation
(b) Leadership
(c) Organisational Vision
(d) Bureaucracy

125. What according to author is leadership?

(a) Process which keeps system of people and technology running smoothly
(b) Planning the future and budgetary resources of the organizations
(c) Inspiring people to realize the vision
(d) Carrying out the crucial functions of management

126. Why, according to author a distinction between management and leadership is crucial?

(a) Leaders are reactive whereas managers are proactive
(b) Organizations are facing problems of not getting good managers
(c) Organizations are pursuing the strategy of status quo
(d) In today's context organizations need leaders much more than managers in transforming them

127. Management education was emphasized in the management programs because

(a) Establishing direction was the main focus of organizations
(b) Motivating employees was thought to be done by managers
(c) Organizations wanted to create powerful guiding coalition.
(d) Management was the main item of agenda in organizations

128. Why does the attention of large organizations turn inward?

(a) Their manager become arrogant
(b) They have to keep themselves under control
(c) Their success creates market dominance
(d) They want to project their predictability

129. Fill in the blank with correct preposition.

Rita drove .... a red light.:

(a) From (b) Through
(c) In (d) Among

130. Given below is the sentence in active voice. Choose the correct sentence given in passive voice among the alternatives.

Ought you not to reveal the truth now?

(a) Ought the truth not to be revealed by you then?
(b) Ought the truth not to be revealed by you now?
(c) Should the truth need not be revealed by you then?
(d) Ought the truth need not be revealed by you now?

131. Identify the incorrect sentence:

(a) It is he who came to see us.
(b) It is I who caught the thief.
(c) Suman is more good than brave.
(d) He is very stronger than I am.

132. Identify the incorrect sentence:

(a) The first two chapters of the novel are dull.
(b) He was elected President of our society
(c) I am looking forward to receive your reply
(d) You and I must finish our work in time.

133. Fill in the gap with proper preposition:

He succeeded ____ perseverance and sheer hard work.

(a) By virtue of (b) By dint of
(c) By means of (d) By way of

134. Identify the incorrect sentence:

(a) The patient had died before the doctor arrived.
(b) He behaves as if he was a king.
(c) Read regularly lest you should fail.
(d) Bread and butter makes a good breakfast.

135. Identify the correct sentence:

(a) Neither he nor his friends are reading.
(b) It is one of the important day in my life.
(c) A great many student has been declared successful.
(d) He not only comes for swimming but also for coaching of learners.

136. Given below are four substitutions for the underlined part. Choose the correct alternative to make the sentence grammatically correct.

Make haste lest you <u>should not be</u> caught in the storm:

(a) You should be (b) You could be
(c) You might be (d) Otherwise you can be

137. In the given alternatives, there are four sentences with underlined word. Identify the sentence with underlined word which is not an adverb:

(a) He is nearly sixty

(b) He gave me a lovely present

(c) He hardly works

(d) Suzy is quite beautiful

138. Identify the correct sentence:

(a) All the students of the class are friendly; they love each other.

(b) Each student must bring their books.

(c) Each of the seven boys in the class has finished his task.

(d) A few men are free from fault.

139. Fill in the gap with proper preposition:

There was no money____him.

(a) With (b) On

(c) To (d) About

140. Identify the correct sentence:

(a) Forty miles are a good distance.

(b) He was turned out to copy answers.

(c) The flowers smelt sweetly.

(d) I say my prayers in the morning.

141. Fill in the gap suitably:

Measles___infectious.

(a) is (b) are

(c) were (d) are being

142. Identify the incorrect plural:

(a) Swine (b) Sheep

(c) Species (d) Aircrafts

143. Identify the incorrect sentence:

(a) They are having lunch.

(b) These grapes taste sour.

(c) He is having a cellular phone.

(d) I am thinking of going to Malaysia.

144. Fill in the blank with correct preposition.

Butter is made.........milk:

(a) Of (b) From

(c) By (d) Through

145. Select the correct meaning of the bold idioms and phrases out of the four choices given below:

He does not understand that he stands in his own light.

(a) Acts against his own interests

(b) Burnt himself

(c) To come in dark

(d) Ignore by others

146. Despite his injuries, he won the match fair and square.

(a) With tricks (b) With an honest way

(c) Illegally (d) Forcefully

147. Fill in the blank using the correct tense:

I .... at five o'clock and .... for two hours. After taking a bath, I .... breakfast. Then, I .... to school on my bike.

(a) Got up, studied, took, went

(b) Get up, will study, take, went

(c) Get up, study, will take, go

(d) Got up, will study, will take, will go

148. Three sentences have been given. You have to find out the sentence in which an adverb modifies the meaning of another adverb:

1. Jalpa sings well
2. He speaks very fluently
3. This soap is delicately perfumed.

(a) 1 (b) 2

(c) 2 and 3 (d) 1 and 3

149. Choose the correct synonym out of the four choices given:

GIMMICK

(a) Stunt (b) To sparkle

(c) Gloomy (d) Glorify

150. Some words are used sometimes as adjectives, sometimes as adverbs. Four sentences have been given where in three sentences are containing words being used as adjectives and one sentence contains word being used as adverb. Identify the odd one:

(a) He knows me better than you.

(b) Rama is our fast bowler

(c) He is the only child of his parents.

(d) The teacher has a high opinion of that boy.

151. Choose the correct synonym out of the four choices given:

SCATHING

(a) Sabotage (b) Perpetuated

(c) Incense (d) Scornful

152. Combine each of the following sentence by using an Adverb Clause.

He was quite tired. He could scarcely stand.

(a) He was tired and he could not stand.

(b) He could scarcely stand because he was quite tired.

(c) Being tired he could scarcely stand.

(d) On account of being tired he could scarcely stand.

153. Identify the correct sentence:
(a) I dislike to see a film late in the night.
(b) I can reproduce this lesson word by word
(c) An Europian lives in my area.
(d) The higher we go, the cooler we feel

154. Identify the incorrect plural:
(a) Thieves (b) Halves
(c) Mice (d) Daughter-in-laws

155. Given below are the words with different spellings. Select the one which you consider to be correctly spelled:
(a) Paraphernalia (b) Paraphrenalia
(c) Parapherenalia (d) Paraphrennalia

156. Choose the correct synonym out of the four choices given:
WINSOME
(a) Captivating (b) Enthusiasm
(c) Instigate (d) Utopian

157. Choose the correct synonym out of the four choices given:
PERCHANCE
(a) Prerogative (b) Fluke
(c) Debauched (d) Sordid

158. In the following question, a word is given in capital letters followed by four words.
Choose the word that is nearly opposite in meaning to the word in capital letters.
SCRIMP:
(a) Lavish (b) Parsimonious
(c) Meticulous (d) Polite

159. In the following question, a word is given in capital letters followed by four words.
Choose the word that is nearly opposite in meaning to the word in capital letters.
DELECTABLE:
(a) Agonising (b) Apetising
(c) Distasteful (d) Laborious

160. In the following question, a word is given in capital letters followed by four words.
Choose the word that is nearly opposite in meaning to the word in capital letters.
BOISTEROUS:
(a) Serene (b) Tumultuous
(c) Brazen (d) Opaque

161. This question consists of an idiom followed by four alternative words or phrases giving the possible meaning of idiom. Select the alternative which gives closest meaning of the idiom.
To eat crow:
(a) To become a non vegetarian
(b) To quieten down
(c) Made to feel small
(d) To swallow one's own word

162. This question consists of an idiom followed by four alternative words or phrases giving the possible meaning of idiom. Select the alternative which gives closest meaning of the idiom.
Feathering one's nest
(a) Enriching oneself at the expense of public
(b) Buying insurance in time
(c) Taking adequate precaution against danger
(d) Making a new home

163. A Group of flies is a .... :
(a) Flock (b) Swarm
(c) Gaggle (d) Nest

164. In the following question, the sentence has a blank space and four words are given after the sentence.
Choose the word from alternative that you consider most appropriate for the blank space.
He was so .... in the music that he did not hear the door open :
(a) Embroiled (b) Engrossed
(c) Engaged (d) Amused

165. In the following question, the sentence has a blank space and four words are given after the sentence. Choose the word from alternative that you consider most appropriate for the blank space.
The television station was .... with letters and phone calls after the announcement:
(a) Drowned (b) Stormed
(c) Absorbed (d) Deluged

166. Change into passive voice.
The boys were playing cricket.
(a) Cricket was being played by the boys.
(b) Cricket was played by the boys.
(c) Cricket will have been played by the boys.
(d) The cricket being played by the boys.

167. In the following question, identify the correct reported speech from the given alternatives.
He said, " I must go home at once.":
(a) He said that he must have gone home at once.
(b) He said that he must go home then and there.
(c) He said that he had to go home then
(d) He said that he had to go home at once

168. In the following question, identify the correct reported speech from the given alternatives.

She said , "oh dear! I have just missed the bus .":

(a) She regretted that she just missed the bus.

(b) She said with regret that she had just missed the bus.

(c) She exclaimed that she has just missed the bus.

(d) She narrated that she just missed the bus.

169. Which of phrases given below in each sentence should replace the phrase in Underlined to make the sentence grammatically correct?

Why did you not threw the bag away?

(a) did you not throw

(b) had you not threw

(c) did you not thrown

(d) you did not thrown

170. Find out the sentence in Passive Voice

(a) The judge convicted the culprit for his offence

(b) The culprit apologised to the judge for his offence

(c) The judge scolded the culprit for his offence

(d) The culprit was punished for his offence by the judge

171. Mark the appropriate words to fill in the blanks.

God helps those .... help themselves.

(a) Who (b) Which

(c) That (d) Whom

172. Identify the complex sentence among the followings:

(a) The mother tiger saw her cubs.

(b) The mother tiger saw her cubs and she became happy

(c) When the mother tiger saw her cubs, she became happy

(d) Both (b) & (c) above

173. Identify the correct tense which has been used in this sentence.

By the end of this year he will have saved a lot of money.

(a) Future continuous tense

(b) Future perfect tense

(c) Future perfect continuous tense

(d) Simple future tense

174. In the following options, identify the sentence with interrogative adverb of manner:

(a) How did you contrive it?

(b) Where is Ram?

(c) When did you come?

(d) How many boys are there in your class?

175. Identify the kind of conjunction in the following sentence:

His bark is worst than his bite.

(a) Copulative coordinating conjunction

(b) Disjunctive Coordinating Conjunction

(c) Subordinating Conjunction

(d) Illative coordinating conjunction

176. Select the word choice that gives the meaning of the sentence or phrase.

A student who absents himself from the class without permission:

(a) Censorious (b) Vagabond

(c) Reticent (d) Truant

177. Select the word choice that gives the meaning of the sentence or phrase.

A person who does not speak much:

(a) Candid (b) Outspoken

(c) Reticent (d) Defiant

178. Select the word choice that gives the meaning of the sentence or phrase.

Anything of little value:

(a) Travail

(b) Primitive

(c) Trivial

(d) Trammels

179. In following question, four spellings are given for a word. Pick the right choice which has the correct spelling.

(a) Fiasco

(b) Faisco

(c) Fiasoco

(d) Faisiaco

180. In following question, four spellings are given for a word. Pick the right choice which has the correct spelling.

(a) Courageus

(b) Couragious

(c) Courageous

(d) Couragous

OMR SHEET

1. (A) (B) (C) (D)
2. (A) (B) (C) (D)
3. (A) (B) (C) (D)
4. (A) (B) (C) (D)
5. (A) (B) (C) (D)
6. (A) (B) (C) (D)
7. (A) (B) (C) (D)
8. (A) (B) (C) (D)
9. (A) (B) (C) (D)
10. (A) (B) (C) (D)
11. (A) (B) (C) (D)
12. (A) (B) (C) (D)
13. (A) (B) (C) (D)
14. (A) (B) (C) (D)
15. (A) (B) (C) (D)
16. (A) (B) (C) (D)
17. (A) (B) (C) (D)
18. (A) (B) (C) (D)
19. (A) (B) (C) (D)
20. (A) (B) (C) (D)
21. (A) (B) (C) (D)
22. (A) (B) (C) (D)
23. (A) (B) (C) (D)
24. (A) (B) (C) (D)
25. (A) (B) (C) (D)
26. (A) (B) (C) (D)
27. (A) (B) (C) (D)
28. (A) (B) (C) (D)
29. (A) (B) (C) (D)
30. (A) (B) (C) (D)
31. (A) (B) (C) (D)
32. (A) (B) (C) (D)
33. (A) (B) (C) (D)
34. (A) (B) (C) (D)
35. (A) (B) (C) (D)
36. (A) (B) (C) (D)
37. (A) (B) (C) (D)
38. (A) (B) (C) (D)
39. (A) (B) (C) (D)
40. (A) (B) (C) (D)
41. (A) (B) (C) (D)
42. (A) (B) (C) (D)
43. (A) (B) (C) (D)
44. (A) (B) (C) (D)
45. (A) (B) (C) (D)
46. (A) (B) (C) (D)
47. (A) (B) (C) (D)
48. (A) (B) (C) (D)
49. (A) (B) (C) (D)
50. (A) (B) (C) (D)
51. (A) (B) (C) (D)
52. (A) (B) (C) (D)
53. (A) (B) (C) (D)
54. (A) (B) (C) (D)
55. (A) (B) (C) (D)
56. (A) (B) (C) (D)
57. (A) (B) (C) (D)
58. (A) (B) (C) (D)
59. (A) (B) (C) (D)
60. (A) (B) (C) (D)
61. (A) (B) (C) (D)
62. (A) (B) (C) (D)
63. (A) (B) (C) (D)
64. (A) (B) (C) (D)
65. (A) (B) (C) (D)
66. (A) (B) (C) (D)
67. (A) (B) (C) (D)
68. (A) (B) (C) (D)
69. (A) (B) (C) (D)
70. (A) (B) (C) (D)
71. (A) (B) (C) (D)
72. (A) (B) (C) (D)
73. (A) (B) (C) (D)
74. (A) (B) (C) (D)
75. (A) (B) (C) (D)
76. (A) (B) (C) (D)
77. (A) (B) (C) (D)
78. (A) (B) (C) (D)
79. (A) (B) (C) (D)
80. (A) (B) (C) (D)
81. (A) (B) (C) (D)
82. (A) (B) (C) (D)
83. (A) (B) (C) (D)
84. (A) (B) (C) (D)
85. (A) (B) (C) (D)
86. (A) (B) (C) (D)
87. (A) (B) (C) (D)
88. (A) (B) (C) (D)
89. (A) (B) (C) (D)
90. (A) (B) (C) (D)
91. (A) (B) (C) (D)
92. (A) (B) (C) (D)
93. (A) (B) (C) (D)
94. (A) (B) (C) (D)
95. (A) (B) (C) (D)
96. (A) (B) (C) (D)
97. (A) (B) (C) (D)
98. (A) (B) (C) (D)
99. (A) (B) (C) (D)
100. (A) (B) (C) (D)

OMR SHEET

101. (A) (B) (C) (D)	121. (A) (B) (C) (D)	141. (A) (B) (C) (D)	161. (A) (B) (C) (D)
102. (A) (B) (C) (D)	122. (A) (B) (C) (D)	142. (A) (B) (C) (D)	162. (A) (B) (C) (D)
103. (A) (B) (C) (D)	123. (A) (B) (C) (D)	143. (A) (B) (C) (D)	163. (A) (B) (C) (D)
104. (A) (B) (C) (D)	124. (A) (B) (C) (D)	144. (A) (B) (C) (D)	164. (A) (B) (C) (D)
105. (A) (B) (C) (D)	125. (A) (B) (C) (D)	145. (A) (B) (C) (D)	165. (A) (B) (C) (D)
106. (A) (B) (C) (D)	126. (A) (B) (C) (D)	146. (A) (B) (C) (D)	166. (A) (B) (C) (D)
107. (A) (B) (C) (D)	127. (A) (B) (C) (D)	147. (A) (B) (C) (D)	167. (A) (B) (C) (D)
108. (A) (B) (C) (D)	128. (A) (B) (C) (D)	148. (A) (B) (C) (D)	168. (A) (B) (C) (D)
109. (A) (B) (C) (D)	129. (A) (B) (C) (D)	149. (A) (B) (C) (D)	169. (A) (B) (C) (D)
110. (A) (B) (C) (D)	130. (A) (B) (C) (D)	150. (A) (B) (C) (D)	170. (A) (B) (C) (D)
111. (A) (B) (C) (D)	131. (A) (B) (C) (D)	151. (A) (B) (C) (D)	171. (A) (B) (C) (D)
112. (A) (B) (C) (D)	132. (A) (B) (C) (D)	152. (A) (B) (C) (D)	172. (A) (B) (C) (D)
113. (A) (B) (C) (D)	133. (A) (B) (C) (D)	153. (A) (B) (C) (D)	173. (A) (B) (C) (D)
114. (A) (B) (C) (D)	134. (A) (B) (C) (D)	154. (A) (B) (C) (D)	174. (A) (B) (C) (D)
115. (A) (B) (C) (D)	135. (A) (B) (C) (D)	155. (A) (B) (C) (D)	175. (A) (B) (C) (D)
116. (A) (B) (C) (D)	136. (A) (B) (C) (D)	156. (A) (B) (C) (D)	176. (A) (B) (C) (D)
117. (A) (B) (C) (D)	137. (A) (B) (C) (D)	157. (A) (B) (C) (D)	177. (A) (B) (C) (D)
118. (A) (B) (C) (D)	138. (A) (B) (C) (D)	158. (A) (B) (C) (D)	178. (A) (B) (C) (D)
119. (A) (B) (C) (D)	139. (A) (B) (C) (D)	159. (A) (B) (C) (D)	179. (A) (B) (C) (D)
120. (A) (B) (C) (D)	140. (A) (B) (C) (D)	160. (A) (B) (C) (D)	180. (A) (B) (C) (D)

ANSWERS

1. (*d*)	**2.** (*b*)	**3.** (*c*)	**4.** (*b*)	**5.** (*d*)	**6.** (*b*)	**7.** (*b*)	**8.** (*)	**9.** (*b*)	**10.** (*c*)
11. (*b*)	**12.** (*d*)	**13.** (*b*)	**14.** (*c*)	**15.** (*a*)	**16.** (*a*)	**17.** (*d*)	**18.** (*b*)	**19.** (*a*)	**20.** (*c*)
21. (*b*)	**22.** (*b*)	**23.** (*c*)	**24.** (*d*)	**25.** (*b*)	**26.** (*d*)	**27.** (*c*)	**28.** (*a*)	**29.** (*c*)	**30.** (*b*)
31. (*d*)	**32.** (*c*)	**33.** (*c*)	**34.** (*b*)	**35.** (*a*)	**36.** (*b*)	**37.** (*a*)	**38.** (*d*)	**39.** (*c*)	**40.** (*b*)
41. (*c*)	**42.** (*d*)	**43.** (*c*)	**44.** (*c*)	**45.** (*b*)	**46.** (*d*)	**47.** (*b*)	**48.** (*a*)	**49.** (*a*)	**50.** (*b*)
51. (*a*)	**52.** (*d*)	**53.** (*a*)	**54.** (*b*)	**55.** (*d*)	**56.** (*c*)	**57.** (*a*)	**58.** (*a*)	**59.** (*c*)	**60.** (*d*)
61. (*c*)	**62.** (*b*)	**63.** (*a*)	**64.** (*a*)	**65.** (*c*)	**66.** (*b*)	**67.** (*c*)	**68.** (*d*)	**69.** (*c*)	**70.** (*a*)
71. (*d*)	**72.** (*d*)	**73.** (*c*)	**74.** (*c*)	**75.** (*c*)	**76.** (*d*)	**77.** (*a*)	**78.** (*c*)	**79.** (*d*)	**80.** (*b*)
81. (*a*)	**82.** (*a*)	**83.** (*c*)	**84.** (*c*)	**85.** (*c*)	**86.** (*a*)	**87.** (*b*)	**88.** (*d*)	**89.** (*d*)	**90.** (*a*)
91. (*b*)	**92.** (*d*)	**93.** (*d*)	**94.** (*b*)	**95.** (*a*)	**96.** (*c*)	**97.** (*d*)	**98.** (*a*)	**99.** (*b*)	**100.** (*c*)
101. (*a*)	**102.** (*b*)	**103.** (*c*)	**104.** (*c*)	**105.** (*c*)	**106.** (*a*)	**107.** (*b*)	**108.** (*c*)	**109.** (*c*)	**110.** (*c*)
111. (*c*)	**112.** (*b*)	**113.** (*b*)	**114.** (*b*)	**115.** (*d*)	**116.** (*a*)	**117.** (*b*)	**118.** (*a*)	**119.** (*d*)	**120.** (*c*)
121. (*c*)	**122.** (*a*)	**123.** (*a*)	**124.** (*d*)	**125.** (*c*)	**126.** (*d*)	**127.** (*d*)	**128.** (*d*)	**129.** (*b*)	**130.** (*b*)
131. (*d*)	**132.** (*c*)	**133.** (*b*)	**134.** (*b*)	**135.** (*a*)	**136.** (*a*)	**137.** (*b*)	**138.** (*c*)	**139.** (*b*)	**140.** (*d*)
141. (*a*)	**142.** (*d*)	**143.** (*c*)	**144.** (*b*)	**145.** (*a*)	**146.** (*b*)	**147.** (*a*)	**148.** (*b*)	**149.** (*a*)	**150.** (*a*)
151. (*d*)	**152.** (*b*)	**153.** (*d*)	**154.** (*d*)	**155.** (*a*)	**156.** (*a*)	**157.** (*b*)	**158.** (*a*)	**159.** (*c*)	**160.** (*a*)
161. (*c*)	**162.** (*a*)	**163.** (*b*)	**164.** (*b*)	**165.** (*d*)	**166.** (*a*)	**167.** (*d*)	**168.** (*b*)	**169.** (*a*)	**170.** (*d*)
171. (*a*)	**172.** (*c*)	**173.** (*b*)	**174.** (*a*)	**175.** (*c*)	**176.** (*d*)	**177.** (*c*)	**178.** (*c*)	**179.** (*a*)	**180.** (*c*)

EXPLANATIONS

1. N < M

O < N

$$\left.\begin{array}{l} O < P \\ P < Q \end{array}\right] \Rightarrow O < P < Q$$

$O < Q^N < M$ $\boxed{O < P < Q < N < M}$

Hence 'O' is lightest.

2. 'Q' lighter than M and N

3. 'M' is heaviest in weight

5. If the S.P of 1 chocolate be ₹ 1

⇒ S.P of 18 chocolate = ₹18

Now C.P = 18 + 2 = ₹ 20

$$\therefore \text{Loss \%} = \left[\frac{2}{20} \times 100\right]\% = 10\%$$

6. $\frac{10}{100} \Rightarrow \frac{1}{10}$

$\frac{20}{100} \Rightarrow \frac{1}{5}$ and $\frac{50}{100} \Rightarrow \frac{1}{2}$

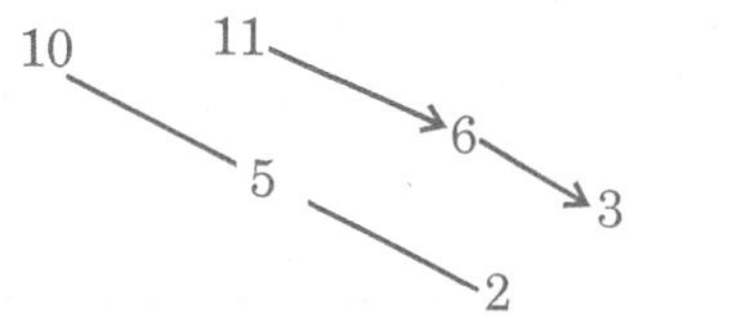

10 × 5 × 2 : 11 × 6 × 3

10 × 5 × 2 : 11 × 6 × 3

∴ 100 : 198

∴ % Change in volume of cuboid = 98%

7. 80 km is travelled in 2 hrs then rest of 20 mins. Again 80 km is travelled in 2 hrs and then rest of again 20 mins

Now final 80 km is travelled in 2 hrs

Hence total time taken = (2 + 2 + 2) hrs + 40 mins

= 6 hrs 40 min

9. Let the Income of Anil be 7x

Income of firoz be 9x and Income of Ketan be 12x

and Spending of Anil be 8 y

Spending of firoz be 9y and Spending of Ketan be 15y

According to question $8y = \frac{1}{4} \times 7x$

$\therefore \quad y = \frac{7x}{32}$

$\therefore$ Ratio of Saving of Anil, Firoz and Ketan is

$(7x - 8y) : (9x - 9y) : (12x - 15y)$

= 56 : 99 : 69

10. Ramesh : Sujeet

2100 × 12 : x × 8

$\therefore \frac{2100 \times 12}{8} = x$

$\therefore$ x = ₹ 3150

11. The boy in the photograph is the only son of the Lokesh mother. i.e Son of Lokesh. Hence, Lokesh is the father of the boy.

12. 'mat' is not animal ⇒ 'mat' is cow

$\therefore$ option (d) is correct

13.

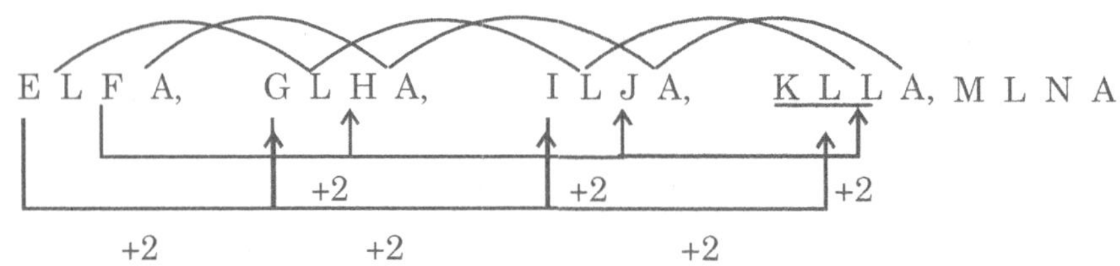

$\therefore$ option (b) is correct

14. In Cars, mirror used is concave

16. TCS has developed super computer of India Param Yuva - II

15. Equinox is a position on equator

18. Name of Prince Khurram was known to be Shahjahan

20. Ramon Magsasay Award for 2014 was given to Chinese Journalist Hu Shuli

22. "Vistar" is brand name of newly launched Airlines.

23. "Rang De Basanti" and Bhaag Mikha Bhaag" was directed by Om Prakash Mehra.

24. Haldia port is in west Bengal. Tuticorin port is Tamilnadu. Kandla port is in Gujarat while paradip port is in Odhisa.

25. "Hitopedesh" Sanskrit Granth is written by Narayan Pandit.

26. Ravi Prabha Arjun
\- - - - - - - - - - - - - - - - - -

1st Statement is true.

2nd Statement is also true

3rd Statement is also true.

Arjun is 14th from left end

27. Mountbatten plan of 3rd june 1947 contains the proporsal for the partition of India and Pakistan.

28. Total there are 13 cards of heart

$\therefore$ Required probability is $\frac{13}{52} = \frac{1}{4}$

29. A candidate for election to the Rajya Sabha must be not less than 30 years of age

30. At curie point, magnetic material undergo a sharp change in their magnetic properties

31. Gross National product (GNP)

= Market values of all goods and service + Net income from abroad.

32. 60% of Students in school are boys ⇒ 40% of students in school are girls

Now according to question, 40% = 300

$\therefore$ 60% = ?

$\therefore$ No of boys $= \frac{60 \times 300}{40} = 450$

33. According to question $\frac{40}{100} \times x = 208$

$\therefore x = \frac{208 \times 100}{40} = 520$

34. According to question $\frac{x}{63} = 24$

$\therefore x = 24 \times 63$

Now $x' = \frac{24 \times 63}{36} = 42$

35.

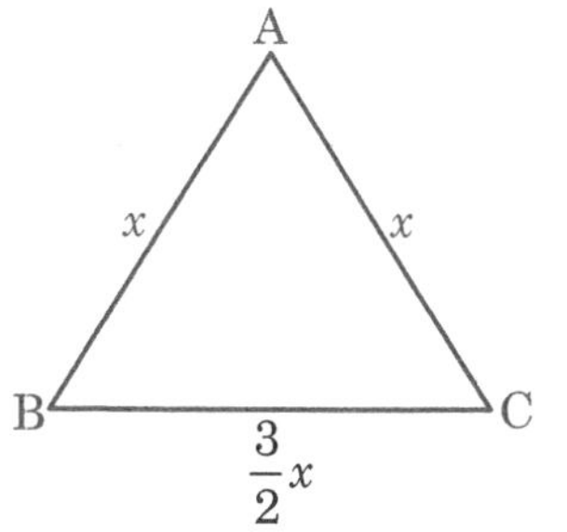

Now $2x + \frac{3}{2}x = 42$

$\therefore \frac{7x}{2} = 42$

$\therefore x = 12$

Hence length of each side of triangle is 12cm, 12cm and 18 cm

36. A quadrilateral with pair of opposite sides parallel is parallelogram

37. r = 14 cm

and h = 30 cm

T.S.A of cylinder = $2\pi r(r + h)$

$= 2 \times \frac{22}{7} \times 14(14 + 30)$

$= 2 \times 22 \times 2\ (44)$

$= 3872\ cm^2$

$\therefore$ option (a) is correct

40. A = 26 $\Rightarrow$ From last

SUN = 27 $\Rightarrow$ S(8) + U(6) + N(13)

= 8 + 6 + 13 = 27

Similarly, CAT = C(24) + A(26) + T(7)

= 24 + 26 +7 = 57

41.

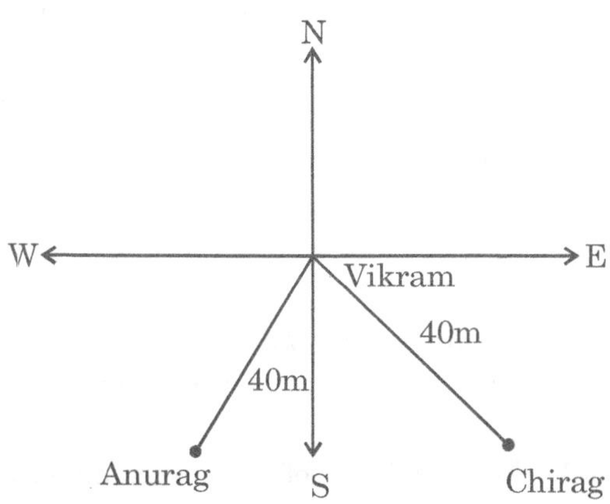

Hence Chirag is in East direction from Anurag

42. BRAIN $\quad$ TEAR

↓ ↓ ↓↓↓ $\quad$ ↓ ↓↓ ↓

* % ÷ # x $\quad$ + $ ÷ %

$\Rightarrow$

B → *

R → %

A → ÷

I → #

N → ×

T → +

E → $

$\therefore$ Code of NEAR is '×$ ÷ %'

43. $\underbrace{3,10}_{+7},\underbrace{17}_{+7},\underbrace{24}_{+7},\underbrace{31}_{+7},\underbrace{38}_{+7},\underbrace{45}_{+7},\underbrace{52}_{+7}$

44. In 12 hrs, the hand of clock are right angle at 22 times

45. Cake and biscuit are product of bakery

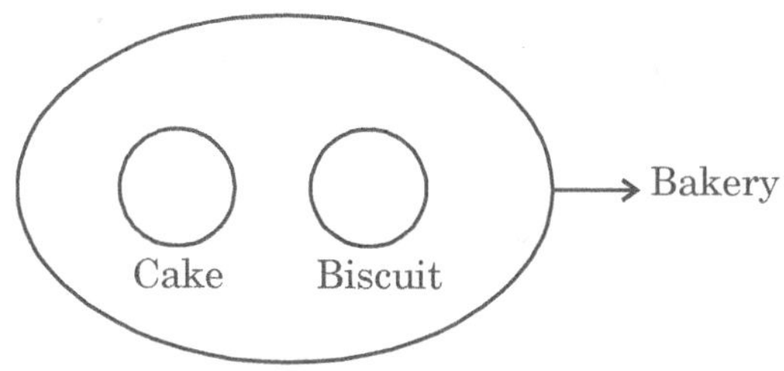

46. Digital signals are least affected by noise and interference.

47. No. of bits = $\log_2 32 = 5$

48. Pulse amplitude modulation is more sensitive to noise

49. The phase velocity of wave travelling in a hollow metal waveguide is more than the velocity of light in free space.

50. Characteristic impedance,

$Z_0 = \sqrt{Z_{oc}.Z_{sc}} = \sqrt{100 \times 25} = 50\Omega$

52.

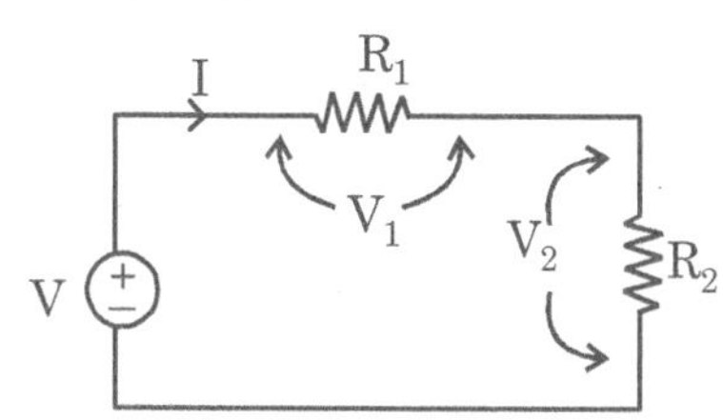

$I = \frac{V}{R_1 + R_2}$

$\therefore V_1 = IR_1, V_2 = IR_2$

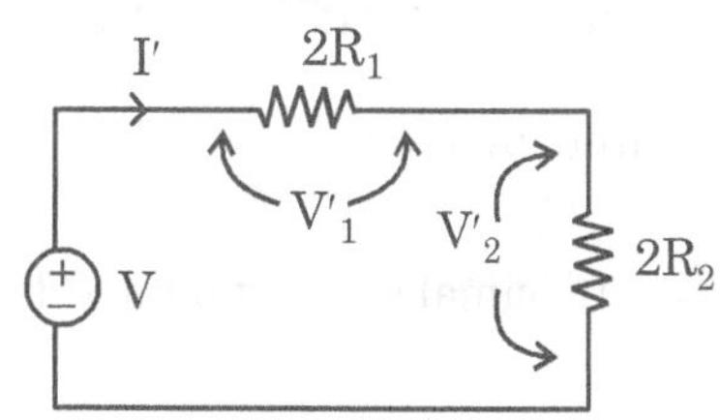

$$I' = \frac{V}{2(R_1 + R_2)} = \frac{I}{2}$$

$$V_1' = I' \times 2R_1 = IR_1$$

$$V_2' = I' \times 2R_2 = IR_2$$

Assuming, $I = \frac{V}{Req}$

53.

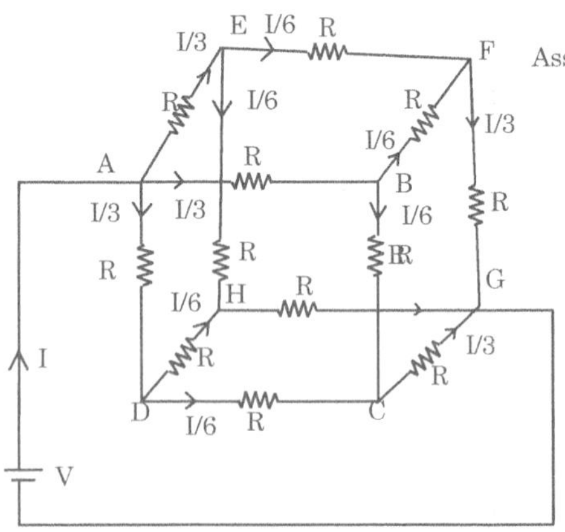

Assuming, $I = \frac{V}{Req}$

Due to symmetry of circuit, current will be equally divide at each node.

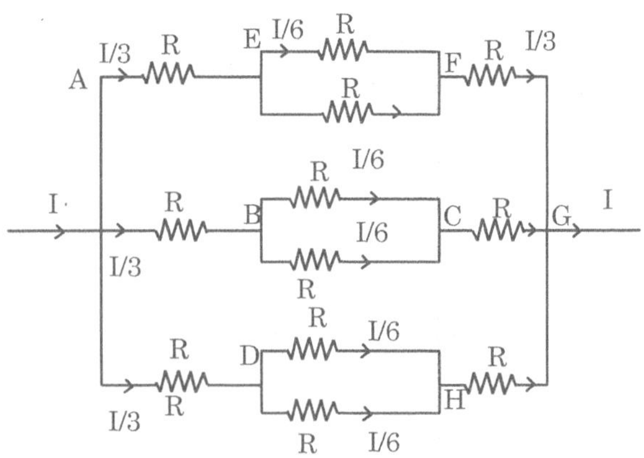

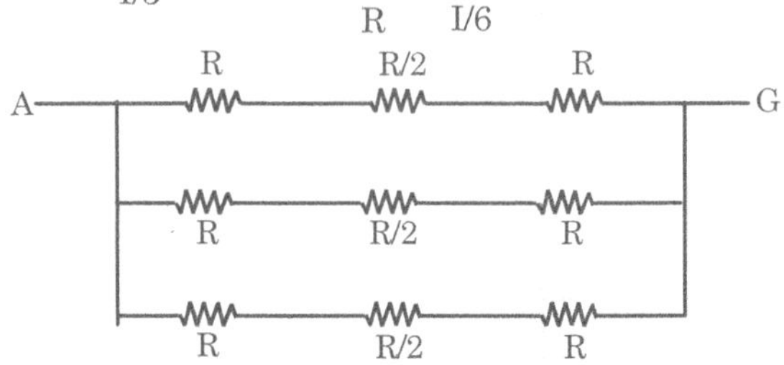

$$\Rightarrow Req = \frac{5R}{6} = \frac{5 \times 1}{6} = \frac{5}{6}\Omega$$

54. Analog output voltage

$$= \frac{V_{ref}}{2^n - 1} \times \text{(decimal value of binary data)}$$

$$= \frac{10}{255} \times 172 = 6.745V$$

55. $f_r = 1$ KHz, Q = 100

$$Q = \frac{\omega L}{R} = \frac{1}{\omega RC} \quad \left| \begin{array}{l} f_r = \frac{1}{2\pi\sqrt{LC}} \\ f_r' = \frac{1}{2\pi\sqrt{2L.2C}} = \frac{f_r}{2} \end{array} \right.$$

$$Q' = \frac{\omega' L'}{R'}$$

$$Q' = \frac{\omega L}{2} = \frac{100}{2} = 50$$

56. Coherent detector rejects quadrature component of noise. Output consists of only in - phase component of noise.

57. In negative resistance region of gunn diode, it acts as both oscillator and amplifier.

58. $n_p.n_e = n_i^2$

$$\Rightarrow n_p = \frac{\left(2.5 \times 10^{16}\right)^2}{5 \times 10^{20}} = 1.25 \times 10^{12} / m^3.$$

59. $[H] \equiv A/m$

$$\overline{\nabla} \times \overline{H} \cong A/m^2$$

60. PIN diode is used in high frequency switching circuits, limiters, modulators etc.

61. In 2's complement representation, MSb represents polarity. If MSb = 0, positive number; MSb = 1, negative number.

2's complement of 11001 = 00110 + 1 = 00111b = 7d

$(11001)_2 \equiv (-7)_{10}$

2' s complement of 1001 = 0110 + 1 = 0111b = 7d

$(1001)_2 = (-7)_{10}$

2's complement of 111001 = 000110 + 1 = 000111b = 7d $(111001)_2 = (-7)_{10}$

62. Balun gives 4 : 1 impedance transformation and is suited for coupling a coaxial line to two wire line.

63. NMOS dominates the LSI field and CMOS is used extensively where least power consumption is necessary.

64. Reflex klystron is one of the most commonly used low power microwave generator.

65. $P_T = P_C\left(1 + m_a^2/2\right) \Rightarrow P_C = P_t/\left(1 + m_a^2/2\right)$

$$\therefore \quad P_C = \frac{10KW}{1 + 0.6^2/2} = 8.47KW$$

66. 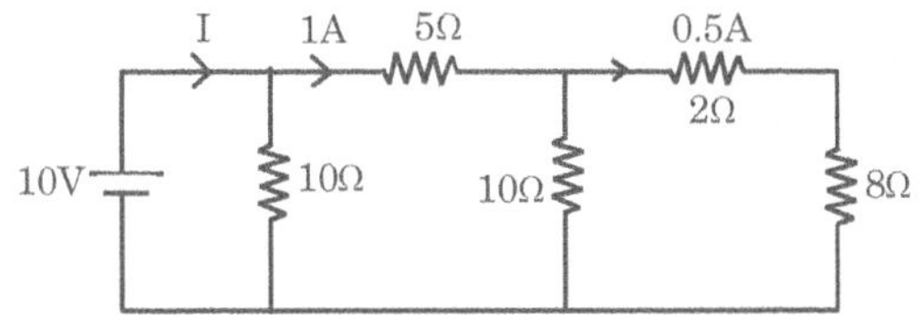

$$I = \frac{10V}{Req} = \frac{10}{5} = 2A$$

Current passing through 8Ω resistor = 0.5 A.

67. JFET is a voltage controlled current device.

69. A cascade amplifier has a high gain, moderately high input impedance, a high output impedance.

70. 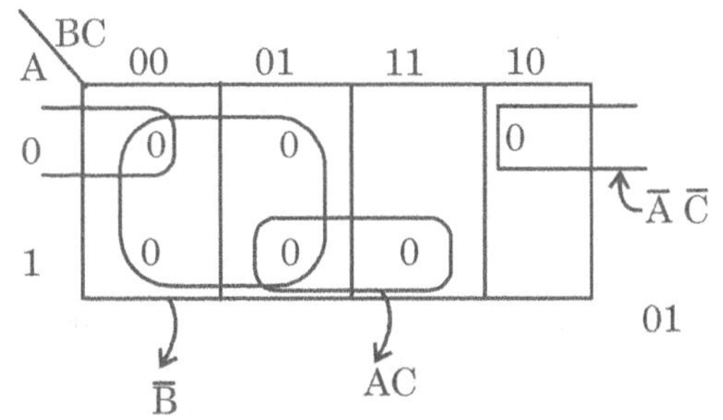

$\overline{f} = \overline{B} + AC + \overline{A}\,\overline{C}$

$\overline{\overline{f}} = \overline{\overline{B} + AC + \overline{A}\,\overline{C}}$

$f = B(\overline{A} + \overline{C})(A + C)$

71. An 8 : 1 MUX is sufficient to provide all functions of 4 variables.

73. Negative feedback in amplifier reduces the effect of disturbances.

74. Spectral density is the average power present in a waveform as a function of frequency.

75. Multiplexer – Data selector

De - Multiplexer – Data distributor

76. For an even signal x(n), it satisfies x(n) = x(– n).

77. y(n) = x(n) * h(n)

↑ ↑ ↑

Output Input convolution

78. In communication system, compander improves the signal to noise ratio.

79. Small variations in cut - off current I_{CBO} is due to temp variations; smaller I_{CB_o} for Si - type as compare to Ge - type transistor.

80. Transmission of signal in optical fiber cable is based on total internal reflection at the core cladding boundary.

81. Complement of AB + BC′ + CD′

= (AB + BC′ + CD′)′

$= (\overline{A} + \overline{B}).(\overline{B} + C).(\overline{C} + D)$

$= \overline{A}\,\overline{B}\,\overline{C} + \overline{A}\,\overline{B}D + \overline{A}C\overline{C} + \overline{A}CD$

$+ \overline{B}C\overline{C} + \overline{B}CD + \overline{B}\,\overline{C} + \overline{B}D$

$= (\overline{A} + 1)\overline{B}\,\overline{C} + (\overline{A} + 1)\overline{B}D + \overline{A}CD + \overline{B}CD$

$= \overline{B}\,\overline{C} + \overline{B}D + \overline{A}CD + \overline{B}CD$

$= \overline{B}\,\overline{C} + \overline{B}D + (1 + C) + \overline{A}CD$

$= \overline{A}CD + \overline{B}\,\overline{C} + \overline{B}D$

82. xy(x′yz + x′y′z′ + xy′z) = x.x′.y.y.z + x.x′.y′y′z′ + xy.y′xz = 0 + 0 + 0 = 0

$\boxed{x.x' = 0}$

83. In PCM, quantization noise $\alpha\, l$ (quantization level).

85. Pulse code modulation is digital modulation which is most immune to noise

86. In crystal microphone, no additional button current, field current or polarizing voltage is required.

87. The large signal bandwidth of op - amp is limited by its slow rate.

88.

S	R	Q		
0	0	Q	←	hold
0	1	0	←	Reset
1	0	1	←	Set
1	1	x	←	Invalid

Truth - table for NOR gate SR FF.

89. 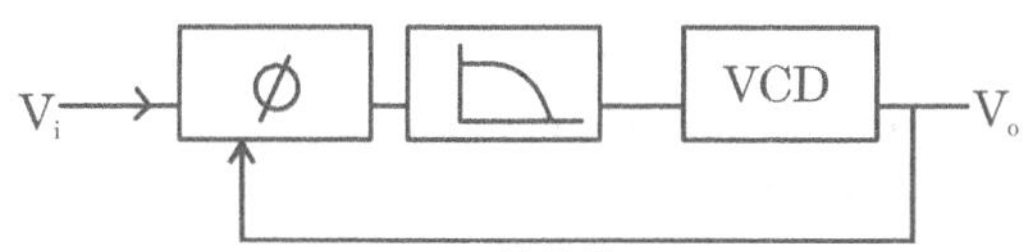

P.L.L

90. LM 339 is a comparator IC with four inbuilt comparators.

92. All given circuits utilise phase - locked - loop circuit

93. Rectification efficiency of full - wave rectifier (without filter)

$$= \frac{P_{dc}}{P_{ac}} = \frac{V_{dc}^2/R_L}{V_{rms}^2/R_L} = \left(\frac{2V_{m/\pi}}{V_{m/\sqrt{2}}}\right)^2 = \frac{8}{\pi^2} = 81.2\%$$

94. Resolution $= \frac{V_r}{2^n} = \frac{V_r}{2^8} = \frac{V_r}{256}$

95. Stable operating point of tunnel diode lies in negative resistance region.

96. Photodiode operates in reverse - biased region.

97. All materials are used in manufacturing of solar cells

98. $y(s) = \frac{s^2+1}{S} = s + \frac{1}{S} = 1s + \frac{1}{1s}$

$\uparrow$ Capacitor $\quad \uparrow$ Inductor

99. Semiconductors are N.T.C. material.

100. Schering bridge – capacitance, dissipation factor, loss tangent.

101.

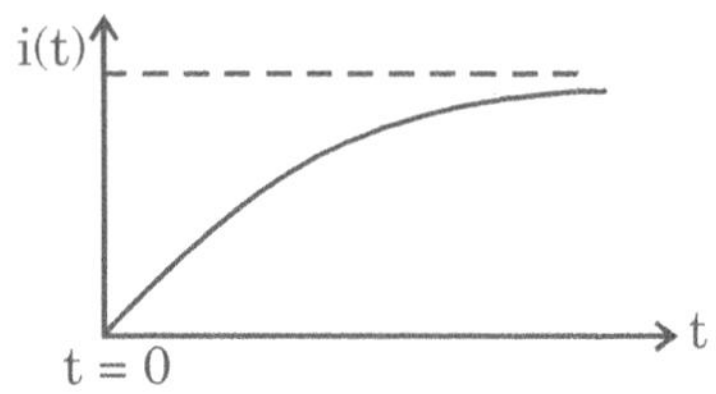

For RL circuit

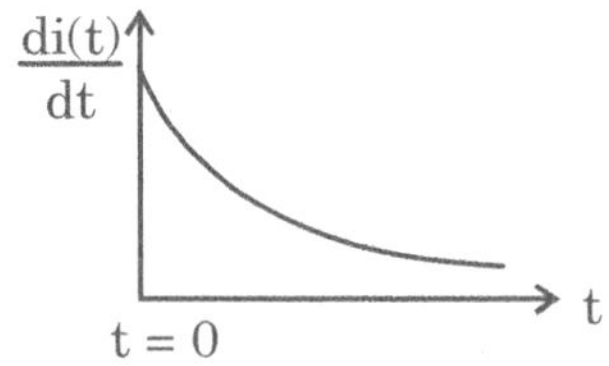

103. Holding current is the minimum forward conduction current.

104.

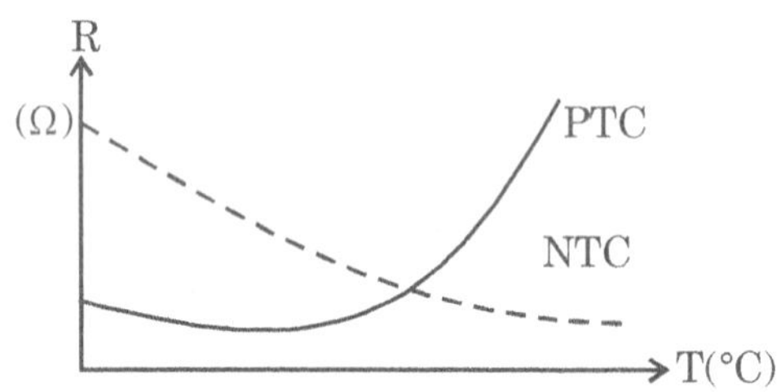

105. One machine cycle consists of 3 to 6 clock cycles or T - states.

106. In memory mapped I/o, devices (I/o) have 16 - bit address lines.

108. Multiplexer : Data selector parallel to serial converter.

109. $M = K\sqrt{L_1 L_2}$

$= 0.45 \times \sqrt{75 \times 105}$ mH = 39.93mH

111. RAL instruction multiplies the number with 2.

112. Energy density spectrum = $G^2(f) = 10^2 = 100$

113. For sinusoidal excitation to begin, the voltage gain of the Wein bridge circuit must be equal or greater than 3, i.e, $A_v \geq 3$

114. For R – 2R DAC, $V_{out} = \frac{V_{ref}}{2^n}$ (decimal equivalent)

$= \frac{5 \times 10}{16} = 3.125$ V.

115. (d)

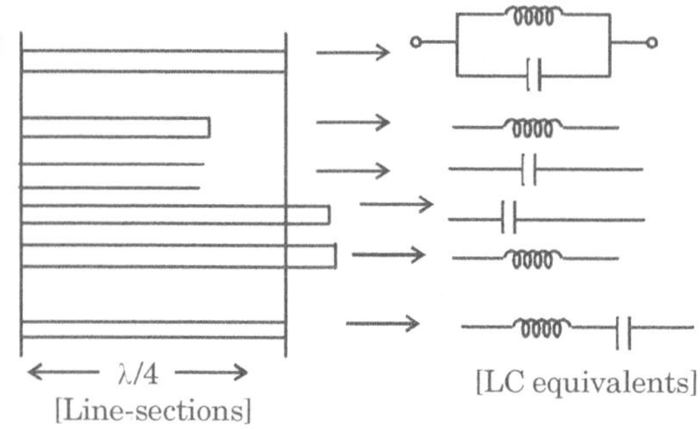

116. Isolation (in dB) = Coupling + Directivity .

117. Dominant mode :

TE_{10} (Rectangular)

TE_{11} (Circular).

118. In a Klystron amplifier, the input is fed at buncher cavity and output is taken at catcher cavity.

119. A pure reactance does not absorb any power. Hence, VSWR is infinite because $|\rho_V| = 1$.

120. Velocity factor is the ratio of the speed at which a wavefront passes through the medium to the speed of light in vaccum. It depends on dielectric medium ($\in_r$) and $\in_r$ depends on humidity.

121. Clearly option (c) is the correct answer as it states exactly the opposite of what is mentioned in the passage. All other options are clearly mentioned in the passage.

122. The word 'smother' as used in the passage means to suffocate or to suppress. Hence (a) is the correct answer.

123. The passage clearly states "Successful transformation is 70-90 percent leadership and only 10-30 management", hence A is the correct answer.

124. The author clearly equates management with bureaucracy. He has equated the two at several places.

125. The passage clearly states "Leadership defines what the future should look like, aligns people with that vision and inspires them to make it happen despite the obstacle". Hence option (a) is correct.

126. The passage clearly states "This distinction is absolutely crucial for our purpose here. Successful transformation is 70-90 percent leadership and only 10-30 management. Therefore, it means that leaders are needed much more than managers.

127. The passage clearly states "So many companies and universities developed management programs and hundreds and thousands people were encouraged to learn management on the job. and they did. But people were taught little about leadership. To some degree, management was emphasized because it is easier to teach them leadership. But even more so management was the main item on the twentieth century agenda because that is what was needed." Hence (d) is the correct answer.

128. All (a), (b) and (c) depict the after effects of inward attention. (d) alone points out the reason for doing the same.

129. The correct answer is 'through' as drive through means to drive past something.

130. Option (b) is correct since it is only in this option that object and subject replace each other perfectly, and the verb remains the same. In all other options, the verb form is changed.

131. Option (d) is the answer as the correct statement should be- "He is much stronger than I am".

132. Option (c) is the answer as the correct statement should be- "I am looking forward to receiving your reply".

133. The correct answer is (b) "by dint of". It is an idiom which means 'due to something'.

134. Statements (a), (c) and (d) are all correct. Though 'lest' should not be followed by any 'auxiliary verb', 'should' is the only one that can follow it. Option (b) is the answer as the correct statement should be- "He behaves as if he is a king."

135. Apart from option (a), all other options are incorrect. (b) is incorrect as it should have been- ".... Important days....." (c) is incorrect as it should have been- "... have been..." Option (d) is constructed in an odd fashion.

136. Though 'lest' should not be followed by any 'auxiliary verb', 'should' is the only one that can follow it. So option (a) is correct.

137. In option (b), 'lovely' is used as an adjective and not as an adverb. Hence it is the correct answer.

138. A is incorrect as the sentence should end at 'friendly' itself. (b) is incorrect as 'his/her' should have been used instead of 'their'. (d) is incorrect as the correct usage is - "Few men are free from fault." Option (c) alone is the correct option.

139. Out of all the options, clearly 'no money on him' is the correct phrase.

140. (a) is incorrect as 'is' should have been used instead of 'are'. B is incorrect as the correct usage is "..for copying answers..." (c) is incorrect as the correct usage is - "....smelt sweet...." Option (d) alone is the correct option.

141. The correct usage is "measles is..." as it is singular and not plural.

142. The answer is (d) 'aircrafts' as the plural of aircraft is aircraft.

143. Sentence (c) is incorrect as the correct usage is 'he has a cell phone'.

144. Option (b) is incorrect as the correct usage is 'from' since it tells what something is made up of.

145. 'To stand in one's own light' is to harm, hinder, or tarnish one's own reputation, opportunities, or advantages, as through foolish or reproachable actions; to be a hindrance or encumbrance to one's own success or ambitions. Hence, (a) is the correct answer.

146. 'Fair and square' means honestly and without cheating.

147. All option other than A, lead to parallelism error. Only through option (a) will the sentence remain grammatically correct. The correct usage is- "I got at five o'clock and studied for two hours. After taking a bath, I took breakfast. Then, I went to school on my bike."

148. It is only in statement 2 that 2 adverbs appear- very and fluently. And the first one modifies the other.

149. GIMMICK refers to stunt. Hence, (a) is the correct answer.

150. Apart from A, 'fast', 'only' and 'high' in B, C A and D respectively are all used as 'adjectives'. 'Better' in A does not describe anything, hence it is used as an adverb and not adjective.

151. Scathing means mocking and sarcastic. Hence, scornful is the correct answer.

152. Clearly option (b) alone uses adverb clause. All other options skip the use of 'quite'.

153. It is evident that apart from option (d), all other options are flawed. In A, it should be 'seeing a film' and not 'to see a film'. B incorrectly uses 'reproduce'. In C, 'a' should have been used instead of 'an'.

154. The correct plural form should be daughters in law, hence (d) is the answer.

155. The correct answer (a) as it spells Paraphernalia correctly.

156. Winsome means something that is charming and pleasant, hence option (a) is correct.

157. Perchance means by chance or perhaps, hence fluke is the correct option.

158. Scrimp is to economize, and hence lavish is its clear opposite. Parsimonious means unwilling to spend money that is stingy or frugal.

159. Delectable is something that is good to taste, hence distasteful is the correct antonym.

160. Boisterous means lively and active, therefore, serene, which means calm is its appropriate antonym.

161. To eat crow means to admit that one was wrong and hence to feel small. To swallow one's own word is an idiom which means to refract from your own words.

162. To feather one's nest means to enrich oneself by using one's influence or position or with other people's money, hence option (a) is correct.

163. A group of flies is called a swarm.

164. Engrossed is to be deeply enthralled and captivated, hence (b) is correct.

165. 'Deluged with' is the correct phrase as it means to be flooded with something. Other options don't make correct phrases.

166. Option (a) is correct since it is only in this option that object and subject replace each other perfectly, and the verb remains the same. In all other options, the verb form is changed.

167. In indirect speech, the sentence should be in the past tense. Hence both 1 and 2 are ruled out as they use 'must' which indicates a future action. 'At once' should remain the same and hence (d) is correct.

168. In indirect speech, the sentence should be in the past tense. Therefore (b) is correct and (c) is ruled out. (a) is faulty and ignores the verbs. She was not 'narrating' and hence (d) is incorrect.

169. The correct usage is 'why did you not throw'? With 'did', the first form of the verb should come and not the past form.

170. In the passive voice, the subject receives the action of the verb. Here, the object of the active verb becomes the subject of the passive verb. This is what happens in option (d), hence it is the answer.

171. The correct saying is 'god helps those who help themselves'.

172. A complex sentence contains a dependent clause and independent clause. Sentence C alone contains both of these. Hence, it is the correct answer.

173. The sentence uses 'will have'. 'Will' indicates future + 'Have' indicates perfect. Therefore, future perfect tense is the correct answer.

174. Clearly option A alone talks about how a thing is made, hence talks about manner.

175. The correct answer is C- subordinating conjunction as it is a conjunction that introduces a subordinating clause, here 'than'.

176. A pupil who stays away from school without leave or explanation is called a truant.

177. A person who does not speak much is called reticent.

178. Trivial is something that is of little value.

179. The correct spelling is 'fiasco', hence (a) is the correct answer.

180. The correct spelling is 'courageous', hence (c) is the correct answer.

LMRC SOLVED PAPER – 2015

ELECTRONICS ENGINEERING

Read the following instructions carefully before you begin to answer the questions. This Booklet contains questions in English.

INSTRUCTIONS

1. This Booklet contains 100 questions in total comprising of the following four parts:

Part – I:	Domain Knowledge	[1 – 75 Questions]
Part – II:	Quantitative Aptitude	[76 – 85 Questions]
Part – III:	Computer	[86 – 95 Questions]
Part – IV:	General Knowledge	[96 – 100 Questions]

2. All questions are compulsory and carry 1 mark each.

3. Before you start to answer the questions you must check this Booklet to ensure that it contains all the pages and see that no page is missing or repeated.

4. For every wrong answer 1/3 marks will be deducted.

5. You are supplied with the OMR-Sheet separately.

6. Answers must be indicated by completely blackening one of the corresponding ovals on OMR-Sheet against the relevant question number, by black/blue ball pen only. Answers which are not indicated by black/blue ball pen will not be awarded any mark.

7. The OMR-Sheet must be handed over to the Invigilator before you leave the Examination Hall.

8. Failure to comply with any of the above instructions will render a candidate liable to such action/penalty as may be deemed fit.

LMRC SOLVED PAPER-2015

DOMAIN KNOWLEDGE

1. Which of the following is a current statement about P type semiconductors ?
(a) They are negatively charged
(b) They are positively charged
(c) They are neutral
(d) Some are positive whereas some are negative.

2. Forbidden energy gap in silicon is:
(a) 1.1 eV (b) 2.4 eV
(c) 0.72 eV (d) 0.97 eV

3. Purest form of semiconductors are called:
(a) Intrinsic (b) Extrinsic
(c) Diodes (d) Transistor

4. N-type semiconductors are formed by adding.
(a) divalent impurities to pure semiconductor
(b) trivalent impurities to pure semiconductor
(c) tetravalent impurities to pure semiconductor
(d) pentavalent impurities to pure semiconductor

5. Zener diodes are used as:
(a) light emitter (b) light absorber
(c) voltage regulator (d) microwave detector

6. What does the diagram given below depict?

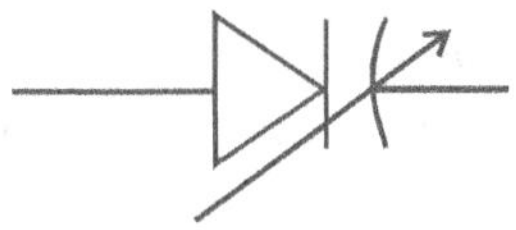

(a) LED (b) Zener diode
(c) Tunnel diode (d) Varactor diode

7. The energy of a photon of light whose wavelength is 620 nm equals the band gap of a semiconducting material. What is the minimum energy required to create a hole electron pair?
(a) 2 eV (b) 0.72 eV
(c) 0.125 eV (d) 0.5 eV

8. If a small amount of phosphorus is added to germanium, then
(a) the conductivity decreases
(b) silicon becomes a P-type semiconductor
(c) phosphorus becomes an acceptor impurity
(d) there will be more free electrons than holes in semiconductor

9. An intrinsic semiconductor at 0 K behaves as:
(a) a semiconductor (b) an insulator
(c) a conductor (d) a superconductor

10. Merging of a free electron and a hole is called:
(a) Recombination (b) Neutralization
(c) Restriking (d) Zeroing

11. A passive network contains
(a) 1 source of emf (b) 2 sources of emf
(c) 3 sources of emf (d) no source of emf

12. Barrier potential of a p-n junction diode does not depend upon
(a) temperature (b) foward bias
(c) doping density (d) diode design

13. Superposition Theorem is based on the concept of:
(a) Reciprocity (b) Duality
(c) Non-linearity (d) Linearity

14. Kirchoff's law is applicable to
(a) Passive networks only
(b) Active networks only
(c) D.C. only
(d) Both A.C. and D.C.

15. Higher the Q of a series circuit
(a) broader its resonance curve
(b) narrower its pass band
(c) greater its band width
(d) sharper its resonance

16. The peak value of a sine wave is 100 V. Its average value is
(a) 100 V (b) 50.5 V
(c) 63.6 V (d) 74.7 V

17. A resonance curve for a series circuit is a plot of frequency versus
(a) Current (b) Voltage
(c) Impedance (d) Reactance

18. Power factor of an electric bulb is
(a) Lagging (b) Leading
(c) Infinity (d) Unity

19. Which of the following is a non linear circuit parameter?
(a) Inductance
(b) Condenser
(c) Wire wound resistor
(d) Transistor

20. The circuit whose properties are the same in either directions is known as
(a) Unilateral circuit (b) Bilateral circuit
(c) Reversible circuit (d) Irreversible circuit

21. The purpose of parallel circuit resonance is to magnify.
(a) Current (b) Voltage
(c) Power (d) Frequency

22. The shunt combination in band pass filter is
(a) Capacitive
(b) Inductive
(c) Parallel combination of L and C
(d) Series combination of L and C

23. The dual of a link is
(a) Twig (b) Node
(c) Loop (d) Tree branch

24. At half power frequency, power drawn in the circuit is:
(a) I^2R (b) $\frac{1}{2}I^2R$
(c) $\frac{1}{2}IR^2$ (d) IR^2

25. According to star connected system the line voltage in a 3-phase circuit is
(a) 1.732 times the phase voltage
(b) 1.414 times the phase voltage
(c) 2.05 times the phase voltage
(d) equal to the phase voltage

26. The sum of two wattmeter readings in a 3-phase circuit gives:
(a) total resistance in the circuit
(b) total voltage drop in the circuit
(c) total power in the circuit
(d) total current in the circuit

27. In a series R-L-C circuit, impedance at resonance is
(a) Unity (b) Maximum
(c) Infinity (d) Minimum

28. Which of the following statements is correct about resonance in parallel circuits?
(a) Admittance of circuit is maximum.
(b) Current drawn is maximum
(c) Phase angle between current and voltage is a right angle.
(d) Power factor is unity

29. In a parallel resonant circuit, effective impedance is given by:
(a) R (b) $\frac{L}{C}$
(c) $\frac{L}{CR}$ (d) $\frac{R}{LC}$

30. A coil with large distributed capacitance has a
(a) High resonant frequency
(b) low resonant frequency
(c) low quality factor
(d) low resistance

31. Which of the following does not measure self inductance?
(a) Maxwell's bridge
(b) Kelvin's double bridge
(c) Hay's bridge
(d) Owen's bridge

32. Attractive type and repulsive type are two types of
(a) Rectifiers
(b) Moving coil instruments
(c) Moving iron instruments
(d) Energy meters

33. Method of control in a thermocouple is
(a) Hair spring (b) Moving coil
(c) Eddy currents (d) Air

34. Which of the following bridges is used to measure high frequency oscillations?
(a) Maxwell's bridge (b) Hay's bridge
(c) wein's bridge (d) Anderson's bridge

35. A measure of reproducibility of measurements is called
(a) Resolution (b) Reproduction
(c) Accuracy (d) Precision

36. Dynamometer is likely to have
(a) Air damping
(b) Eddy current damping
(c) Fluid friction damping
(d) Moving coil damping

37. A piezoelectric pick-up is a transducer of the type
(a) Variable resistance
(b) Variable inductance
(c) Voltage generating
(d) Voltage divider

38. Which of the following does not result in conditioning of signal?
(a) Signal converter
(b) Amplifier
(c) Equalizing network
(d) Damping network

39. Which of the following represents Bellows?

(a) 

(b) 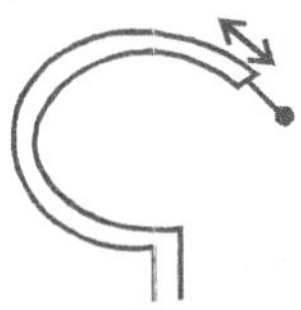

(c) 

(d) 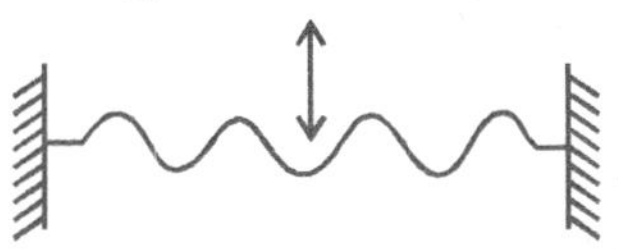

40. Temperature coefficient of a thermistor is
(a) low and positive (b) high and positive
(c) low and negative (d) high and negative

41. Which of the following is a digital transducer?
(a) Encoder
(b) Photovoltaic
(c) Thermocouple
(d) Piezoelectric transducer

42. LVDTs give output in the form of
(a) rotary movement of core
(b) pulses
(c) high frequency signals
(d) linear displacement of core

43. A multimeter consists of
(a) Voltmeter and current meter
(b) Voltmeter and ohm meter
(c) current meter and ohm meter
(d) voltmeter, current meter and ohm meter

44. Transducers are not known as:
(a) Gauges (b) Transistors
(c) Pickups (d) Signal generators

45. In self generating type transducers, voltage is generated because of
(a) relative motion between conductor and magnetic field
(b) change in electric field
(c) motion of charges
(d) residual magnetism

46. Transducers that make use of a variable core are referred to as
(a) Reluctance type (b) Permeance type
(c) Resistance type (d) Inductance type

47. In LVDT when the core is in the centre, the induced voltages are:
(a) equal and in the same direction
(b) unequal but in the same direction
(c) equal and opposite
(d) unequal and opposite

48. What does the following figure represent?

(a) Capacitive transducer
(b) Stream gauge
(c) Piezoelectic transducer
(d) LVDT

49. Semiconductor strain gauges are used where
(a) high gauge factor is required
(b) low gauge factor is required
(c) large envelopes are required
(d) gauge factor is not required

50. When measurement provides impulse for remote automatic action, then the result is called:
(a) Process electronics
(b) Control system
(c) Sensing
(d) Conversion

51. For converting a number in decimal number system to its binary equivalent, remainders are recorded after successive division of the number by
(a) 2 (b) 10
(c) 8 (d) 16

52. The decimal equivalent of $(404011)_2$ is:
(a) 48 (b) 54
(c) 44 (d) 43

53. "A + (B + C) = (A + B)+ C" is
(a) Associative law (b) Distributive law
(c) Commutative law (d) Absorption law

54. On multiplying $(4011)_2$ by $(1401)_2$, we get:

(a) 40001400 (b) 40001111
(c) 40001140 (d) 40004011

55. Which of the following refers to NOR gate?

(a) 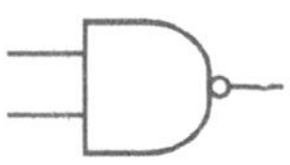 (b) 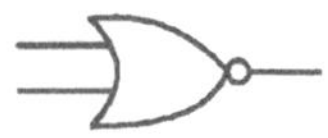
(c)  (d) 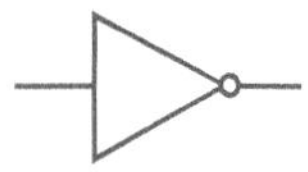

56. Consider the figure below,

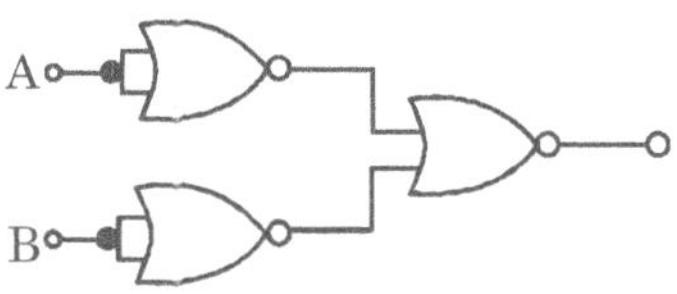

The figure is an electric equivalent of :

(a) NAND gate (b) XOR gate
(c) AND gate (d) OR gate

57. What is the output of the figure given below?

(a) A.B (b) A + B
(c) (A + B).B (d) (A.B).A

58. The following picture is a logical diagram of:

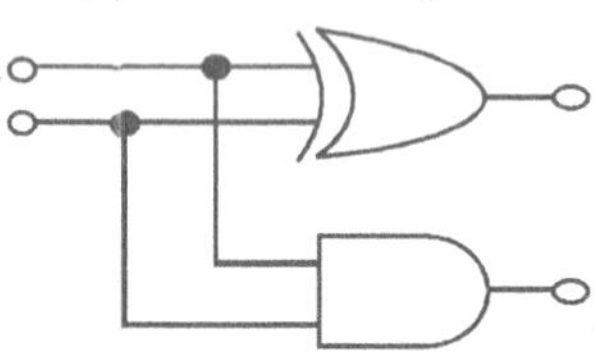

(a) Full adder (b) Full subtractor
(c) Half adder (d) Half subtractor

59. According to algebra logic (A + A') generates:

(a) O (b) 1
(c) A (d) A + A'

60. In general, which of the following flip flops are used as shift registers?

(a) D (b) JK
(c) SR (d) T

61. A full adder consists of

(a) two inputs and one output
(b) one input and two ouputs
(c) three inputs and two outputs
(d) two inputs and three outputs

62. The flip flop circuit is

(a) unstable (b) multistable
(c) monostable (d) bistable

63. Octal coding involves grouping of bits in

(a) 3's (b) 4's
(c) 5's (d) 6's

64. Which of the following is the fastest unsaturated logic gate?

(a) RTL (b) HTL
(c) TTL d) ECL

65. The range of output voltage when circuit changes from 1 state to another is.

(a) 0.1-0.5 ns (b) 0.5-50 ns
(c) 50-400 ns (d) 0.5-5.0 ns

66. As compared to parallel adder, serial adder:

(a) is less faster
(b) needs more components
(c) uses registers with parallel load
(d) is a combinational circuit

67. Look ahead carry generators have

(a) low propagation delay time
(b) high propagation delay time
(c) no additional circuitry
(d) speed dependent on number of bits

68. A decoder is a:

(a) synchronous circuit
(b) combinational circuit
(c) sequential circuit
(d) digital circuit

69. A multiplexer is a logic circuit that gets

(a) single input to several outputs
(b) single input to single output
(c) several inputs to single output
(d) several inputs to no output

70. Race condition always arise in

(a) digital circuits
(b) synchronous circuits
(c) asynchronous circuits
(d) combinational circuits

71. Which of the following instruments present signal wave forms visually?

(a) CRO (b) Wattmeter
(c) Thermocouple (d) LVDT

72. direct measurement of pressure can be done using

(a) Rota meter (b) Borden tube
(c) LVDT (d) Strain gauge

73. At the resonant frequency in LC circuit,

(a) current is maximum
(b) current is minimum
(c) impedance is maximum
(d) voltage across C is minimum

74. Form factor is the ratio of
(a) average value to rms value
(b) rms value to average value
(c) average value to peak value
(d) rms value to peak value

75. Free electrons exist in
(a) Valence band (b) Conduction band
(c) Free band (d) Excited band

QUANTITATIVE APTITUDE

76. Find out the word-pair in which the two words hold the same relationship as the two words in the word-pair given below.
Orange : Apple
(a) House : Family (b) Pen : Ink
(c) Tree : Flower (d) Table : Chair

77. The first two figures on the left side of the sign '::' are related in a certain way. The same relationship holds for the second pair of figures on the right side of the sign '::' of which one is missing. Which of the following is the missing one?

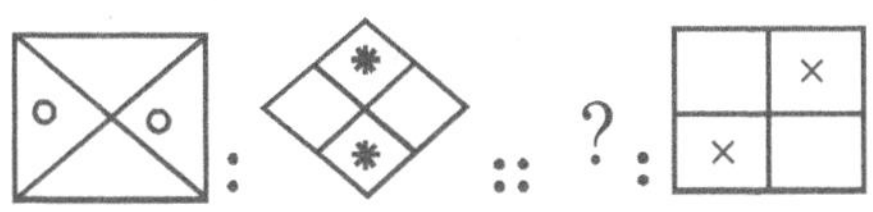

(a) 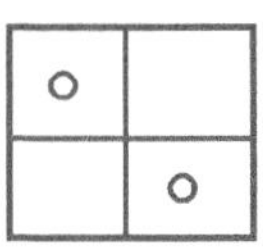 (b) 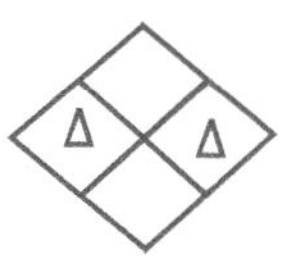
(c) 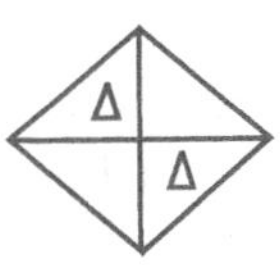 (d) 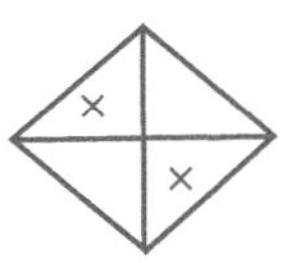

78. Which number will replace the question mark in the number series below?
3, 4, 13, 38, ?, 168
(a) 63 (b) 87
(c) 77 (d) 102

79. If 12% of 40% of X = 84, then what is the value of X?
(a) 1600 (b) 1680
(c) 1750 (d) 1860

80. P is the brother of Q and son of R. V is the sister of T, who is the daughter of Q. If R is the husband of S, then which of the following is true?
(a) V is the granddaughter of S
(b) R is the grandfather of Q
(c) S is the mother of T
(d) V is the nephew of P

81. Four months are given below out of which three are alike in some manner and one is different. Which of the given months is different from the other three?
(a) October (b) January
(c) June (d) December

82. Four figures are given below out of which three are similar in certain way and one is different. Which is of the given figures is different from the other three?

(a)  (b) 
(c)  (d) 

83. If '+' means '÷', '÷' means '–', '–' means '×' and '×' means '+', then
$(12 + 4 \times 2) \times 6 - 4 \div 2 = ?$
(a) 27 (b) 24
(c) 17 (d) 14

84. If 'DAUGHTER' is coded as 'SFUIHVBE' in a certain language, then how would 'PROBLEM' be coded as in the same language?
(a) LDKANQO (b) NFMCPSQ
(c) QSPCMIN (d) NFMCSPQ

85. Each of the following two rows contains 3 numbers. Some rules are given below to be used for getting the resultant for each row separately. Apply the rules for each row from left to right and answer the question.
Rules :
(i) If an even number is followed by an odd but not prime number, both are to be multiplied.
(ii) If an even number is followed by a prime number, both are to be added.
(iii) If an odd number is followed by a composite odd number, the first number is to be subtracted from the square of the second number.
(iv) If an odd number is followed by an even number, the odd one is to be subtracted from the even number.
Row I: 16, 9, 23
Row II: 13, 38, 25
What is the difference between the resultants of the first row and the second row?
(a) 193 (b) 26
(3) 767 (d) 433

COMPUTER

86. The operating system for mobile phones developed by Google is:

(a) Amoeba (b) Magellan

(c) Android (d) Ubuntu

87. The CPU of a computer consists of

(a) ALU control unit and registers

(b) ALU and control unit

(c) ALU, control unit and hard disk

(d) ALU, control unit and monitor

88. Which of the following is both an input and an output device?

(a) Touch screen (b) LCD projector panel

(c) Audio cards (d) Modem

89. A program in execution is called:

(a) Process (b) Instruction

(c) Procedure (d) Function

90. Criterion used for judging appropriateness of tool software is:

(a) Scalability (b) Compatibility

(c) Functionality (d) Security

91. Which one of the following decimal number is equivalent to $(10000010)_2$?

(a) 130_{10} (b) 200_{10}

(c) 1010_{10} (d) 304_{10}

92. In a spreadsheet, one function inside another is called :

(a) Text (b) Nested

(c) Sum (d) Round

93. Which of the following is not a part of the office Suite?

(a) Database (b) Image editor

(c) File manager (d) Word processor

94. Which of the following is not a Search Engine?

(a) Yahoo (b) Alta Vista

(c) Google (d) Facebook

95. A shared network within an organisation to provide connectivity to the staff is called.

(a) Internet (b) Intranet

(c) Delnet (d) Extranet

GENERAL KNOWLEDGE

96. Which Mughal ruler was on the throne of India when East India company was permitted to establish a factory at Surat?

(a) Aurangzeb (b) Humayun

(c) Jehangir (d) Shahjahan

97. Who advises the centre on legal matters?

(a) Advocate General of India

(b) President of India

(c) Attorney General of India

(d) chief Justice of India

98. Which of the following payment instruments is known as plastic money?

(a) Bearer Cheques (b) Gift Cheques

(c) Demand Drafts (d) Credit Cards

99. EEG is taken out for diagnosing ailments of which of the following?

(a) Brain (b) Heart

(c) Lungs (d) Stomach

100. Prime Minister Narender Modi has recently launched the "Give it Up" campaign for voluntarily giving up which of the following?

(a) Drugs (b) Plastics

(c) Ration (d) LPG Subsidy

OMR SHEET

1. Ⓐ Ⓑ Ⓒ Ⓓ	26. Ⓐ Ⓑ Ⓒ Ⓓ	51. Ⓐ Ⓑ Ⓒ Ⓓ	76. Ⓐ Ⓑ Ⓒ Ⓓ
2. Ⓐ Ⓑ Ⓒ Ⓓ	27. Ⓐ Ⓑ Ⓒ Ⓓ	52. Ⓐ Ⓑ Ⓒ Ⓓ	77. Ⓐ Ⓑ Ⓒ Ⓓ
3. Ⓐ Ⓑ Ⓒ Ⓓ	28. Ⓐ Ⓑ Ⓒ Ⓓ	53. Ⓐ Ⓑ Ⓒ Ⓓ	78. Ⓐ Ⓑ Ⓒ Ⓓ
4. Ⓐ Ⓑ Ⓒ Ⓓ	29. Ⓐ Ⓑ Ⓒ Ⓓ	54. Ⓐ Ⓑ Ⓒ Ⓓ	79. Ⓐ Ⓑ Ⓒ Ⓓ
5. Ⓐ Ⓑ Ⓒ Ⓓ	30. Ⓐ Ⓑ Ⓒ Ⓓ	55. Ⓐ Ⓑ Ⓒ Ⓓ	80. Ⓐ Ⓑ Ⓒ Ⓓ
6. Ⓐ Ⓑ Ⓒ Ⓓ	31. Ⓐ Ⓑ Ⓒ Ⓓ	56. Ⓐ Ⓑ Ⓒ Ⓓ	81. Ⓐ Ⓑ Ⓒ Ⓓ
7. Ⓐ Ⓑ Ⓒ Ⓓ	32. Ⓐ Ⓑ Ⓒ Ⓓ	57. Ⓐ Ⓑ Ⓒ Ⓓ	82. Ⓐ Ⓑ Ⓒ Ⓓ
8. Ⓐ Ⓑ Ⓒ Ⓓ	33. Ⓐ Ⓑ Ⓒ Ⓓ	58. Ⓐ Ⓑ Ⓒ Ⓓ	83. Ⓐ Ⓑ Ⓒ Ⓓ
9. Ⓐ Ⓑ Ⓒ Ⓓ	34. Ⓐ Ⓑ Ⓒ Ⓓ	59. Ⓐ Ⓑ Ⓒ Ⓓ	84. Ⓐ Ⓑ Ⓒ Ⓓ
10. Ⓐ Ⓑ Ⓒ Ⓓ	35. Ⓐ Ⓑ Ⓒ Ⓓ	60. Ⓐ Ⓑ Ⓒ Ⓓ	85. Ⓐ Ⓑ Ⓒ Ⓓ
11. Ⓐ Ⓑ Ⓒ Ⓓ	36. Ⓐ Ⓑ Ⓒ Ⓓ	61. Ⓐ Ⓑ Ⓒ Ⓓ	86. Ⓐ Ⓑ Ⓒ Ⓓ
12. Ⓐ Ⓑ Ⓒ Ⓓ	37. Ⓐ Ⓑ Ⓒ Ⓓ	62. Ⓐ Ⓑ Ⓒ Ⓓ	87. Ⓐ Ⓑ Ⓒ Ⓓ
13. Ⓐ Ⓑ Ⓒ Ⓓ	38. Ⓐ Ⓑ Ⓒ Ⓓ	63. Ⓐ Ⓑ Ⓒ Ⓓ	88. Ⓐ Ⓑ Ⓒ Ⓓ
14. Ⓐ Ⓑ Ⓒ Ⓓ	39. Ⓐ Ⓑ Ⓒ Ⓓ	64. Ⓐ Ⓑ Ⓒ Ⓓ	89. Ⓐ Ⓑ Ⓒ Ⓓ
15. Ⓐ Ⓑ Ⓒ Ⓓ	40. Ⓐ Ⓑ Ⓒ Ⓓ	65. Ⓐ Ⓑ Ⓒ Ⓓ	90. Ⓐ Ⓑ Ⓒ Ⓓ
16. Ⓐ Ⓑ Ⓒ Ⓓ	41. Ⓐ Ⓑ Ⓒ Ⓓ	66. Ⓐ Ⓑ Ⓒ Ⓓ	91. Ⓐ Ⓑ Ⓒ Ⓓ
17. Ⓐ Ⓑ Ⓒ Ⓓ	42. Ⓐ Ⓑ Ⓒ Ⓓ	67. Ⓐ Ⓑ Ⓒ Ⓓ	92. Ⓐ Ⓑ Ⓒ Ⓓ
18. Ⓐ Ⓑ Ⓒ Ⓓ	43. Ⓐ Ⓑ Ⓒ Ⓓ	68. Ⓐ Ⓑ Ⓒ Ⓓ	93. Ⓐ Ⓑ Ⓒ Ⓓ
19. Ⓐ Ⓑ Ⓒ Ⓓ	44. Ⓐ Ⓑ Ⓒ Ⓓ	69. Ⓐ Ⓑ Ⓒ Ⓓ	94. Ⓐ Ⓑ Ⓒ Ⓓ
20. Ⓐ Ⓑ Ⓒ Ⓓ	45. Ⓐ Ⓑ Ⓒ Ⓓ	70. Ⓐ Ⓑ Ⓒ Ⓓ	95. Ⓐ Ⓑ Ⓒ Ⓓ
21. Ⓐ Ⓑ Ⓒ Ⓓ	46. Ⓐ Ⓑ Ⓒ Ⓓ	71. Ⓐ Ⓑ Ⓒ Ⓓ	96. Ⓐ Ⓑ Ⓒ Ⓓ
22. Ⓐ Ⓑ Ⓒ Ⓓ	47. Ⓐ Ⓑ Ⓒ Ⓓ	72. Ⓐ Ⓑ Ⓒ Ⓓ	97. Ⓐ Ⓑ Ⓒ Ⓓ
23. Ⓐ Ⓑ Ⓒ Ⓓ	48. Ⓐ Ⓑ Ⓒ Ⓓ	73. Ⓐ Ⓑ Ⓒ Ⓓ	98. Ⓐ Ⓑ Ⓒ Ⓓ
24. Ⓐ Ⓑ Ⓒ Ⓓ	49. Ⓐ Ⓑ Ⓒ Ⓓ	74. Ⓐ Ⓑ Ⓒ Ⓓ	99. Ⓐ Ⓑ Ⓒ Ⓓ
25. Ⓐ Ⓑ Ⓒ Ⓓ	50. Ⓐ Ⓑ Ⓒ Ⓓ	75. Ⓐ Ⓑ Ⓒ Ⓓ	100. Ⓐ Ⓑ Ⓒ Ⓓ

ANSWERS

1. (*c*)	**2.** (*a*)	**3.** (*a*)	**4.** (*d*)	**5.** (*c*)	**6.** (*d*)	**7.** (*a*)	**8.** (*d*)	**9.** (*b*)	**10.** (*a*)
11. (*d*)	**12.** (*d*)	**13.** (*d*)	**14.** (*d*)	**15.** (*b*)	**16.** (*c*)	**17.** (*a*)	**18.** (*d*)	**19.** (*a*)	**20.** (*b*)
21. (*b*)	**22.** (*c*)	**23.** (*d*)	**24.** (*b*)	**25.** (*a*)	**26.** (*c*)	**27.** (*d*)	**28.** (*d*)	**29.** (*a*)	**30.** (*a*)
31. (*b*)	**32.** (*c*)	**33.** (*b*)	**34.** (*c*)	**35.** (*d*)	**36.** (*a*)	**37.** (*c*)	**38.** (*b*)	**39.** (*c*)	**40.** (*d*)
41. (*a*)	**42.** (*d*)	**43.** (*d*)	**44.** (*b*)	**45.** (*a*)	**46.** (*b*)	**47.** (*c*)	**48.** (*c*)	**49.** (*a*)	**50.** (*b*)
51. (*a*)	**52.** (*d*)	**53.** (*a*)	**54.** (*b*)	**55.** (*b*)	**56.** (*c*)	**57.** (*b*)	**58.** (*c*)	**59.** (*b*)	**60.** (*a*)
61. (*c*)	**62.** (*a*)	**63.** (*a*)	**64.** (*d*)	**65.** (*b*)	**66.** (*a*)	**67.** (*a*)	**68.** (*c*)	**69.** (*c*)	**70.** (*c*)
71. (*a*)	**72.** (*b*)	**73.** (*a*)	**74.** (*b*)	**75.** (*b*)	**76.** (*d*)	**77.** (*c*)	**78.** (*b*)	**79.** (*c*)	**80.** (*a*)
81. (*c*)	**82.** (*b*)	**83.** (*a*)	**84.** (*b*)	**85.** (*d*)	**86.** (*c*)	**87.** (*a*)	**88.** (*d*)	**89.** (*a*)	**90.** (*c*)
91. (*a*)	**92.** (*b*)	**93.** (*c*)	**94.** (*d*)	**95.** (*b*)	**96.** (*c*)	**97.** (*c*)	**98.** (*d*)	**99.** (*a*)	**100.** (*d*)

EXPLANATIONS

1. P-type or N-type semiconductors are electrically neutral since ideally number of positively charged protons in the nuclei is still equal to the number of free and orbiting negatively charged electrons in the structure.

2. Forbidden energy gap,

$$E_g = \begin{cases} 0.67 \text{ eV (Ge)} \\ 1.1 \text{ eV (Si)} \\ 1.43 \text{ eV (GaAs)} \end{cases}$$

3. Undoped or purest form of semiconductors are called intrinsic semiconductors.

4. An n-type matrial is created by introducing impurity elements that have five valence electrons (pentavalent) like, Sb, As, P etc.

5. Zener diode (in reverse biased region can be used as voltage regulator.

6.

Varactor diode

7. $\lambda = 620\text{nm}$.

$$E_g = \frac{hc}{\lambda} = \frac{6.6 \times 10^{-34} \times 3 \times 10^8}{620 \times 10^{-9}} \text{J}$$

$$= \frac{6.6 \times 3 \times 10^{-17}}{620 \times 10^{-19} \times 1.6} \text{eV}$$

Eg = 2 eV.

8. Phosphorus is an pentavalent impurity and when added, it produces more free electrons than holes.

9. At absolute zero temperature (zero), intrinsic semiconductor behaves as an insulator.

10. The process of merging of a free electron and a hole in a semiconductor is called recombination.

11. In passive networks, there is no source of energy (or emf)

12. Barrier potential of a p-n junction diode does not depend upon diode design.

13. Superposition theorem is based on the concept of linearity homogeneity and additivity

14. Kirchoff's law are applicable to any lumped AC and DC networks.

15.

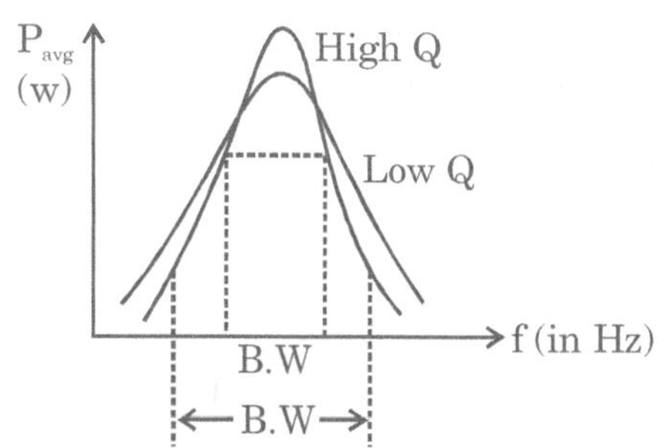

16. Average value of sine wave $= \frac{2V_{peak}}{\pi}$

$$= \frac{2 \times 100}{\pi} = 63.66 \text{ V}.$$

17. At resonance in series RLC circuit,

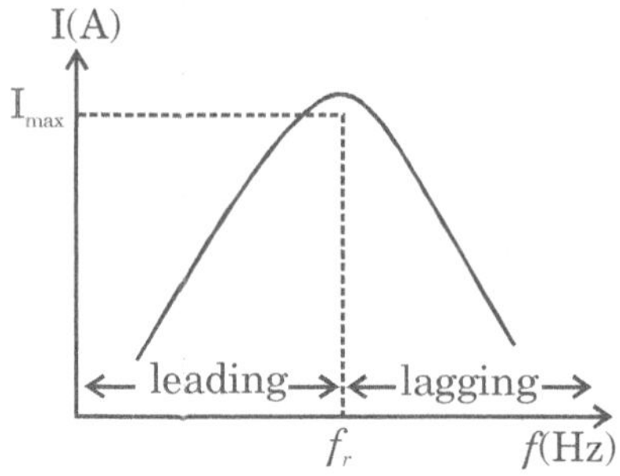

18. Electric bulb has resistive filament material, hence, its power factor is unity.

19. Condenser, wire-wound resistor and transistor are elements not circuit parameters. Inductance is the property or circuit parameter of a coil, which can be non-linear when it has some energy stored in it.

20. Bilateral circuit properties do not change with change in direction of input voltage or current.

21. In parallel RLC circuit at resonance, $V_{out} = Q.V_{in}$ where Q can be made higher than 1.

22.

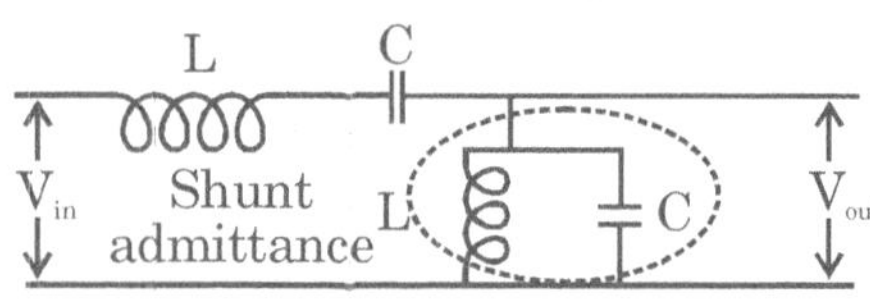

24. At half-power frequency, power drawn in circuit is half i.e., $\frac{1}{2}I^2R$ or 3 dB.

25. In star connected system, $\tilde{V}_L = \sqrt{3}\tilde{V}_{ph}\angle +30°$

$= 1.732\ \tilde{V}_{ph}\angle +30°$

26. For balanced 3 - ϕ circuit, sum of reading of two wattmeter (in 2-wattmeter method) = power consumed in circuit.

27. At resonance, for any ac circuit, its impedance is minimum, ie., $\tilde{Z} = R$.

28. At resonance, for any ac circuit, power factor is unity.

29. In a parallel resonant circuit, effective impedance, $\tilde{Z} = R$.

30. Distributed capacitance, $C_{eff} = \dfrac{C}{1-\left(\dfrac{f}{f_r}\right)^2}$

$$f_r = \frac{f}{\sqrt{1-\left(\dfrac{C}{C_{eff}}\right)}}$$

31. Kelvin's double bridge is used to measure low resistance.

32. Moving iron instruments are categorized as attractive type and repulsive type instruments on the basis of their working.

33. Method of control in thermocouple is moving coil type.

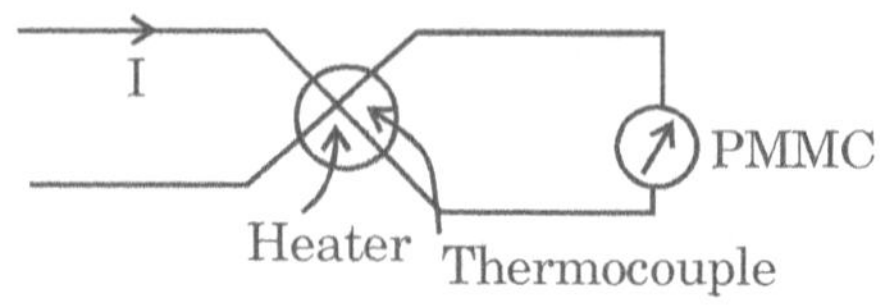

34. Wein's bridge is used to measure high frequency oscillations.

35. A measure of reproducibility of measurements is called precision.

36. In dynamometer type instruments, air-damping is employed.

37. A piezo-electric pick-up is a transducer which produces electrical voltage by applying time varying force.

40. Temperature coefficient of thermistor is high and negative.

41. Encoder is a digital transducer.

42. LVDT gives electrical output in the form of displacement of core.

43. A multimeter can measure voltage (voltmeter), current (ammeter) and resistance (ohm-meter).

44. Transducers are also known as pick-up or signal generators. e.g. strain gauges.

45. Self generating type transducer works on Faraday's law of electromagnetic inductin.

46. Variable air-gap-reluctance-type transducer.

Variable core type - permeance type transducer

47. In LVDT, when core is in the centre, the induced voltages are equal and opposite.

48. It is a piezo-electric transducer.

49. Semiconductor strain gauges are used when high gauge factor is required.

52. $(101011)_2 = 1\times2^5 + 0\times2^4 + 1\times2^3 + 0\times2^2 + 1\times2^1 + 1\times2^0 = (43)_{10}$.

53. Associative law : A + (B + C) = (A + B) + C

54.

$$\begin{array}{cccccccc}
 & & & & 1 & 0 & 1 & 1 \\
 & & & \times & 1 & 1 & 0 & 1 \\
\hline
 & & & & 1 & 0 & 1 & 1 \\
 & & & 0 & 0 & 0 & 0 & \times \\
 & & 1 & 0 & 1 & 1 & \times & \times \\
 & 1 & 0 & 1 & 1 & \times & \times & \times \\
\hline
1 & 0 & 0 & 0 & 1 & 1 & 1 & 1 \\
\hline
\end{array}$$

55.

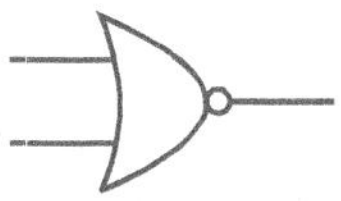

NOR Gate

56.

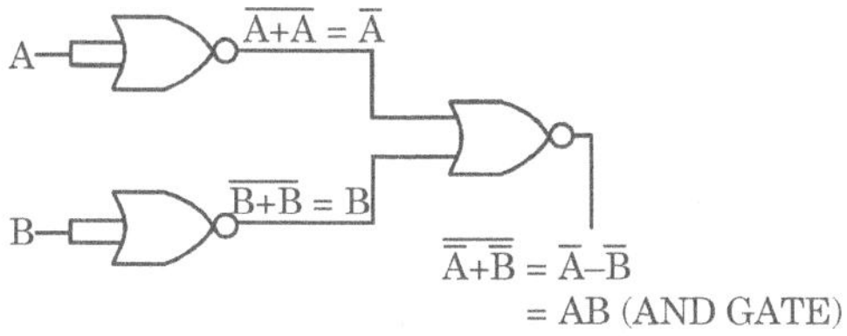

$\overline{\overline{A}+\overline{B}} = \overline{\overline{A}}\cdot\overline{\overline{B}} \Rightarrow$ AND Gate

57.

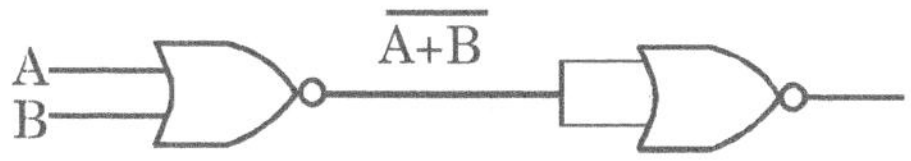

$\overline{\overline{A+B}+\overline{A+B}} = \overline{\overline{A+B}} = A+B$

58.

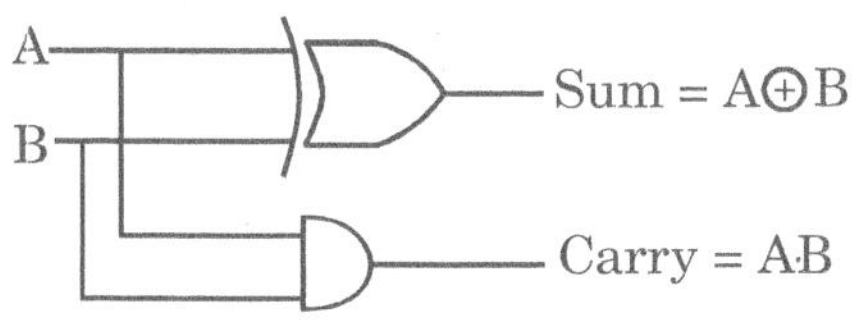

Half-Adder circuit

59. $A + \overline{A} = 1$

60. D-flip flop is used in shift registers.

61. A full adder adds 3 inputs (two significant bits and a previous carry) and gives 2 outputs (sum and carry).

62. Flip-flop is a bistable circuit.

63. 3-bits are used to represent a digital in octal numbers.

64. Speed of emitter-coupled logic family is highest among others.

65. State transition period in logic families is about 0.5-50 ns.

66. Serial circuits are slower than parallel circuits.

67. Look ahead carry generaters have low propagation time as compared to parallel adders.

68. Decoder is a combination circuit (no memory device).

69.

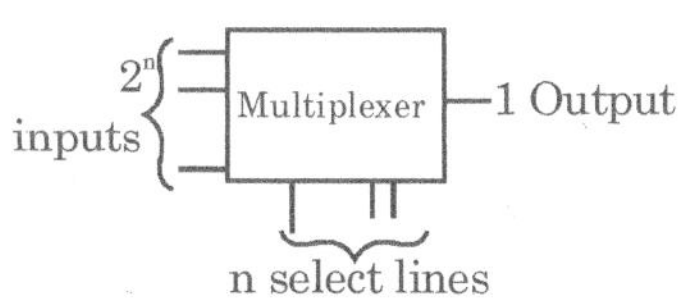

70. Race-around condition always arises in asynchronous sequential circuits.

71. C.R.O. has a display to present signal waveform.

72. Direct measurement of pressure can be done using circular or twisted Bourdon tube.

73. At the resonance in LC circuit, current is maximum and impedance is minimum.

74. Form factor $= \dfrac{\text{RMS value}}{\text{Average value}}$

75. Conduction band only consists of free electrons.

76. Orange and Apple are fruits, Similarly table and chair are furniture

77. Option (c) satisfies the pattern

78. 3, 4, 13, 38, ?, 168

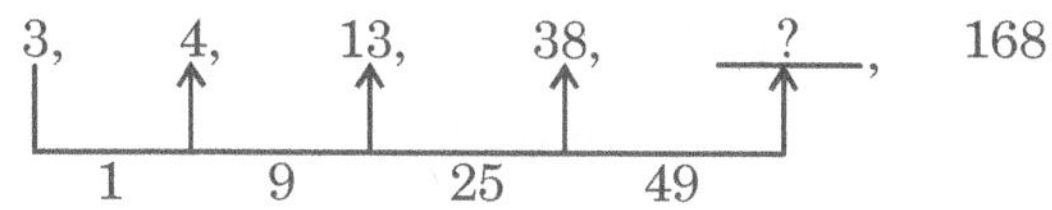

as $(1)^2 = 1$

$(3)^2 = 9$

$(5)^2 = 25$

$\therefore\ (7)^2 = 49$

$\Rightarrow 38 + 49 = 87$

$\therefore$ Option (b) is correct

79. 12% of 40% of x = 84

$$\therefore \frac{12}{100} \times \frac{40}{100} \times x = 84$$

$$\therefore x = \frac{840000}{12 \times 40}$$

$\therefore x = 1750$

80. R $\downarrow$ + P $\rightarrow$ Q   Q $\downarrow$ $\overline{V} \rightarrow \overline{T}$

$\therefore$ +

$R \longleftrightarrow \overline{S}$

$\downarrow$

+

$P \rightarrow Q$

$\downarrow$

$\overline{V} \rightarrow \overline{T}$

'V' is grand daughter of 'S'

81. October, January and December has 31 days while June has 30 days

82. Option (b) satisfies the given pattern

83. $+ \Rightarrow \div$

$\div \Rightarrow -$

$- \Rightarrow \times$

$\times \Rightarrow +$

$(12 + 4 \times 2) \times 6 - 4 \div 2$

$(12 \div 4 + 2) + 6 \times 4 - 2$

$= 3 + 2 + (6 \times 4) - 2$

$= 3 + 2 + 24 - 2 = 27$

$\therefore$ Option (a) is correct

84. 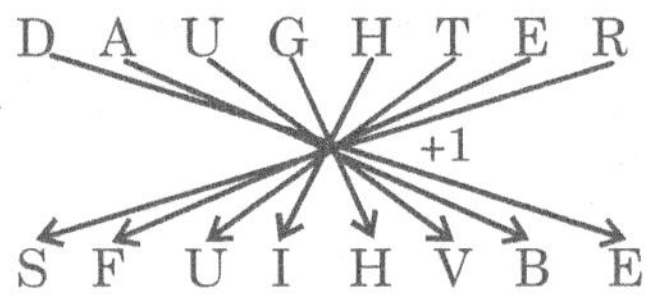

Similarly,

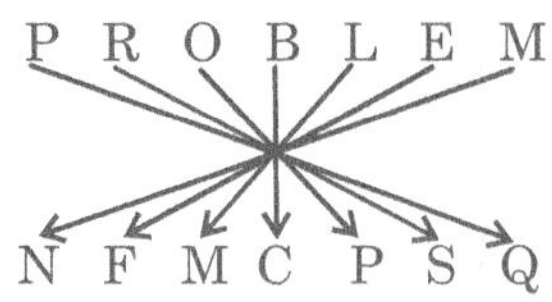

85. Row I $\rightarrow (16 \times 9) + 23 = 144 + 23 = 167$

Row II $\rightarrow (38 - 13) \times 25 = 25 \times 25 = 625$

Required difference $= 625 - 167$

$= 458$

86. Operating system for mobile phones developed by Google is Android

87. C.P.U of computer consists of ALU, control units and registers

88. Modem act as both input as well as output device

89. A program in execution is called process

91. $(10000010)_2 = (1 \times 2^7 + 1 \times 2^1)_{10}$

$= (128 + 2)_{10}$

$= (130)_{10}$

92. In Spreadsheet, one function inside the other is called nested function

93. Database, Image Editor and word processor part of office suit.

Hence file manger is not the part of office suit.

94. Facebook is not a Search engine.

95. A Shared network within an organization to provide connectivity to the Staff is called Intranet

SOLVED PAPER 2016

ELECTRONICS ENGINEERING

Read the following instructions carefully before you begin to answer the questions. This Booklet contains questions in English.

INSTRUCTIONS

1. This Booklet contains 120 questions in total comprising of the following two parts:
   Part – I: Aptitude [1 – 45 Questions]
   Part – II: Technical [46 – 120 Questions]

2. All questions are compulsory and carry 1 mark each.

3. Before you start to answer the questions you must check this Booklet to ensure that it contains all the pages and see that no page is missing or repeated.

4. For every wrong answer 1/3 marks will be deducted.

5. You are supplied with the OMR-Sheet separately.

6. Answers must be indicated by completely blackening one of the corresponding ovals on OMR-Sheet against the relevant question number, by black/blue ball pen only. Answers which are not indicated by black/blue ball pen will not be awarded any mark.

7. The OMR-Sheet must be handed over to the Invigilator before you leave the Examination Hall.

8. Failure to comply with any of the above instructions will render a candidate liable to such action/penalty as may be deemed fit.

SOLVED PAPER 2016

APTITUDE

1. A car travels first half distance between two places with a speed of 40 km/hr and the rest half distance with a speed of 60 km/hr. The average speed of the car is:
   (a) 37 km/hr (b) 44 km/hr
   (c) 48 km/hr (d) None of these
2. In how many different ways can the letters of the word 'RIDDLED' be arranged?
   (a) 1680 (b) 840
   (c) 2520 (d) 5040

Directions : In the following question, a series is given with one term missing. Choose the correct alternative from the given ones that will complete the series.

3. AGMSY, CIOUA, EKQWC, ___, IOUAG, KQWCI.
   (a) GNTYE (b) GLRYE
   (c) GMSYE (d) GMTXE
4. A sphere and a cube have equal surface areas. The ratio of the volume of the sphere to that of the cube is:
   (a) $\sqrt{\pi}:\sqrt{6}$ (b) $\sqrt{6}:\sqrt{\pi}$
   (c) $\sqrt{2}:\sqrt{\pi}$ (d) None of these

Directions (Question nos. 5 to 7) : There are 3 types of industries, paper, drug and toy industry. Six workers A, B, C, D, E and F work in these industries. Only two of them can work in one industry. No one can work in more than one industry. 'A' does not work in paper industry. 'B' and 'E' do not work in toy industry. 'A' and 'B' do not work in the same industry. 'D' does not work in drug industry. 'A' and 'D' do not work in toy industry. 'E' and 'F' are not in the same industry.

5. Which two workers work in the drug industry?
   (a) AE (b) EB
   (c) CF (d) None of these
6. Which two work in the paper industry?
   (a) AE (b) EB
   (c) BD359 (d) None of these
7. Paper, Drug and Toy are represented by which of the following respectively?
   (a) ACD (b) EFD
   (c) BAC (d) None of these
8. Which of the following occurred first:
   (a) Jallianwallah Massacre
   (b) Dandi March
   (c) Champaran Satyagraha
   (d) Khilafat Movement
9. Which of the following represents the statement that some of the psychologists are philosophers. Some philosophers are writers. But no psychologist is a writer.

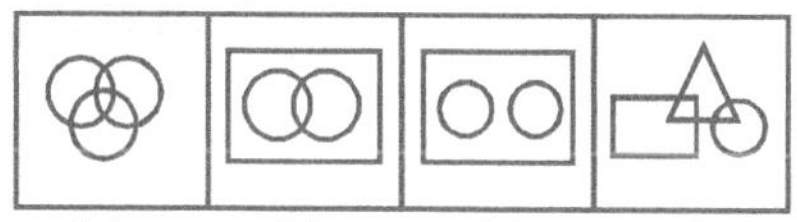

(a) (b) (c) (d)

   (a) B (b) A
   (c) D (d) C
10. If $(243)^{\frac{3}{5}} \times (729)^{\frac{-2}{3}} = 3^x$ then the value of x is
   (a) – 2 (b) – 1
   (c) 1 (d) None of these
11. The velocity of a shell leaving the gun is 300 m/sec. Two shots are fired, first with elevation of barrel at 30° and second with elevation 60°. The range achieved by the first will be:
   (a) More than the second
   (b) Less than the second
   (c) Same as the second
   (d) Half the second
12. Count the number of rectangles in the following figure:

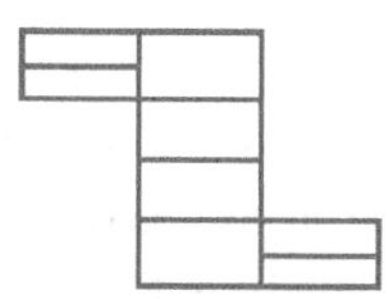

   (a) 8 (b) 17
   (c) 18 (d) None of these
13. The circumference of a semicircle of area 1925 sq. cm is equal to the breadth of a rectangle. If the length of the rectangle is equal to the perimeter of a square of side 48 cm. What is the perimeter of the rectangle?
   (a) 734 cm (b) 744 cm
   (c) 755 cm (d) None of these
14. A cloth merchant sold half of his cloth at 40% profit, half of remaining at 40% loss and rest was sold at cost price. In total transaction his gain or loss will be?
   (a) 20% gain (b) 25% gain
   (c) 10% gain (d) 15% loss
15. Khaira disease of paddy is due to soil deficiency of:
   (a) Zinc (b) Iron
   (c) Potassium (d) Boron

16. "The State shall endeavor to provide early childhood care and education of all children until they complete the age of six years". This is a provision under:

(a) Article 21A of the constitution of India relating to fundamental right enumerated in part III.

(b) Article 51A relating to Fundamental duties enumerated in Part IV A of The Constitution of India

(c) Article 45 relating to directive principles of state policy enumerated in part IV of the constitution of India

(d) Article 38 of the directive principles of State policy to secure a social order for the promotion of welfare of the people

17. If INDIA is codified as XYZXE and GLAD is codified PMEZ then LANDING is codified as:

(a) MEPZXYP (b) PZMEXYZ

(c) MEYZXYP (d) MEXYZPX

Directions : Anil is the son of Bina. Chitra, who is Bina's sister has a son Deepak and a daughter Ela. Fateh is the maternal uncle of Deepak.

18. How is Ela related to Fateh:

(a) Sister (b) Wife

(c) Daughter (d) Niece

19. Find the wrong number in the following series.
10, 14, 28, 32, 64, 68, 132.

(a) 14 (b) 32

(c) 132 (d) 68

20. Which has become the first airport of the world to fully operate on solar power?

(a) Jaipur International Airport

(b) Sardar Vallabh Bhai Patel International Airport

(c) Cochin International Airport Ltd.

(d) Raja Bhoj International Airport

21. Which one of the following European countries borders Atlantic ocean?

(a) Portugal (b) Austria

(c) Finland (d) Romania

22. From the following diagram find out correct answer for the given question.

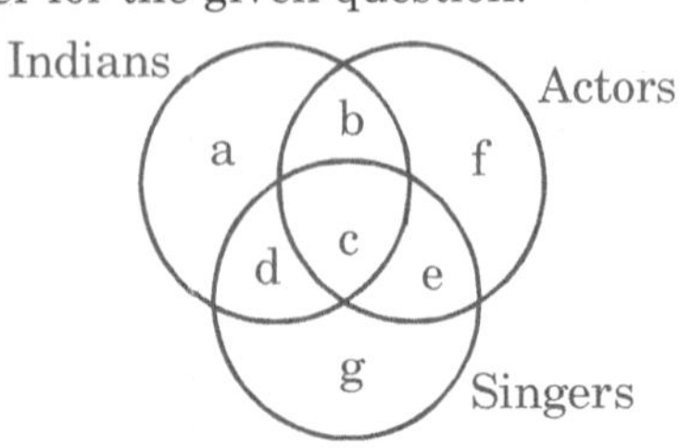

Indians who are actors but not singers:

(a) b (b) c

(c) f (d) g

23. New Horizons became the first space probe to reach closest to Pluto recently. This is a mission by:

(a) ISRO (b) ESA

(c) NASA (d) FKA & RKA

Directions : Select the suitable alternative to satisfy the relationship in the following question.

24. Question figure

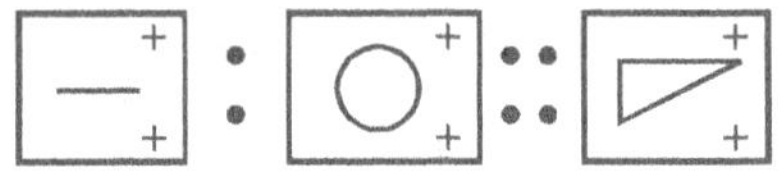

Answer figure

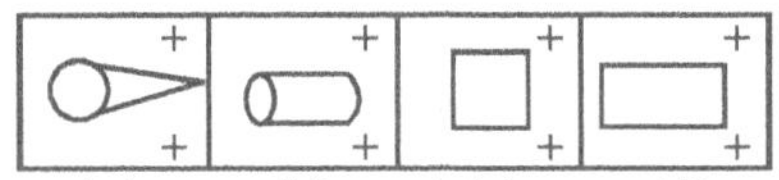

(a) (b) (c) (d)

Directions : Select the suitable alternative to satisfy the relationship in the following question.

25. Question figure

Answer figure

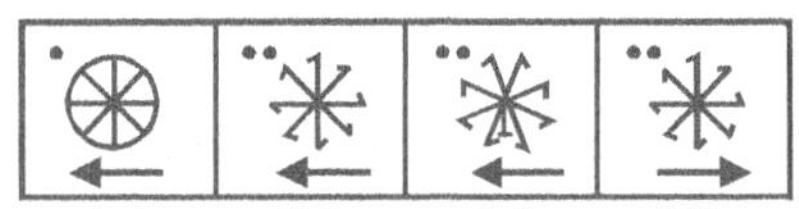

(a) (b) (c) (d)

26. Aman and Babu can complete the work in 4 days. Aman alone starts working and leaves it after working for 3 days, completing only half of the work. In how many days, it can be completed if the remaining job is undertaken by Babu:

(a) 4 days (b) 6 days

(c) 5 days (d) None of these

27. In three vessels, the ratio of water and milk is 6:7, 5:9 and 8:7 respectively. If the mixtures of the three vessels are mixed then what is the ratio of water and milk?

(a) 3691:4499 (b) 2431:3781

(c) 4381:5469 (d) None of these

28. 'Indian Standard Meridian' passes through the states of:

(a) Uttar Pradesh, Madhya Pradesh, Chhatishgarh, Odisha, and Andhra Pradesh

(b) Uttar Pradesh, Bihar, Jharkhand, Odisha and West Bengal

(c) Uttar Pradesh, Rajasthan, Maharashtra, Gujarat and Madhya Pradesh

(d) Tamilnadu, Karnataka, Andhra Pradesh, Odisha and West Bengal

29. If interest payments are subtracted from gross fiscal deficit, the remainder will be:

(a) Revenue deficit (b) Gross primary deficit

(c) Capital deficit (d) Budgetary deficit

30. Who was the famous poet who was deeply affected by Tagore's Gitanjali?

(a) Eliot (b) Auden
(c) Yeats (d) Pound

31. A person mixes three varieties of wheat costing Rs. 20/- kg, Rs. 24/- kg and Rs. 30/- kg in the ratio of 1:5:2 and sells the mixture at a profit of 20%. Find the selling price of the mixture when it is sold at 33% profit:

(a) Rs. 30/- kg (b) Rs. 33.25/- kg
(c) Rs. 39.9/- kg (d) None of these

32. Match the trophies associated with the respective games:

TROPHIES	GAMES
A. Agha Khan Cup	1. Cricket
B. Duleep Trophy	2. Football
C. Welligton Trophy	3. Hockey
D. Ezra Cup	4. Polo
E. Subroto Cup	5. Rowing

(a) A4, B5, C3, D1, E2
(b) A3, B1, C4, D5, E2
(c) A2, B3, C4, D5, E1
(d) A3, B1, C5, D4, E2

33. A triangle has two of its angles in the ratio of 1:2. If the measure of one of its angles is 30 degrees. What is the measure of the largest angle of the triangle in degree?

(a) 100
(b) 90
(c) 135
(d) Can not be determined

34. The alloy containing copper, tin and zinc in the ratio of 87:10:3 is called:

(a) Bronze (b) Brass
(c) Solder (d) Gun metal

35. Which portion of the heart receives oxygenated blood:

(a) Left auricle (b) Right auricle
(c) Left ventricle (d) Right ventricle

36. A form of Hindustani music known as 'Khayal' evolved during the 13th & 14th centruries. The proponent of the style is known as:

(a) Ustad Bismillah Khan
(b) Ustad Amir Khan
(c) Amir Khusro
(d) Thyagaraja

37. The first Indian to receive Nobel Prize:

(a) Dr. C.V. Raman
(b) Rabindra Nath Tagore
(c) Mother Teresa
(d) Dr. Hargobind Khorana

38. When we consider 15° meridian on a world map or globe and count them in an eastward direction starting with Greenwich meridian (0°), we find that the time of this meridian is:

(a) Same as Greenwich (b) 1 hour fast
(c) 1 hour slow (d) 12 hour fast

39. Out of a total of 85 children playing Badminton or Table Tennis or both. Total number of girls in the group is 70% of the total number of boys in the group. The number of boys playing only Badminton is 50% of the number of boys and total number of boys playing Badminton is 60% of the total number of boys. Number of children only playing Table Tennis is 40% of the total number of children and a total of 12 children play Badminton and Table Tennis both. What is the number of girls playing only Badminton?

(a) 16 (b) 14
(c) 17 (d) None of these

40. How many squares can be formed by joining the center of the circles by horizontal and vertical lines:

(a) 6 (b) 8
(c) 10 (d) None of these

41. A cistern has two pipes attached to it, one to supply and one to draw off. If both the pipes are opened together, the cistern is filled in 9 hours but if the waste pipe is opened one hour after the supply pipe, the cistern is filled in 7 hours. The time that will be taken by the supply pipe to fill the empty cistern is:

(a) 2 hrs. 47 minutes (b) 3 hrs. 30 minutes
(c) 3 hrs. 32 minutes (d) 3 hrs.

42. The first woman in NASA history to command space flight:

(a) Kalpana Chawla (b) Sally Ride
(c) Eilean Collins (d) None of these

43. In an examination, pass marks are 36% of maximum marks. If an examine gets 17 marks and fails by 10 marks in this examination, what are the maximum marks?

(a) 50 (b) 75
(c) 85 (d) 100

44. If $x = \frac{\sqrt{3}-\sqrt{2}}{\sqrt{3}+\sqrt{2}}$ and $y = \frac{\sqrt{3}+\sqrt{2}}{\sqrt{3}-\sqrt{2}}$, then find the value of $x^2 + xy + y^2$.

(a) 97 (b) 98

(c) 99 (d) None fo these

45. Rhine valley in France is known for mineral deposits of:

(a) Bauxite (b) Copper

(c) Nickel (d) Zinc

TECHNICAL

46. An FM signal with modulation index mf is passed through a frequency tripler. The modulation index of the output signal will be

(a) m_f (b) $3m_f$

(c) $9m_f$ (d) $27m_f$

47. What is the correct sequence of the following step in the fabrication of a monolithic, bipolar junction transistor?

1. Emitter diffusion
2. Base diffusion
3. Buried layer formation
4. Epi–layer formation

Select the correct answer using the codes given below

(a) 3, 4, 1, 2 (b) 4, 3, 1, 2

(c) 3, 4, 2, 1 (d) 4, 3, 2, 1

48. Two sinusoidal signals having the same amplitude and frequency are applied to the X and Y inputs of a CRO. The observed Lissajous figure is straight line. The phase shift between the two signals would be

(a) Zero

(b) 90 degrees

(c) Either zero or 180 degrees

(d) Either 90 degrees or 270 degrees

49. In a BJT circuit a pnp transistor is replaced by npn transistor. To analyse the new circuit

(a) All calculations done earlier have to be repeated

(b) Replace all calculated voltages by reverse values

(c) Replace all calculated currents by reverse values

(d) Replace all calculated voltages and currents by reverse values

50. In 8085 microprocessor, the value of the most significant bit of the result following the execution of any arithmetic of Boolean instruction is stored in the

(a) Carry status flag

(b) Auxiliary carry status flag

(c) Sign status flag

(d) Zero status flag

51. The analog signal m(t) is given below m(t) = 4 cos 100 pt + 8 sin 200 pt + cos 300 pt, the Nyquist sampling rate will be:

(a) $\frac{1}{100}$ (b) $\frac{1}{200}$

(c) $\frac{1}{300}$ (d) $\frac{1}{600}$

52. Find the noise factor for an antenna at 27°C with equivalent noise temperature 30°K:

(a) 20 (b) 100

(c) 300 (d) 1.1

53. Inverse Laplace transform of $\frac{2s+5}{(s+3)(s+2)}$ is:

(a) 2 exp (– 2.5 t) cos h (0.5 t)

(b) exp (– 2 t) + exp (– 3 t)

(c) 2 exp (– 2.5 t) sin h (0.5 t)

(d) 2 exp (– 2.5 t) cos 0.5 t

54. The analog signal given below is sampled by 600 samples per second for m(t) = 3 sin 500 pt + 2 sin 700 pt then folding frequency is:

(a) 500 Hz (b) 700 Hz

(c) 300 Hz (d) 1400 Hz

55. Which one most appropriate dynamic system?

(a) y(n) + y(n – 1) + y(n + 1)

(b) y(n) + y(n – 1)

(c) y(n) = x(n)

(d) y(n) + y(n – 1) + y(n + 3) = 0

56. In which of these is reverse recovery time nearly zero?

(a) Zener diode (b) Tunnel diode

(c) Schottky diode (d) PIN diode

57. Frequency shift keying is used mostly in:

(a) Telegraphy

(b) Telephony

(c) Satellite communication

(d) Radio transmission

58. A parallel polarized wave is incident from air into paraffin having relative permittivity 3, the value of its Brewster angle is:

(a) 0° (b) 30°

(c) 45° (d) 60°

59. The phase angle corresponding to ë/4 in a standing–wave pattern is:

(a) 30° (b) 90°

(c) 135° (d) 180°

60. A 10 km long line has a characteristic impedance of 400 ohms. If line length is 100 km, the characteristic impedance is:

(a) 4000 Ω (b) 400 Ω

(c) 40 Ω (d) 4 Ω

61. The output of an exclusive–NOR gate is 1. Which input combination is correct?

(a) A = 1, B = 0 (b) A = 0, B = 1

(c) A = 0, B = 0 (d) None of these

62. In figure v1 = 8 V and v2 = 4 V. Which diode will conduct?

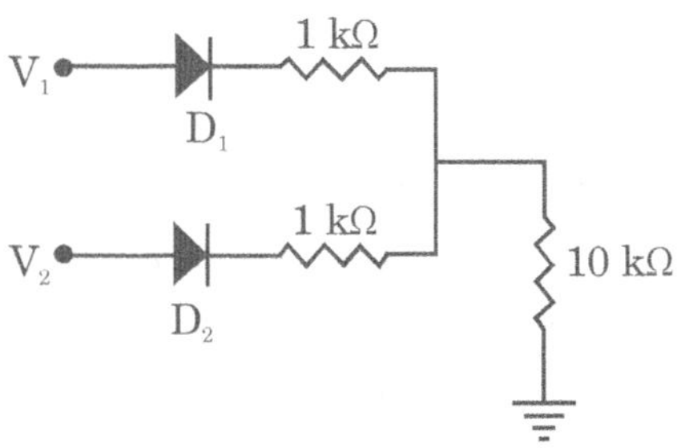

(a) D2 only (b) D1 only

(c) Both D1 and D2 (d) Neither D1 nor D2

63. Determine the output frequency for a frequency division circuit that contains 12 flip–flops with an input clock frequency of 20.48 MHz:

(a) 10.24 kHz (b) 5 kHz

(c) 30.24 kHz (d) 15 kHz

64. Initially the number decimal 8 is stored. If instruction RAL is executed twice, the final number stored will be:

(a) Decimals 8 (b) Decimal 16

(c) Decimal 32 (d) Decimal 2

65. The output of an AND gate with three inputs, A, B, and C, is HIGH when ________:

(a) A = 1, B = 1, C = 0 (b) A = 0, B = 0, C = 0

(c) A = 1, B = 1, C = 1 (d) A = 1, B = 0, C = 1

66. The gain of an FET amplifier can be changed by changing:

(a) rm (b) gm

(c) Rd (d) None of these

67. In the case of a 70–MHz IF carrier for a transponder bandwidth of 36 MHz, energy must lie between MHz:

(a) 34 and 106 (b) 52 and 88

(c) 106 and 142 (d) 34 and 142

68. In a full wave rectifier circuit using centre tapped transformer, the peak voltage across half of the secondary winding is 30 V. Then PIV is:

(a) 30 V (b) 60 V

(c) 15 V (d) 10 V

69. The Lissajous pattern observed on screen of CRO is a straight line inclined at 45° to x axis. If X–plate input is 2 sin ωt, the Y–plate input is

(a) 2 sin ωt (b) 2 sin (ωt + 45°)

(c) 2 sin (ωt – 45°) (d) 22 sin ωt + 45°)

70. The input impedance of op–amp circuit of figure is:

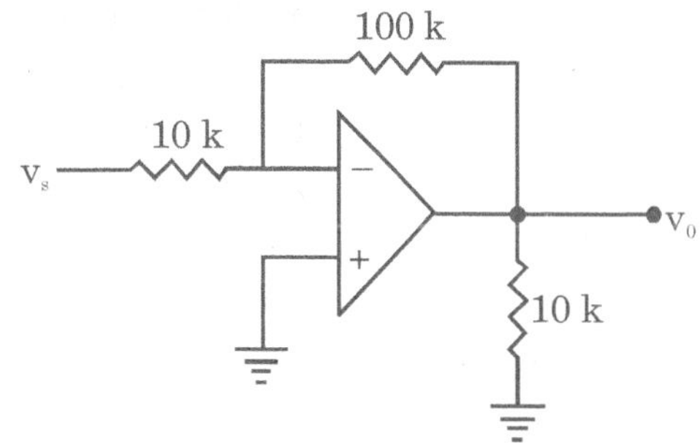

(a) 120 k ohm (b) 110 k ohm

(c) Infinity (d) 10 k ohm

71. In a CRO which of the following is not a part of electron gun:

(a) Cathode (b) Grid

(c) Accelerating anode (d) X–Y plates

72. A 1 μF capacitor is connected to 12 V battery. The energy stored in the capacitor is:

(a) 12×10^{-6} J (b) 24×10^{-6} J

(c) 48×10^{-6} J (d) 72×10^{-6} J

73. For a transmission line open circuit and short circuit impedances are 20 Ω and 5 Ω. Then characteristic impedance is:

(a) 100 Ω (b) 50 Ω

(c) 25 Ω (d) 10 Ω

74. Find i(t) if $I(s) = \dfrac{s+1}{(s+4)(s+3)}$:

(a) $3e^{-4t} - e^{-5t}$ (b) $3e^{-4t} - 2e^{-3t}$

(c) $2e^{-4t} - 5e^{-2t}$ (d) None of these

75. A capacitor used for power factor correction in single phase circuit decreases:

(a) Power factor

(b) Line current

(c) Both line current and p.f.

(d) The line current and increases p.f.

76. The negative sign in the expression e = – N (dØ)/dt is due to:

(a) The Fleming's Rule (b) Thumb's Rule

(c) Faraday's Law (d) Lenz's Law

77. Form factor is the ratio of:

(a) Maximum value to r.m.s. value

(b) Maximum value to average value

(c) r.m.s value to average value

(d) r.m.s. value to instantaneous value

78. The number of valence electrons in pentavalent impurity is:

(a) 5 (b) 4

(c) 3 (d) 1

79. Material which lack permanent magnetic dipoles are known as:

(a) Paramagnetic (b) Diamagnetic

(c) Ferromagnetic (d) Ferrimagnetic

80. The primary function of the bias circuit is to:

(a) Hold the circuit stable at VCC

(b) Hold the circuit stable at vin

(c) Ensure proper gain is achieved

(d) Hold the circuit stable at the designed Q-point

81. The transconductance of a JFET is computed at constant V_{DS} by:

(a) Ratio of change in I_d to change of V_{gs}

(b) Ratio of change inV_{gs} to change of I_d

(c) Product of change in V_{gs} to change of I_d

(d) Ratio of change in V_{ds} to change of I_d

82. To find current in a resistance connected in a network, Thevenin's theorem is used V_{TH} = 20 V and R_{TH} = 5 Ω.

The current through the resistance:

(a) is 4 A

(b) is 4 A or less

(c) is less than 4 A

(d) May be 4 A or less or more than 4 A

83. Superposition theorem is not applicable to network containing:

(a) Non linear element

(b) Dependent voltage source

(c) Dependent current source

(d) Transformer.

84. In a microwave test bench, why is the microwave signal amplitude modulated at 1 kHz?

(a) To increase the sensitivity of measurement

(b) To transmit the signal to a far-off place

(c) To study amplitude modulation

(d) Because crystal detector fails at microwave frequencies

85. The directive gain of an antenna:

(a) is less than or equal to 1

(b) is greater than or equal to 1

(c) lies between 0 and directivity of antenna

(d) is greater than directivity of antenna

86. An amplifier without feedback has a voltage gain of 50, input resistance 1kΩ and output resistance of 2.5kΩ. The input resistance of the current-shunt negative feedback amplifier using the above amplifier with a feedback factor of 0.2 would be:

(a) $\frac{1}{11}$ kΩ (b) $\frac{1}{5}$ kΩ

(c) 5 kΩ (d) 11 kΩ

87. In amplitude modulation AM, if modulation index is more than 100% then:

(a) Power of the wave increases.

(b) Efficiency of transmission increases.

(c) The wave gets distorted.

(d) Bandwidth increases

88. The use of non-uniform quantization leads to:

(a) Reduction in transmission bandwidth.

(b) Increase in maximum SNR.

(c) Increase in SNR for signal levels.

(d) Simplification of quantization process.

89. A PD controller is used to compensate a system. Compared to the uncompensated system, the compensated system has:

(a) A higher type number.

(b) Reduced damping.

(c) Higher noise amplification.

(d) Larger transient overshoot.

90. Which of the following is an inherent property of an optical signal and cannot be eliminated even in principle?

(a) Thermal noise (b) Shot noise

(c) Environmental noise (d) Background Noise

91. The disadvantage of CDMA system is:

(a) Higher MAI with more user count.

(b) That it needs least power consumption.

(c) That it needs coherent signal.

(d) That it is not suitable with PSK systems

92. The Ku frequency band used in satellite communication (for uplink and downlink) is:

(a) 14 GHz and 11 GHz (b) 30 GHz and 20 GHz

(c) 70 GHz and 50 GHz (d) 60 GHz and 45 GHz

93. Following transducer is an active transducer:

(a) Carbon Microphone

(b) Piezoelectric Microphone

(c) LVDT

(d) Strain Gauge

94. A NOR gate is equivalent to a bubbled AND gate. This statement is an outcome of:

(a) De Morgan's Law (b) Involution Law
(c) Law of Absorption (d) Idempotent Law

95. The circuit shown in the figure has initial current $i_L(0)$ = 1A through the inductor and an initial voltage $v_c(0)$ = 1 V across the capacitor. For input v(t) = U(t) the Laplace transform of the current i(t) for t ≥ 0 is:

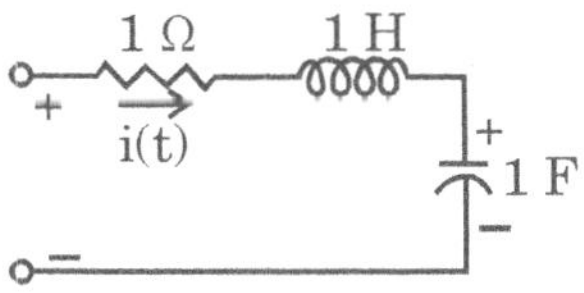

(a) $\frac{s}{s^2+s+1}$ (b) $\frac{s+2}{s^2+s+1}$

(c) $\frac{s-2}{s^2+s+1}$ (d) $\frac{s-2}{s^2+s+2}$

96. Which of the following noise must be considered at high frequencies (microwave frequencies)?

(a) Shot noise (b) Random noise
(c) Impulse noise (d) Transit time noise

97. In a broadcast communication receiver, most of the receiver selectivity is achieved in:

(a) RF section (b) IF section
(c) Mixer (d) Local oscillator

98. Slew rate is defined by:

(a) di/dt(max) (b) dv/dt(max)
(c) di/dt(min) (d) dv/dt(min)

99. The ideal operational amplifier does not have:

(a) Infinite input resistance
(b) Infinite output resistance
(c) Infinite voltage gain
(d) Infinite bandwidth

100. The following property of semiconductors cannot be determined from Hall effect:

(a) Semiconductor is n–type or p–type
(b) The carrier concentration
(c) The mobility of semiconductor
(d) The atomic concentration of semiconductor

101. The ABCD parameters of ideal n:1 transformer shown in the figure are $\begin{bmatrix} n & o \\ o & x \end{bmatrix}$. The value of X will be:

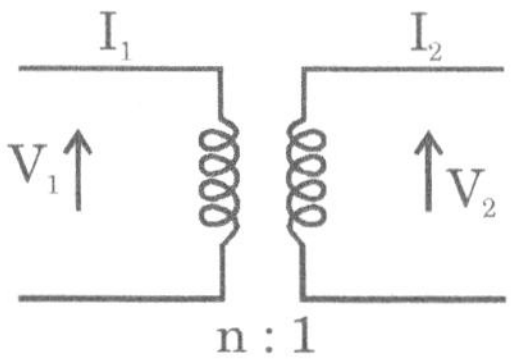

(a) n (b) $\frac{1}{n}$

(c) n^2 (d) $\frac{1}{n^2}$

102. In the circuit shown below, the voltmeter will read:

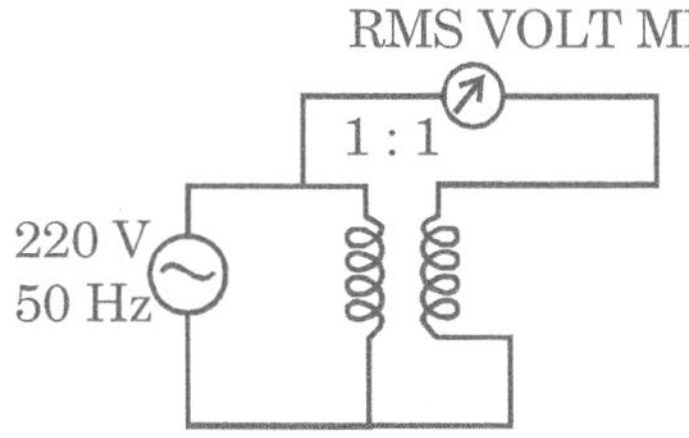

(a) Zero (b) 110 V
(c) 220 V (d) 440 V

103. This figure is a block diagram of ________:

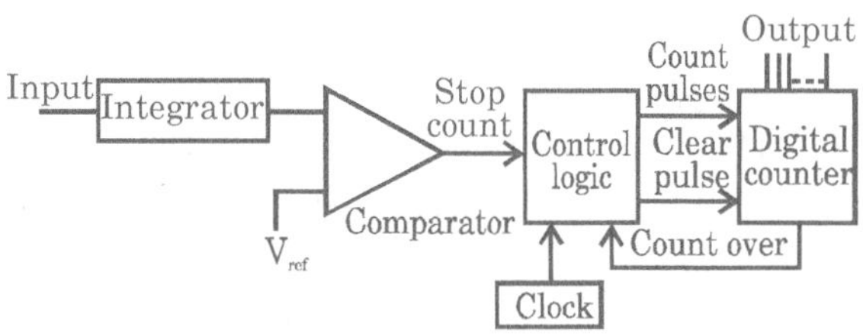

(a) ADC
(b) DAC
(c) Comparator
(d) 555 timer

104. Theoretical bandwidth requirement for FM system is:

(a) Same as that of the modulating signal bandwidth
(b) Twice of modulating signal Bandwidth
(c) Thrice as of modulating signal Bandwidth
(d) Infinite

105. Waveguides are used mainly for microwave signals because:

(a) They depend on straight-line propagation which applies to microwaves only
(b) Losses would be too heavy at lower frequencies
(c) There are no generators powerful enough to excite them at lower frequencies
(d) They would be too bulky at lower frequencies

106. Indicate which one of the following is NOT an advantage of FM over AM?
(a) Better noise immunity is provided
(b) Lower bandwidth is required
(c) The transmitted power is more useful
(d) Less modulating power is required

107. Which of the following statement is false? Modulation is used to:
(a) Reduce the bandwidth used
(b) Separate differing transmissions
(c) Ensure that intelligence may be transmitted over long distances
(d) Allow the use of practicable antennas

108. In an n-type semiconductor, as the donor concentration ND increases, the Fermi level EF:
(a) Remains unaltered
(b) Moves towards the conduction band
(c) Move towards the center of forbidden energy gap
(d) May or may not move depending on temperature

109. The term "white noise" refers to the following:
(a) A random signal with flat power spectral density
(b) A random signal generated by beat signal
(c) A random signal with long range correlation
(d) All of the above are true

110. Which of the following relation is true for two digital signals A and B?
(a) $\overline{A+B} = \bar{A} \cdot \bar{B}$
(b) $\overline{A+B} = A \cdot B$
(c) $\overline{A+B} = \bar{A} \cdot \bar{B} + A \cdot B$
(d) $\overline{A+B} = \bar{A} \cdot B$

111. The output waveform of a 555 Timer is:
(a) Sinusoidal (b) Triangular
(c) Rectangular (d) Elliptical

112. The phase correcting circuit is:
(a) All–pass filter (b) Low–pass filter
(c) High–pass filter (d) Band–pass filter

113. The light emitted diodes consist of:
(a) Si (b) GaAs
(c) Ge (d) Diamond

114. When the initial slope of input sine wave is greater than the slew rate of an OPAMP, the output:
(a) Has no offset
(b) Approaches to triangular waveshape
(c) Is pure sinusoidal
(d) Is square wave

115. A circuit is said to be free from phase distortion if its phase response shows:
(a) Quadrature variation with frequency
(b) Independent of frequency
(c) Linear variation with frequency
(d) Proportional to frequency

116. In a full wave rectifier with input frequency of 50 Hz, the frequency of the output is:
(a) 50 Hz (b) 100 Hz
(c) 150 Hz (d) 200 Hz

117. In the common–emitter transistor circuit, if the current gain is 100 and the collector current is 10 mA, the base current is:
(a) 10 μA (b) 100 μA
(c) 1 A (d) 10 A

118. The basic Hartley oscillator uses:
(a) One inductor and two capacitors
(b) A centre tapped inductor and a capacitor
(c) Two capacitors
(d) Tickler coil

119. The correct order of electromagnetic spectrum with decreasing frequency is:
(a) Microwaves, Radiowaves, Infrared rays, Ultraviolet rays, X–rays
(b) Radiowaves, Infrared rays, Ultraviolet rays, Microwaves, X–rays
(c) X–rays, Infrared rays, Microwaves, Radiowaves, Ultraviolet rays
(d) X–rays, Ultraviolet rays, Infrared rays, Microwaves, Radiowaves

120. In the given figure reflection coefficient at load is:

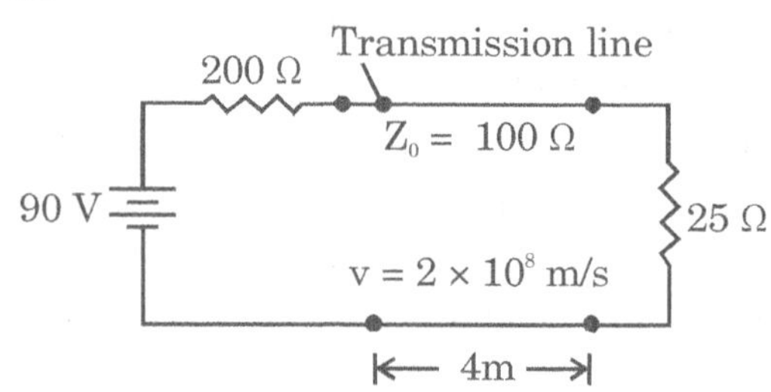

(a) 0.6 (b) – 0.6
(c) 0.4 (d) None of these

OMR SHEET

1.	Ⓐ Ⓑ Ⓒ Ⓓ	31.	Ⓐ Ⓑ Ⓒ Ⓓ	61.	Ⓐ Ⓑ Ⓒ Ⓓ	91.	Ⓐ Ⓑ Ⓒ Ⓓ
2.	Ⓐ Ⓑ Ⓒ Ⓓ	32.	Ⓐ Ⓑ Ⓒ Ⓓ	62.	Ⓐ Ⓑ Ⓒ Ⓓ	92.	Ⓐ Ⓑ Ⓒ Ⓓ
3.	Ⓐ Ⓑ Ⓒ Ⓓ	33.	Ⓐ Ⓑ Ⓒ Ⓓ	63.	Ⓐ Ⓑ Ⓒ Ⓓ	93.	Ⓐ Ⓑ Ⓒ Ⓓ
4.	Ⓐ Ⓑ Ⓒ Ⓓ	34.	Ⓐ Ⓑ Ⓒ Ⓓ	64.	Ⓐ Ⓑ Ⓒ Ⓓ	94.	Ⓐ Ⓑ Ⓒ Ⓓ
5.	Ⓐ Ⓑ Ⓒ Ⓓ	35.	Ⓐ Ⓑ Ⓒ Ⓓ	65.	Ⓐ Ⓑ Ⓒ Ⓓ	95.	Ⓐ Ⓑ Ⓒ Ⓓ
6.	Ⓐ Ⓑ Ⓒ Ⓓ	36.	Ⓐ Ⓑ Ⓒ Ⓓ	66.	Ⓐ Ⓑ Ⓒ Ⓓ	96.	Ⓐ Ⓑ Ⓒ Ⓓ
7.	Ⓐ Ⓑ Ⓒ Ⓓ	37.	Ⓐ Ⓑ Ⓒ Ⓓ	67.	Ⓐ Ⓑ Ⓒ Ⓓ	97.	Ⓐ Ⓑ Ⓒ Ⓓ
8.	Ⓐ Ⓑ Ⓒ Ⓓ	38.	Ⓐ Ⓑ Ⓒ Ⓓ	68.	Ⓐ Ⓑ Ⓒ Ⓓ	98.	Ⓐ Ⓑ Ⓒ Ⓓ
9.	Ⓐ Ⓑ Ⓒ Ⓓ	39.	Ⓐ Ⓑ Ⓒ Ⓓ	69.	Ⓐ Ⓑ Ⓒ Ⓓ	99.	Ⓐ Ⓑ Ⓒ Ⓓ
10.	Ⓐ Ⓑ Ⓒ Ⓓ	40.	Ⓐ Ⓑ Ⓒ Ⓓ	70.	Ⓐ Ⓑ Ⓒ Ⓓ	100.	Ⓐ Ⓑ Ⓒ Ⓓ
11.	Ⓐ Ⓑ Ⓒ Ⓓ	41.	Ⓐ Ⓑ Ⓒ Ⓓ	71.	Ⓐ Ⓑ Ⓒ Ⓓ	101.	Ⓐ Ⓑ Ⓒ Ⓓ
12.	Ⓐ Ⓑ Ⓒ Ⓓ	42.	Ⓐ Ⓑ Ⓒ Ⓓ	72.	Ⓐ Ⓑ Ⓒ Ⓓ	102.	Ⓐ Ⓑ Ⓒ Ⓓ
13.	Ⓐ Ⓑ Ⓒ Ⓓ	43.	Ⓐ Ⓑ Ⓒ Ⓓ	73.	Ⓐ Ⓑ Ⓒ Ⓓ	103.	Ⓐ Ⓑ Ⓒ Ⓓ
14.	Ⓐ Ⓑ Ⓒ Ⓓ	44.	Ⓐ Ⓑ Ⓒ Ⓓ	74.	Ⓐ Ⓑ Ⓒ Ⓓ	104.	Ⓐ Ⓑ Ⓒ Ⓓ
15.	Ⓐ Ⓑ Ⓒ Ⓓ	45.	Ⓐ Ⓑ Ⓒ Ⓓ	75.	Ⓐ Ⓑ Ⓒ Ⓓ	105.	Ⓐ Ⓑ Ⓒ Ⓓ
16.	Ⓐ Ⓑ Ⓒ Ⓓ	46.	Ⓐ Ⓑ Ⓒ Ⓓ	76.	Ⓐ Ⓑ Ⓒ Ⓓ	106.	Ⓐ Ⓑ Ⓒ Ⓓ
17.	Ⓐ Ⓑ Ⓒ Ⓓ	47.	Ⓐ Ⓑ Ⓒ Ⓓ	77.	Ⓐ Ⓑ Ⓒ Ⓓ	107.	Ⓐ Ⓑ Ⓒ Ⓓ
18.	Ⓐ Ⓑ Ⓒ Ⓓ	48.	Ⓐ Ⓑ Ⓒ Ⓓ	78.	Ⓐ Ⓑ Ⓒ Ⓓ	108.	Ⓐ Ⓑ Ⓒ Ⓓ
19.	Ⓐ Ⓑ Ⓒ Ⓓ	49.	Ⓐ Ⓑ Ⓒ Ⓓ	79.	Ⓐ Ⓑ Ⓒ Ⓓ	109.	Ⓐ Ⓑ Ⓒ Ⓓ
20.	Ⓐ Ⓑ Ⓒ Ⓓ	50.	Ⓐ Ⓑ Ⓒ Ⓓ	80.	Ⓐ Ⓑ Ⓒ Ⓓ	110.	Ⓐ Ⓑ Ⓒ Ⓓ
21.	Ⓐ Ⓑ Ⓒ Ⓓ	51.	Ⓐ Ⓑ Ⓒ Ⓓ	81.	Ⓐ Ⓑ Ⓒ Ⓓ	111.	Ⓐ Ⓑ Ⓒ Ⓓ
22.	Ⓐ Ⓑ Ⓒ Ⓓ	52.	Ⓐ Ⓑ Ⓒ Ⓓ	82.	Ⓐ Ⓑ Ⓒ Ⓓ	112.	Ⓐ Ⓑ Ⓒ Ⓓ
23.	Ⓐ Ⓑ Ⓒ Ⓓ	53.	Ⓐ Ⓑ Ⓒ Ⓓ	83.	Ⓐ Ⓑ Ⓒ Ⓓ	113.	Ⓐ Ⓑ Ⓒ Ⓓ
24.	Ⓐ Ⓑ Ⓒ Ⓓ	54.	Ⓐ Ⓑ Ⓒ Ⓓ	84.	Ⓐ Ⓑ Ⓒ Ⓓ	114.	Ⓐ Ⓑ Ⓒ Ⓓ
25.	Ⓐ Ⓑ Ⓒ Ⓓ	55.	Ⓐ Ⓑ Ⓒ Ⓓ	85.	Ⓐ Ⓑ Ⓒ Ⓓ	115.	Ⓐ Ⓑ Ⓒ Ⓓ
26.	Ⓐ Ⓑ Ⓒ Ⓓ	56.	Ⓐ Ⓑ Ⓒ Ⓓ	86.	Ⓐ Ⓑ Ⓒ Ⓓ	116.	Ⓐ Ⓑ Ⓒ Ⓓ
27.	Ⓐ Ⓑ Ⓒ Ⓓ	57.	Ⓐ Ⓑ Ⓒ Ⓓ	87.	Ⓐ Ⓑ Ⓒ Ⓓ	117.	Ⓐ Ⓑ Ⓒ Ⓓ
28.	Ⓐ Ⓑ Ⓒ Ⓓ	58.	Ⓐ Ⓑ Ⓒ Ⓓ	88.	Ⓐ Ⓑ Ⓒ Ⓓ	118.	Ⓐ Ⓑ Ⓒ Ⓓ
29.	Ⓐ Ⓑ Ⓒ Ⓓ	59.	Ⓐ Ⓑ Ⓒ Ⓓ	89.	Ⓐ Ⓑ Ⓒ Ⓓ	119.	Ⓐ Ⓑ Ⓒ Ⓓ
30.	Ⓐ Ⓑ Ⓒ Ⓓ	60.	Ⓐ Ⓑ Ⓒ Ⓓ	90.	Ⓐ Ⓑ Ⓒ Ⓓ	120.	Ⓐ Ⓑ Ⓒ Ⓓ

ANSWERS

1. (*c*)	**2.** (*b*)	**3.** (*c*)	**4.** (*b*)	**5.** (*a*)	**6.** (*c*)	**7.** (*c*)	**8.** (*c*)	**9.** (*d*)	**10.** (*b*)
11. (*c*)	**12.** (*c*)	**13.** (*b*)	**14.** (*c*)	**15.** (*a*)	**16.** (*c*)	**17.** (*c*)	**18.** (*d*)	**19.** (*c*)	**20.** (*c*)
21. (*a*)	**22.** (*a*)	**23.** (*c*)	**24.** (*a*)	**25.** (*b*)	**26.** (*b*)	**27.** (*a*)	**28.** (*a*)	**29.** (*b*)	**30.** (*c*)
31. (*b*)	**32.** (*d*)	**33.** (*d*)	**34.** (*d*)	**35.** (*a*)	**36.** (*c*)	**37.** (*b*)	**38.** (*b*)	**39.** (*b*)	**40.** (*b*)
41. (*d*)	**42.** (*c*)	**43.** (*b*)	**44.** (*c*)	**45.** (*a*)	**46.** (*b*)	**47.** (*d*)	**48.** (*c*)	**49.** (*d*)	**50.** (*c*)
51. (*c*)	**52.** (*d*)	**53.** (*b*)	**54.** (*c*)	**55.** (*a*)	**56.** (*c*)	**57.** (*a*)	**58.** (*d*)	**59.** (*b*)	**60.** (*b*)
61. (*c*)	**62.** (*b*)	**63.** (*b*)	**64.** (*c*)	**65.** (*c*)	**66.** (*b*)	**67.** (*b*)	**68.** (*b*)	**69.** (*a*)	**70.** (*d*)
71. (*d*)	**72.** (*d*)	**73.** (*d*)	**74.** (*b*)	**75.** (*d*)	**76.** (*d*)	**77.** (*c*)	**78.** (*a*)	**79.** (*b*)	**80.** (*d*)
81. (*a*)	**82.** (*c*)	**83.** (*a*)	**84.** (*d*)	**85.** (*c*)	**86.** (*a*)	**87.** (*c*)	**88.** (*c*)	**89.** (*c*)	**90.** (*b*)
91. (*a*)	**92.** (*a*)	**93.** (*b*)	**94.** (*a*)	**95.** (*b*)	**96.** (*d*)	**97.** (*a*)	**98.** (*b*)	**99.** (*b*)	**100.** (*d*)
101. (*b*)	**102.** (*a*)	**103.** (*a*)	**104.** (*d*)	**105.** (*d*)	**106.** (*b*)	**107.** (*a*)	**108.** (*b*)	**109.** (*a*)	**110.** (*a*)
111. (*c*)	**112.** (*a*)	**113.** (*b*)	**114.** (*b*)	**115.** (*d*)	**116.** (*b*)	**117.** (*b*)	**118.** (*d*)	**119.** (*d*)	**120.** (*b*)

EXPLANATIONS

1. Average speed = $\frac{2xy}{x+y}$

$= \frac{2(40)(60)}{(40+60)}$

$= \frac{4800}{100}$

= 48 Km/hr

$\therefore$ Option (c) is correct

2. R I D D L E D

No of ways = $\frac{7!}{3!}$

$= \frac{7 \times 6 \times 5 \times 4 \times 3!}{3!}$

$= 7 \times 6 \times 5 \times 4$

= 840

$\therefore$ Option (b) is correct

3.

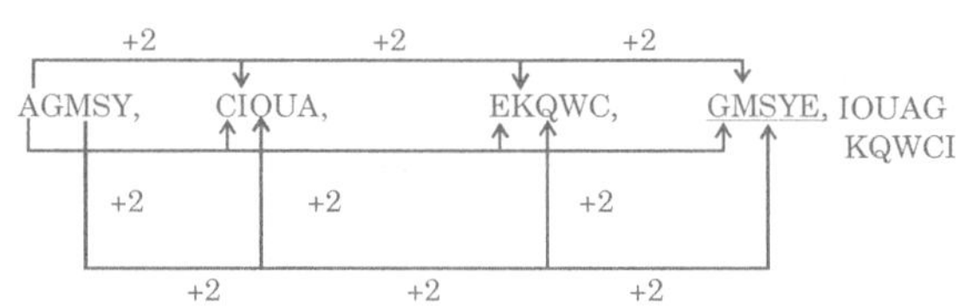

$\therefore$ Option (c) is correct

4. Surface area of cube = $6a^2$.

and surface area of sphere $4\pi r^2$.

According to question

$6a^2 = 4\pi r^2$.

$\therefore a^2 = \frac{4\pi r^2}{6}$

$= \frac{2\pi r^2}{3}$

Now $\frac{\text{volume of sphere}}{\text{volume of cube}}$

$$= \frac{\frac{4}{3}\pi r^3}{a^3}$$

$$= \frac{\frac{4}{3}\pi r^3}{\frac{2\pi r^2}{3}\sqrt{\frac{2\pi}{3}r}}$$

$$= \frac{\frac{4}{3}}{\frac{2}{3}\times\frac{\sqrt{2\pi}}{\sqrt{3}}}$$

$$= \frac{4\times 3\sqrt{3}}{6\times\sqrt{2\pi}}$$

$$= \frac{2\sqrt{3}}{\sqrt{2}\sqrt{\pi}}$$

$$= \frac{\sqrt{6}}{\sqrt{\pi}}$$

$\therefore$ Option (b) is correct

According to question 5 to 7

Industry	Worker
Paper	D,B
Drug	A, E
Toy	C, F

5. A and E work in the drug industry

$\therefore$ Option (a) is correct

6. D and B, work in paper industry

$\therefore$ Option (c) is correct

7. B, A and C represent paper, drugs and toy respectively.

8. Among the following events given in option Champaran styagrah occurred first

9. $\therefore$

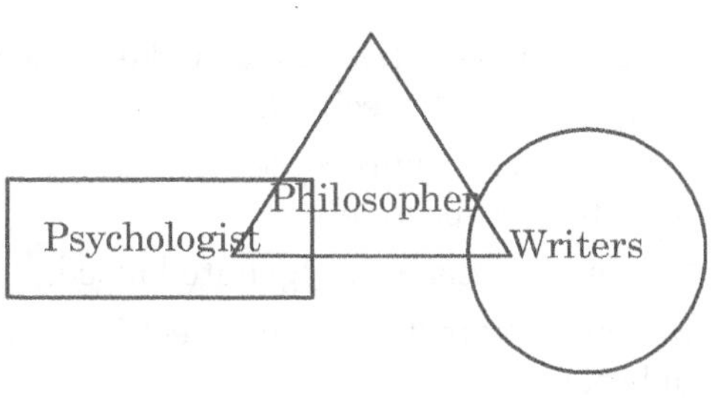

$\therefore$ Option (c) is correct

10. $(243)^{\frac{3}{5}} \times (729)^{\frac{-2}{3}} = 3^x$

$\therefore\ \left(3^5\right)^{\frac{3}{5}} \times \left(9^3\right)^{\frac{-2}{3}} = 3^x$

$\therefore\ 3^3 \times 9^{-2} = 3^x$

$\therefore\ 3^3 \times 3^{-4} = 3^x$

$\therefore\ 3^{-1} = 3^x$

$\therefore\ x = 1$

$\therefore$ Option (b) is correct

12.

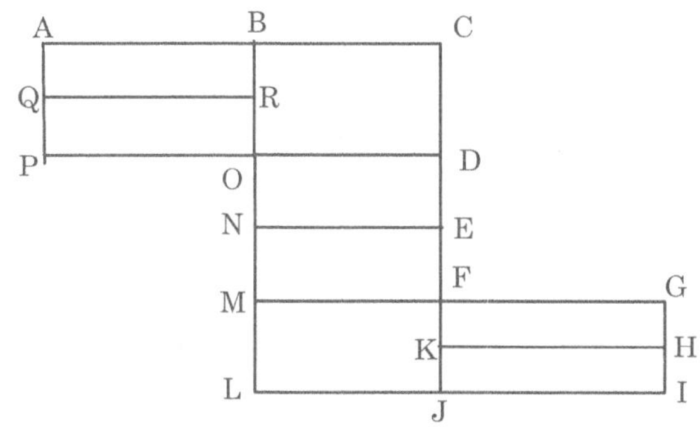

ABQR, ABPO, QROP, BCDO, BCNE, ODNE, BCFM, ODFM, NEFM BCJL, ODLJ, NEJC, MFLJ, FGHK, FGIJ, KHIJ, MFGIJL, ABCDOP

$\therefore$ Total 18 rectangles are there

$\therefore$ Option (c) is correct

13. $1925 = \frac{\pi r^2}{1}$ ....(1)

$\therefore\ \frac{1925\times 2\times 7}{22} = r^2$

$\therefore\ r^2 = 1225$

$r = 35$ cm

Now $\pi r = w$ (where 'w' is breadth of a rectangle)

$\therefore w = \frac{22}{7}\times 35$

$\therefore\ w = 22 \times 5 = 110 + 70 = 180$

$L = 48 \times 4 = 192$

$\therefore$ Perimeter of rectangle = 2 (l + w)

= 2(192 + 180)

= 2(372)

= 744 cm

$\therefore$ Option (b) is correct.

14. Let CP of 1 cloth be ₹100 and let total cloth be 100

S.P of 50 cloth = 140 × 50 = 7000

and S.P of remaining half = 60 × 25 1500

and Remaining S.P of 25 cloth = 25 × 100 = 2500

∴ C.P of 100 cloth = 100 × 100 = ₹10000

and S.P of 100 cloth = 7000 + 1500 + 2500

= 110000

$$\therefore \text{Pr ofit} = \left(\frac{10000}{10000} \times 100\right)\%$$

= 10%

∴ Option (c) is correct

15. Khaira disease found in paddy is due to deficiency of zinc in the soil.

∴ Option (a) is correct

16. Article 45 - directive principle of state policy

∴ Option (c) is correct

17. $I \to X$ and $G \to P$

$N \to Y$ $L \to M$

$D \to Z$ $A \to E$

$I \to X$ $D \to Z$

$A \to E$

$\Rightarrow I \to X, N \to Y, D \to Z, A \to E, G \to P, L \to M$

∴ Code LANDING will be "MEYZXYD"

∴ Option (c) is correct.

18.

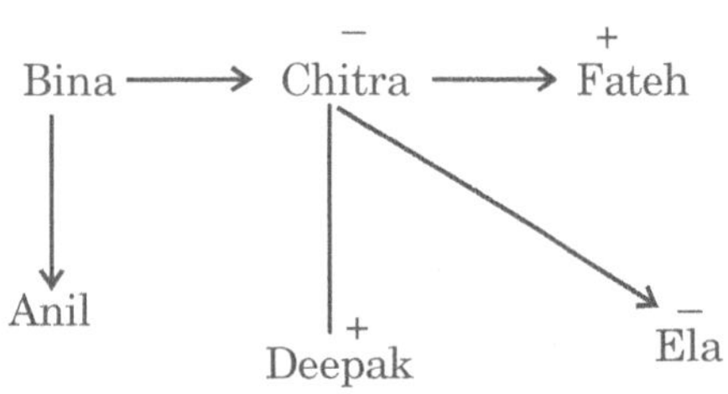

Ela is niece of fateh

∴ Option (d) is correct.

19.

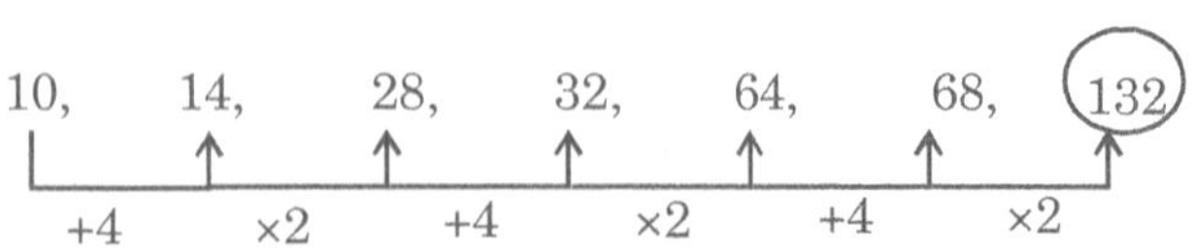

∵ 68 × 2 = 136 ≠ 132

∴ 132 is wrong number in series

∴ Option (c) is correct.

20. Cochin International airport Ltd is the first airport of the world to fully operate on solar power.

∴ Option (c) is correct.

21. Portugal is an European countries among options whose border is with Atlantic ocean

∴ Option (a) is correct.

22. Indian = I = {a, b, c, d}

Actors = A {b, f, c, e}

singers = S = {c, d, e, g}

$I \cap A = \{b, c\}$

Now $(I \cap A) - S = \{b\}$

∴ Option (a) is correct.

23. A mission of NASA in which new Horizons became the first space probe to reach closest to pluto.

24. Option (a) satisfies the condition

25. Option (b) satisfies the condition

26.

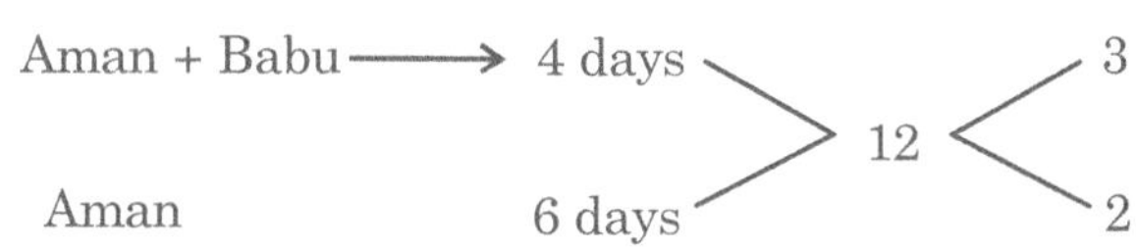

∴ Babu does 1 unit work in a day

⇒ Remaining '6' unit work will be done by Babu in 6 days

∴ Option (b) is correct

28. Indian standard Meridian passes through UP, MP, Chhattisgarh, Odisha and Andhrapradesh

29. (Gross fiscal deficit) – Interest

= Gross primary deficit

∴ Option (b) is correct

30. Yeats was famous poet who was deeply affected by Tagores Gitanjali

32. Agha Khan Cup trophy is given in Hockey Daleep Trophy is given in Cricket welligntos Tropy is given in Rowing Ezra Cup is given in Polo and Subrato cup is given in football

33.

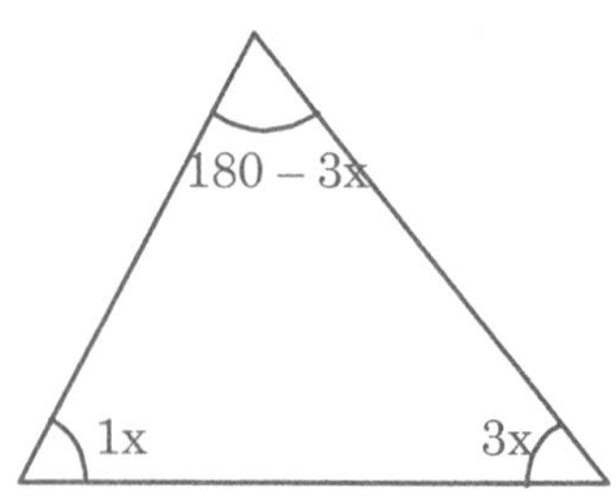

Largest angle of triangle can't be determined as 30° angle is not specified.

34. Gun metal contains copper, tin and zinc in ratio 87 : 10 : 3 respectively.

35. Left circle receives oxygenated blood.

37. Rabindra Nath Tagore received first Nobel prize in India.

39. Let the number of boys be 'x'

then $x + \frac{7x}{10} = 85$

$\therefore$ x = 50

$\Rightarrow$ No of girls = 85 – 50 = 35

Now no of boys playing only badminton

= 50% of boys

$= \frac{50}{100} \times 50 = 25$

and no of children playing only tennis

= 40% of total no of children

$= \frac{40}{100} \times 85 = 34$

$\therefore$ Total no of children playing badminton and tennis both = 12

$\therefore$ No of girls playing only badminton

= 85 – (25 + 34 + 12)

= 14

$\therefore$ Option (b) is correct

40.

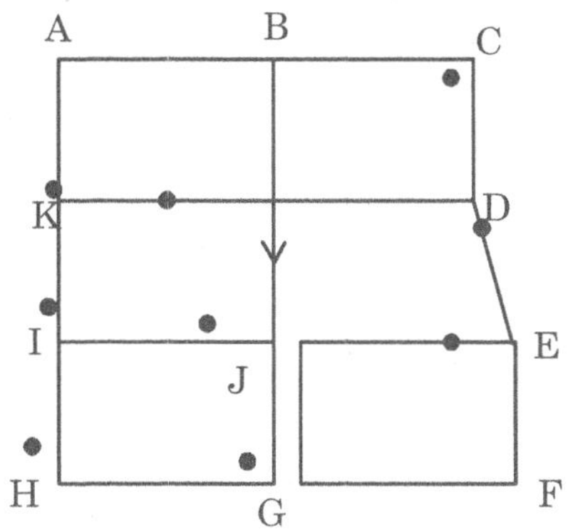

ABLK, BCDL, KLJI, LDJE, ABCDEJI, IJHG, JEFG, ABCDEFGHIK

$\therefore$ Option (b) is correct

42. Eilean Collins was first woman in NASA to command space flight.

$\therefore$ Option (c) is correct

43. Let the maximum marks be x

$\therefore$ Pass mark $= \frac{36}{100} \times x$

Now according to question,

$\frac{36x}{100} = 17 + 10$

$\therefore \frac{36x}{100} = 27$

$\therefore x = \frac{2700}{36}$

$\therefore$ x = 75

$\therefore$ Option (b) is correct

44. $x = \frac{\sqrt{3} - \sqrt{2}}{\sqrt{3} + \sqrt{2}} \times \frac{\sqrt{3} - \sqrt{2}}{\sqrt{3} - \sqrt{2}}$

$\therefore x = 3 + 2 - 2\sqrt{6}$

$= 5 - 2\sqrt{6}$

and $y = \frac{\sqrt{3} + \sqrt{2}}{\sqrt{3} - \sqrt{2}} \times \frac{\sqrt{3} + \sqrt{2}}{\sqrt{3} + \sqrt{2}}$

$= 3 + 2 + 2\sqrt{6}$

Now $x^2 + y^2 + xy$

$= (5 - 2\sqrt{6})^2 + (5 + 2\sqrt{6})^2 + (5 - 2\sqrt{6})(5 + 2\sqrt{6})$

$= 25 + 24 - 20\sqrt{6} + 25 + 24 + 20\sqrt{6} + 25 - 24$

= 50 + 48 + 1

= 99

$\therefore$ Option (c) is correct

45. Rhine valley in France is known for Bauxite

46. Frequency multiplier increases the deviation

$\left(\beta = \frac{\Delta f}{f_m}\right)$

After passing through frequency multiplier,

$\beta' = \frac{3\Delta f}{f_m} = 3\beta$

48. (c)

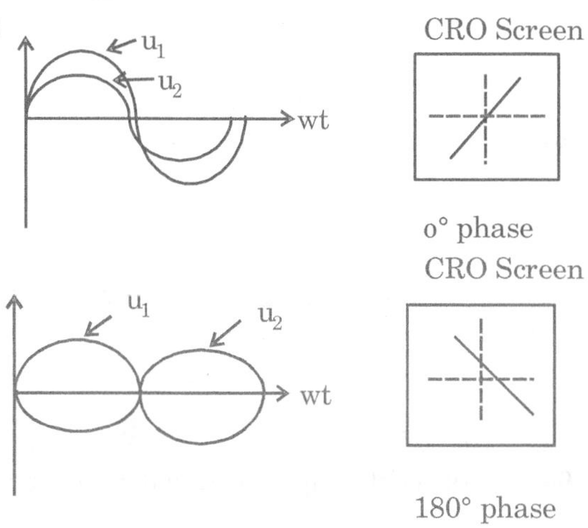

49. In this case, the new voltage and currents are reversed of previous values

50. In 8-bit 8085 microprocessor, 9th bit (or MSb) of result is stored in carry status flag.

51. Max. frequency component is signal, f_{max} = 300 Hz.

Nyquist sampling freq; $f_s = 2f_{max} = 600$ Hz

Nyquist sampling rate, $T_s = \frac{1}{f_s} = \frac{1}{600}$s.

53. $\frac{2s+5}{(s+3)(s+2)} = \frac{A}{s+3} + \frac{B}{s+2}$

$$A = \left.\frac{2s+5}{s+2}\right|_{s=-3} = \frac{-1}{-1} = 1;$$

$$B = \left.\frac{2s+5}{s+3}\right|_{s=-2} = \frac{1}{1} = 1$$

$$L-1\left[\frac{1}{s+3} + \frac{1}{s+2}\right] = e^{-3t} + e^{-2t}$$

52. Noise factor, $F_n = \frac{T_e}{T_o} = \frac{30}{300} + 1 = 1.1$

54. Folding freq,

$$f_{max} = \frac{f_{sampling}}{2} = \frac{600}{2} = 300\text{Hz}$$

55. Output of dynamic system depends upon present as well as past & future values.

$$\underbrace{y(n)}_{\substack{\text{present}\\\text{value}}} + \underbrace{y(n-1)}_{\substack{\text{past}\\\text{value}}} + \underbrace{y(n+3)}_{\substack{\text{future}\\\text{value}}} = 0$$

56. In schottky diode, there is no charge storage and hence almost zero reverse recovery time.

57. frequency shift keying is used mostly is the high frequency radio spectrum and telephone etc.

58. Brewster angle

$$\theta = \tan^{-1}\left(\sqrt{\frac{\varepsilon_{r_2}}{\varepsilon_{r_1}}}\right) = \tan^{-1}\sqrt{3} = 60°$$

59.

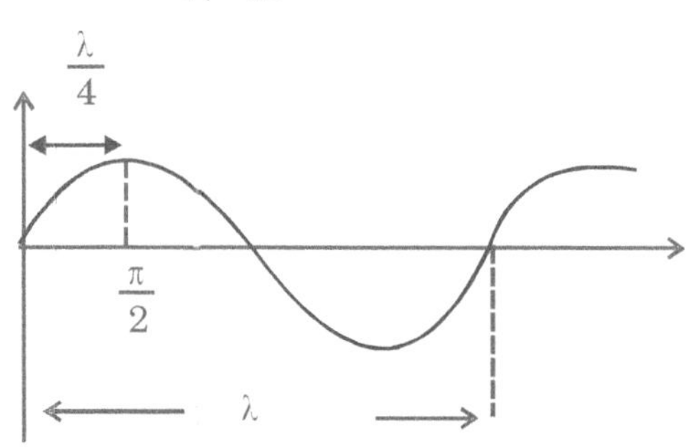

60. Characteristic impedence is independent of length.

61. For NOR output, y = 1,

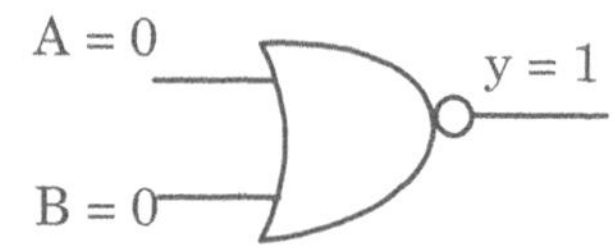

62. D_1 will conduct and the output voltage will be about 7V.

Therefore, D_2 will be reverse biased & will not conduct

63. f = 20.48 MHz, n = 12

output frequency

$$= \frac{f}{2^n} = \frac{20.48 \times 10^6}{2^{12}} = 5 \text{ KHz}$$

64. RAL will rotate the content of Accumulator by one shift and content is doubled the previous value. For 2 RAL operation, 8 will become 8 × 2 × 2 = 32d.

65. The output of an AND with three inputs is high when all the three inputs are high.

66. Gain of FET amplifier $\propto g_m$.

67. B.W. = 36 MHz.

New freq. range will be $70 - \frac{B.W.}{2}$ to $70 + \frac{B.W.}{2}$ = 52 to 88 MHz.

68. For centre tapped full wave rectifier,

PIV = $2V_m$ = 60V.

69. For straight line inclined at 45° to x-axis CRO screen, both X-input & y-input are either in phase or out of phase. Hence, $V_y(t) = 2\sin(\omega t)$

70. Using Miller's theorem,

$$Z_i = Z(1-A) = \frac{R_f.R_i}{(R_i + R_f)} = \frac{10 \times 100}{10 + 100}$$

Z_i = 9.09 KΩ.

72. Energy stored in capacitor

$$W = \frac{1}{2}CV^2 = \frac{1}{2} \times 10^{-6} \times 12 \times 12$$

$= 72 \times 10^{-6}$ J.

73. Charecter impedance,

$$Z_o = \sqrt{Z_{oc}.Z_{sc}}$$

$$= \sqrt{20 \times 5} = 10\Omega$$

74. $I(s) = \frac{s+1}{(s+4)(s+3)} = \frac{3}{s+4} - \frac{2}{s+4}$

$$i(t) = 3e^{-4t} - 2e^{-3t}$$

75. Static capacitor decreases the line current & increases power factor.

76. Negative sign in the expression,

$$e = -N\frac{d\phi}{dt}(\text{faraday's law}) \text{ is due to Lenz's law.}$$

77. $\text{form factor} = \frac{\text{RMS value}}{\text{Average value}}$

78. A dopant with valence electrons is also called a pentavalent impurity.

79. Diamagnetic materials are magnetic insulators, i.e., they do not magnetic dipoles.

81. Transconductance,

$$g_m = \frac{\Delta I_D}{\Delta V_{Gs}}\bigg|_{V_{DS}=\text{constant}}$$

82. $\text{Current} = \frac{V_{th}}{R_{th}+R} = \frac{20}{5+R} < 4\text{A}.$

83. Superposition theorem is applicable only to linear networks.

85. Directive gain, $D_a = \frac{V}{V_o}$

V = radiation intensity due to test antenna (Watt/solid angle)

V_o = radiation intensity due to isotropic antenna (watt/solid angle)

$o \le D_a \le \text{Directivity}$

86. Input resistance with feedback for shunt,

$$Ri' = \frac{Ri'}{1+A\beta}$$

$$Ri' = \frac{1\times 10^3}{1+50\times 0.2} = \frac{1}{11}K\Omega$$

88. Due to non-uniform quantization, signal to noise ratio of signal levels is increased.

89. PD controller is used to increases the bandwidth and also increases the signal to noise ratio by reducing the noise and increasing the signal.

91. In CDMA, as the number of users increases, the overall quality of service decreases (multiple access interference).

92.

Ku Band	Uplink	Downlink
Europe	14 GHz	11GHz
America	14 GHz	11 GHz

93. Piezo-electric transducer is an active transducer.

94.

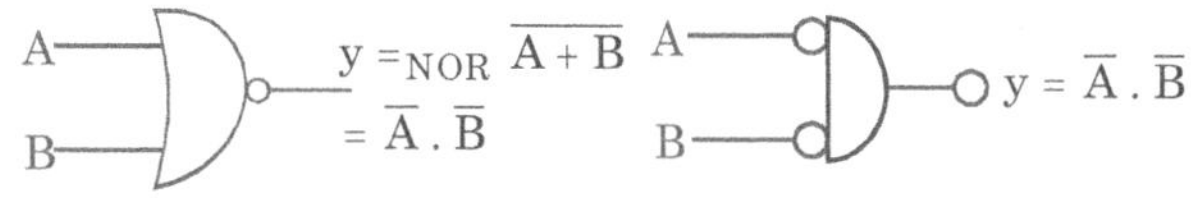

(De-Morgan's theorem)

95. Applying KVL

$$1.\ i(t) + 1.\frac{di(t)}{dt} + \frac{1}{1}\int i(t)dt = u(t)$$

$$I(S) + SI(S) - i(o) + \frac{I(S)}{S} + \frac{v(o)}{S} = \frac{1}{S}$$

$$I(S) + SI(S) - 1 + \frac{I(S)}{S} + \frac{1}{S} = \frac{1}{S}$$

$$I(S) = \frac{S}{S^2+S+1}; t \ge 0$$

96. At high frequencies, time period of wave and transit time become comparable.

97. In broadcast communication receiver, most of the receiver selectivity is achieved is Radio-frequency section.

98. $\text{Slew-rate} = \left|\frac{dv_{out}(t)}{dt}\right|_{max} \text{v}/\mu\text{s}.$

99. Input impedance of ideal amplifier is infinite.

100. Atomic concentration can't be determine using Hall-effect.

101.

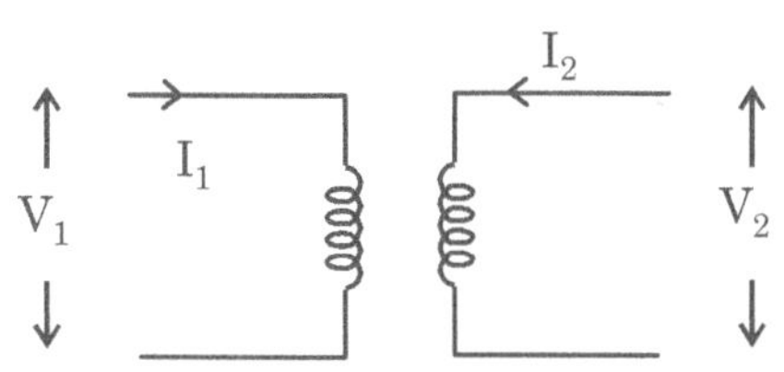

n : 1

$$V_1 = AV_2 + BI_2$$

$$I_1 = CV_2 + DI_2$$

$$\frac{V_1}{V_2} = n = A$$

$$\frac{I_1}{I_2} = \frac{1}{n} = D$$

$$\begin{bmatrix} A & B \\ C & D \end{bmatrix} = \begin{bmatrix} n & o \\ o & \frac{1}{n} \end{bmatrix}$$

102.

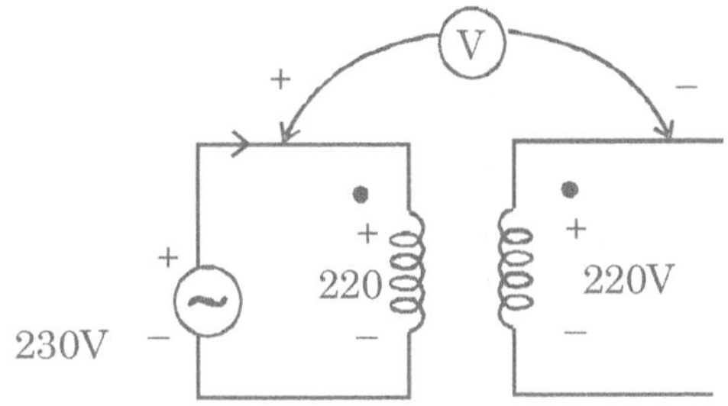

Applying KVL,

$-220 + V + 220 = 0$

$V = 0$ volts.

103. Given block diagram is of dual slope ADC.

104. Theoretical bandwidth requirement for FM system is infinite.

105. Size of waveguide $\propto \frac{1}{\text{frequency}}$

106. Bandwidth of F.M. is larger than A.M.

107. Modulation does not reduce the bandwidth.

108. In n-type semiconductor, on increasing the donor concentration the fermi-level moves towards the conduction band.

109. White noise is the random signal with constant magnitude for all frequencies.

110. $\overline{A+B} = \overline{A}.\overline{B}$ (De-Morgan's Theorem)

111. The output of 555 timer is rectangular pulse train.

112. All pass filter improves the phase of signal with changing the magnitude.

113. GaAs is used to manufacture LEDs.

114. When the initial slope of input sine wave is greater than the slow rate of OP-AMP, the output approaches to rectangular waveshape.

115. For distortionless circuit, phase response is proportional to frequency.

116. For full-wave rectifier,

$f_o = 2f_i = 2 \times 50 = 100$ Hz.

117. $\beta = \frac{I_C}{I_B} \Rightarrow I_B = \frac{10 \times 10^{-3}}{100} = 100\mu A.$

118.

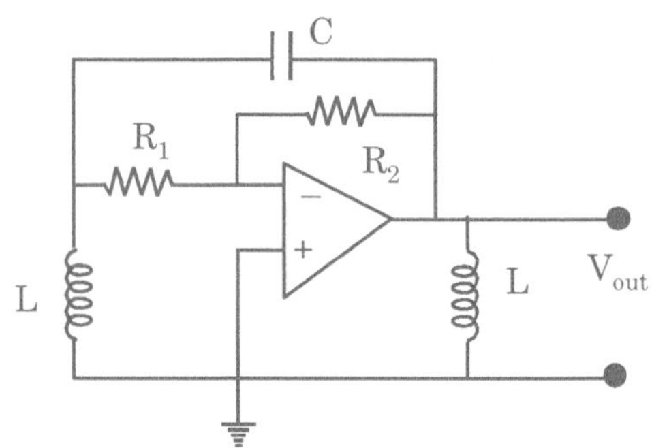

Hartley Oscillator

120. Reflection coefficient,

$$\rho_V = \frac{25-100}{25+100} = -0.6.$$

PRACTICE PAPER-1

ELECTRONICS ENGINEERING

Time Allowed : 2:15 Hours **Maximum Marks : 180**

Read the following instructions carefully before you begin to answer the questions. This Booklet contains questions in English.

INSTRUCTIONS

1. This Booklet contains 180 questions in total comprising of the following five parts:

Part – I:	Domain Knowledge	[1 – 75 Questions]
Part – II:	General Intelligence and Reasoning	[76 – 90 Questions]
Part – III:	General Awareness	[91 – 105 Questions]
Part – IV:	Quantitative Aptitude	[106 – 120 Questions]
Part – V:	General English	[121 – 180 Questions]

2. All questions are compulsory and carry 1 mark each.

3. Before you start to answer the questions you must check this Booklet to ensure that it contains all the pages and see that no page is missing or repeated.

4. For every wrong answer 1/3 marks will be deducted.

5. You are supplied with the OMR-Sheet separately.

6. Answers must be indicated by completely blackening one of the corresponding ovals on OMR-Sheet against the relevant question number, by black/blue ball pen only. Answers which are not indicated by black/blue ball pen will not be awarded any mark.

7. The OMR-Sheet must be handed over to the Invigilator before you leave the Examination Hall.

8. Failure to comply with any of the above instructions will render a candidate liable to such action/penalty as may be deemed fit.

PRACTICE PAPER – 1

DOMAIN KNOWLEDGE

1. Which semiconductor power device out of the following is not a current triggered device?

(*a*) Thyristor (*b*) G.T.O

(*c*) Triac (*d*) MOSFET

2. A switched mode power supply operating at 20 kHz to 100 kHz range uses as the main switching element.

(*a*) Thyristor (*b*) MOSFET

(*c*) Triac (*d*) UJT

3. In a commutation circuit employed to turn-off an SCR, satisfactory turn-off is obtained when

(*a*) circuit turn-off time < device turn-off time

(*b*) circuit turn-off time > device turn-off time

(*c*) circuit time constant > device turn-off time

(*d*) circuit time constant < device turn-off time

4. The conduction loss versus device current characteristic of a power MOSFET is best approximated by

(*a*) a parabola

(*b*) a straight line

(*c*) a rectangular hyperbola

(*d*) an exponentially decaying function

5. An SCR is considered to be a semi-controlled device because

(*a*) it can be turned OFF but not ON with a gate pulse

(*b*) it conducts only during one half-cycle of an alternating current wave

(*c*) it can be turned ON but not OFF with a gate pulse

(*d*) it can be turned ON only during one half-cycle of an alternating voltage wave

6. A certain oscilloscope with 4 cm screen has its own sweep output fed to its input. If the x and y sensitivities are same, the oscilloscope will display a

(*a*) Triangular wave

(*b*) Diagonal line

(*c*) Sine wave

(*d*) Circle

7. The Thevenin's equivalent resistance R_{th} for the given network is

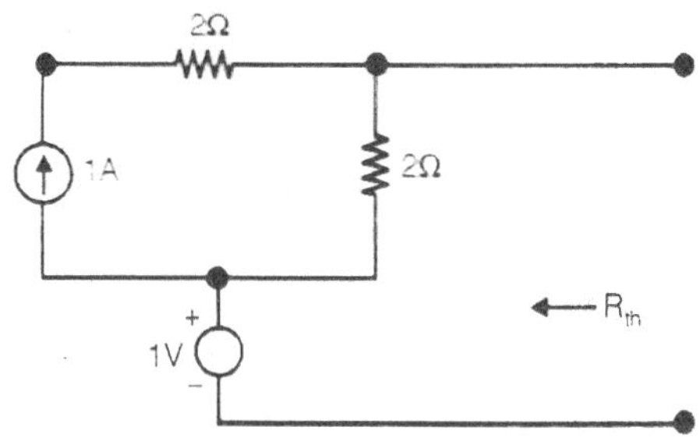

(*a*) 1 Ω (*b*) 2Ω

(*c*) 4 Ω (*d*) infinity

8. An average reading digital multimeter reads 10 V when fed with a triangular wave, symmetric about the time-axis. For the same input an rms-reading meter will read

(*a*) $\frac{20}{\sqrt{3}}$ V (*b*) $\frac{10}{\sqrt{3}}$ V

(*c*) $20\sqrt{3}$ V (*d*) $10\sqrt{3}$ V

9. In an oscilloscope screen, linear sweep is applied at the

(*a*) Vertical axis

(*b*) Horizontal axis

(*c*) Origin

(*d*) Both horizontal and vertical axis

10. A diode whose terminal characteristics are related as $i_D = I_S\, e^{V/V_T}$, where I_S is the reverse saturation current and V_T is thermal voltage (= 25 mV), is biased at i_D = 2 mA. Its dynamic resistance is

(*a*) 25 Ω (*b*) 12.5 Ω

(*c*) 50 Ω (*d*) 100 Ω

11. The most dominant parameter with respect to temperature is

(*a*) I_{CO} (*b*) β

(*c*) V_{BE} (*d*) α

12. Radio broadcasts are generally

(*a*) amplitude modulation

(*b*) frequency modulation

(*c*) both amplitude and frequency modulation

(*d*) neither amplitude nor frequency modulation

13. If the frequency of a wave 20 Hz, the time period is
(*a*) 20 secs (*b*) 2 secs
(*c*) 0.2 secs (*d*) 0.05 secs

14. Push pull amplifier is
(*a*) voltage amplifier (*b*) current amplifier
(*c*) power amplifier (*d*) none of the above

15. If the modulation index of an AM wave is changed from 0 to 1, the transmitted power
(*a*) increases by 50% (*b*) increases by 75%
(*c*) increases by 100% (*d*) remains unaffected

16. Which of the following is true as regards photo emission?
(*a*) Velocity of emitted electrons is dependent on light intensity
(*b*) Rate of photo emission is inversely proportional to light intensity
(*c*) Maximum velocity of electron increases with decreasing wave length
(*d*) Both holes and electrons are produced

17. If the height of transmitting and receiving antenna in a LOS system are 49 meter and a 9 meter respectively then the distance up to which communication may be possible is about
(*a*) 40 km (*b*) 110 km
(*c*) 400 km (*d*) 1100 km

18. In a line VSWR of a load is 6 d(*b*) The reflection coefficient will be
(*a*) 0.033 (*b*) 0.33
(*c*) 0.66 (*d*) 3.3

19. $\nabla.\vec{J} = 0$ is frequently known as
(*a*) continuity equation for steady currents
(*b*) Laplace equation
(*c*) Poisson's equation
(*d*) none of the above

20. **Assertion (A):** If the emitter bypass capacitor of an R-C coupled CE amplifier gets disconnected, its voltage gain increases.
Reason (R): The unbypassed emitter resistor gives negative feedback.
(*a*) Both A and R are true and R is the correct explanation of A
(*b*) Both A and R are true but R is NOT the correct explanation of A
(*c*) A is true but R is false
(*d*) A is false but R is true

21. The input impedance of a λ/8 long short circuited section of a loss less transmission line is
(*a*) zero (*b*) inductive
(*c*) capacitance (*d*) infinite

22. Which of the following relations is correct?
(*a*) $\nabla.\vec{B} = \rho$ (*b*) $\nabla.\vec{D} = \rho$
(*c*) $\nabla.\vec{D} = -\rho$ (*d*) $\nabla.\vec{B} = \vec{J}$

23. A loop antenna may be of
(*a*) square on triangular shape
(*b*) hexagonal shape
(*c*) circular shape
(*d*) any of the above

24. Velocity of the wave when it passes from air into sphere
(*a*) decreases (*b*) increases
(*c*) same (*d*) no change

25. Microwave signals transmitted towards the sky are
(*a*) strongly absorbed by the ionosphere
(*b*) strongly reflected by the ionosphere
(*c*) transmitted through the ionosphere
(*d*) unable to reach the ionosphere because of strong absorption in the lower atmosphere

26. The $\sqrt{L/C}$ has the dimension of
(*a*) time (*b*) capacitance
(*c*) velocity (*d*) resistance

27. The value of VSWR is taken as infinite when the line is
(*a*) short-circuited
(*b*) open-circuited
(*c*) terminated with full load
(*d*) terminated with load equal to the characteristic impedance of the load

28. The wavelength of an electromagnetic wave after reflection at angle on a surface
(*a*) remains same as the wavelength perpendicular to the surface
(*b*) remains same in the free space
(*c*) increases in actual direction of propagation
(*d*) decreases in actual direction of propagation

29. For a reflection co-efficient of 0.5, the value of SWR is
(*a*) -1 (*b*) 1
(*c*) 2 (*d*) 3

30. A 50π characteristic impedance line is connected to a load which shows a reflection coefficients of 0.268. If V_{in} = 15 volt, then the net power delivered to the load will be

(*a*) 0.139 W (*b*) 1.39 W

(*c*) 0.278 W (*d*) 2.78 W

31. Indicate the false statement: If SWR on a transmission line is infinite, the line is terminated in

(*a*) open circuit (*b*) short circuit

(*c*) pure reactance (*d*) complex impedance

32. The resonant frequency of an RF amplifier is 1 MHz and its bandwidth is 10 kHz. The Q factor will be

(*a*) 10 (*b*) 100

(*c*) 0.01 (*d*) 0.1

33. A transmitter power of 5 W is increased by 20 dB The effective radiated power is

(*a*) 5 W (*b*) 100 W

(*c*) 25 W (*d*) 500 W

34. The percentage saving in power of 100% modulated suppressed carrier AM signal is

(*a*) 100 (*b*) 75

(*c*) 66.6 (*d*) 50

35. The maximum distance between television transmission and receiving antennas for reasonably good reception should be

(*a*) less than 20 km (*b*) less than 100 km

(*c*) less than 500 km (*d*) less than 1000 km

36. In 8085, the DAA instruction is used for

(*a*) Direct Address Accumulator

(*b*) Double Add Accumulator

(*c*) Decimal Adjust Accumulator

(*d*) Direct Access Accumulator

37. The addressing mode used in the instruction JMP F347 H. In case of an Intel 8085A microprocessor is which one of the following?

(*a*) Direct (*b*) Register-indirect

(*c*) Implicit (*d*) Immediate

38. What do you call an attenuation that occurs over many different wavelengths of the carrier?

(*a*) Rayleigh fading

(*b*) Rician fading

(*c*) Wavelength fading

(*d*) Slow fading

39. In microwave transmission using digital radio, what causes most inter symbol interference?

(*a*) Delayed spreading

(*b*) Rayleigh fading

(*c*) Random Doppler shift

(*d*) Slow fading

40. The highest frequency which a conventional vacuum-tube oscillator can generate is not limited by the

(*a*) Electron transit time

(*b*) Distributed lead inductance

(*c*) Inter-electrode capacitance

(*d*) Degree of emission from the cathode

41. As the electron beam moves through a klystron's intercavity drift space

(*a*) Frequency modulation at the input cavity creates velocity modulation at the output cavity

(*b*) Velocity modulation at the input cavity creates density modulation at the output cavity

(*c*) Density modulation at the input cavity creates velocity modulation at the output cavity

(*d*) Phase modulation at the input cavity creates velocity modulation at the output cavity

42. Which one of the following photo-detector does not provide gain?

(*a*) Photo-transistor

(*b*) Photo conductor

(*c*) Avalanche photodiode

(*d*) p-i-n photodiode

43. Light travels along the optical fibres by which mechanism?

(*a*) Refraction

(*b*) Reflection

(*c*) Scattering

(*d*) Total internal reflection

44. The code which provides for parity check is

(*a*) Baudot

(*b*) ASCII

(*c*) EBCDIC

(*d*) Excess-3

45. Which of the following can be used to improve range resolution in a Radar?

(*a*) Short duration pulse.

(*b*) Long duration pulse

(*c*) High frequency of operating signal

(*d*) Increasing pulsewidth

46. Doppler effect is employed in which of the following?
1. Moving target plotting on PPI
2. The MTI system
3. FM radar
4. CW radar

Code

(a) 2, 3 and 4 (b) 1,2 and 3
(c) 1,2 and 4 (d) 1, 3 and 4

47. In free space line of sight propagation case, the transmission losses between transmitter and receiver increase with frequency (f) as

(a) f (b) f^2
(c) f^4 (d) $f^{1/2}$

48. In asynchronous TDM for n signal sources, each frame contains m slots, where m is usually

(a) less than n (b) 2 n
(c) n (d) greater than 2n

49. Which of the following pulse modulations is analog?

(a) PCM (b) Differential PCM
(c) PWM (d) Delta

50. For a given data rate, the bandwidth β_p of a BPSK signal and the bandwidth β_0 of the OQPSK signal are related as

(a) $\beta_p = \beta_0/2$ (b) $\beta_p = 2\beta_0$
(c) $\beta_p = \beta_0/4$ (d) $\beta_p = \beta_0$

51. For an AM wave, the maximum voltage was found to be 10 V and the minimum voltage was found to be 5 V. The modulation index of the wave would be

(a) 0.33 (b) 0.52
(c) 0.40 (d) 0.1

52. In an amplitude modulated system, if the total power is 600 W and the power in carrier is 400 W, then the modulation index is

(a) 0.5
(b) 0.75
(c) 0.9
(d) 1

53. Time division multiplexing requires

(a) constant data transmission
(b) transmission of data samples
(c) transmission of data at random
(d) transmission of data of only one measured

54. Which one of the following statement is correct? In a transistor

(a) I_{CBO} is greater than I_{CEO}, and does not depend upon temperature
(b) I_{CBO} is greater than I_{CO}, and doubles for every ten degrees rise in temperature
(c) I_{CBO} is equal to I_{CO} and double for every ten degrees rise in temperature
(d) I_{CBO} is equal to I_{CO} and doubles for every ten degrees rise in temperature.

55. In a p-type silicon sample, the hole concentration is 2.25×10^{15}/cc. If the intrinsic carrier concentration is 1.5×10^{10}/cc, what is the electron concentration in the p-type silicon sample

(a) zero (b) 10^{10}/cc
(c) 10^5/cc (d) 1.5×10^{25}/cc

56. For a voltage follower circuit using an op- amp, which of the following is true?

(a) both input and output impedance are very large
(b) input impedance is very large and the output impedance is very small
(c) input impedance is very small and the output impedance is very large
(d) both input and output impedance are very small

57. The electron density profile in a piece of semiconductor at equilibrium is such that $n(x_1) = 10n(x_2)$. The hole density profile will be such that

(a) $p(x_1) = 10p(x_2)$
(b) $p(x_2) = 10p(x_1)$
(c) $p(x_1) = 100p(x_2)$
(d) insufficient information to answer

58. A bar of certain material having rectangular cross section is one metre in length and has a cross section of 0.0001 square metres. The value of its longitudinal resistance is 100 ohms. The value of its resistivity would be

(a) 1 ohm-metre (b) 0.01 ohm-metre
(c) 0.0001 ohm-metre (d) 100 ohm-metre

59. Two capacitors of equal value, 1 microfarad, are connected in parallel. The effective capacitance of this combination would be

(a) 2 microfarad
(b) 0.5 microfarad
(c) 4 microfarad
(d) 0.25 microfarad

60. Two capacitors $C_1 = 20\ \mu F$ and $C_2 = 25\ \mu F$ with respective breakdown voltages of $V_1 = 50$ V and $V_2 = 40$ V are given,

(*a*) Both capacitors can stores the same maximum energy.

(*b*) When connected up in series and charged, C_1 will fail first.

(*c*) Both capacitors can store the same maximum charge.

(*d*) When connected up in series and charged, C_2 will fail first.

61. The output voltage V_o of the circuit is

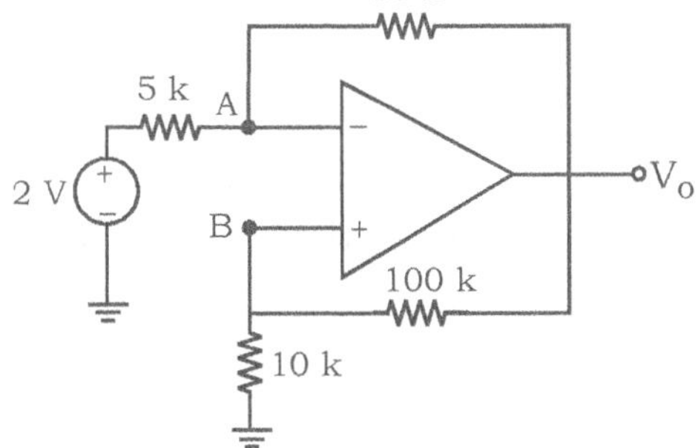

(*a*) –5.5 V (*b*) – 1 V
(*c*) – 10 V (*d*) 2 V

62. In the given circuit, V_- and V_+ are to be amplifier by the same amplification, the value of R should be

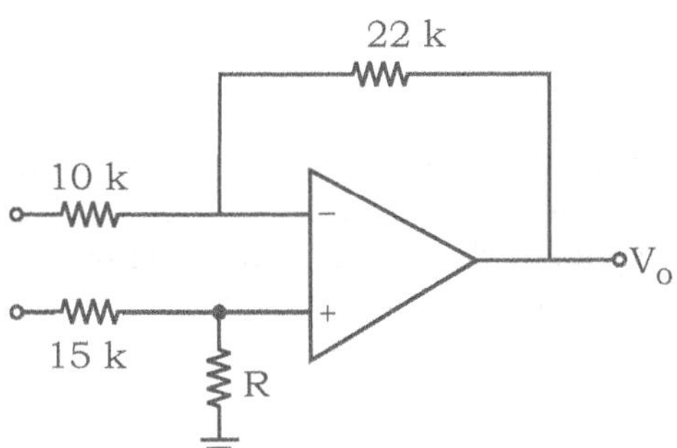

(*a*) 55 kΩ (*b*) 33 kΩ
(*c*) 10 kΩ (*d*) 5 kΩ

63. Given that the op-amp is ideal, the output voltage V_o is

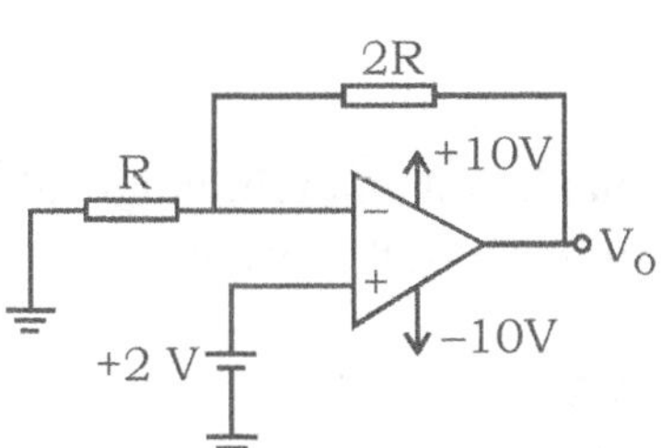

(*a*) 10 V (*b*) 15 V
(*c*) 6 V (*d*) 5 V

64. An ideal sawtooth voltage waveform of frequency 500 Hz and amplitude 3 V is generated by charging a capacitor of 2 pF in every cycle. The charging current is

(*a*) 3 mA (*b*) 4 mA
(*c*) 5 mA (*d*) 10 mA

65. Consider the following circuit using an ideal op-amp. The I-V characteristics of the diode is described by the relation $I = \left(e^{\frac{V}{V_T}} - 1\right)$ where $V_T = 25$ mV, $I_D = 1\ \mu A$ and V is the voltage across the diode (taken as positive for forward bias). For an input voltage $V_i = -1$ V the output voltage is?

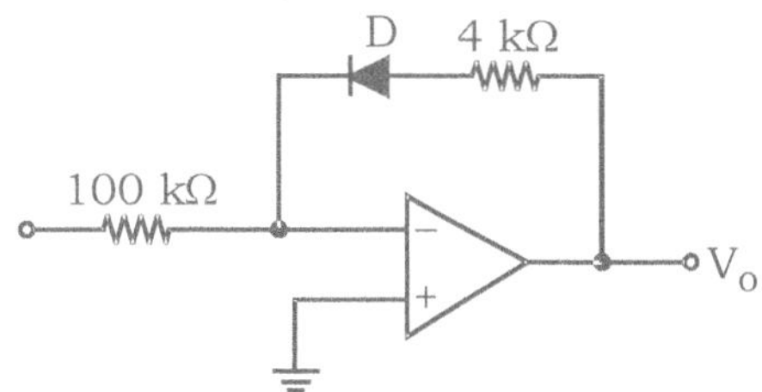

(*a*) 2.0 V (*b*) 0.1 V
(*c*) 1.5 V (*d*) 2.7 V

66. The op-amp in the figure has an input offset voltage of 5 mV and an open loop voltage gain of 10000, then V_o will be

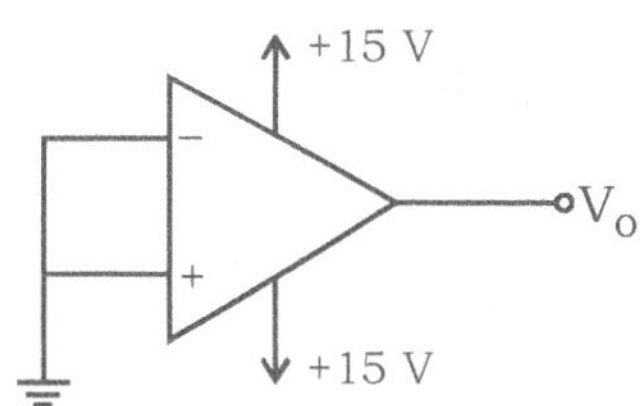

(*a*) ± 15 V (*b*) ± 50 V
(*c*) ± 60 V (*d*) ± 10 V

67. Assume the op-amp to be ideal. The current through 1 kΩ resistor is

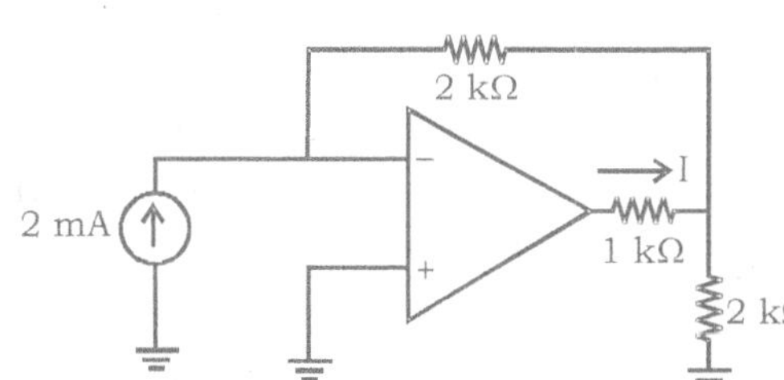

(*a*) – 1 mA
(*b*) – 5 mA
(*c*) – 4 mA
(*d*) – 10 mA

68. In the following circuit the op-amp is ideal. If $\beta_f = 60$ then the total current supplied by 15 V source is

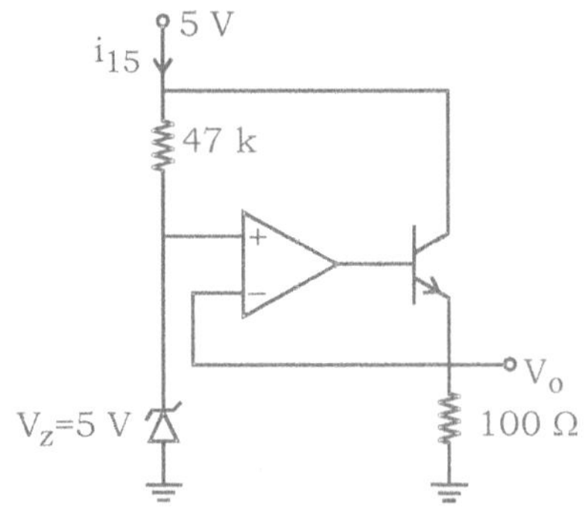

(*a*) 69.4 mA (*b*) 49.4 mA
(*c*) 20.4 mA (*d*) 30.4 mA

69. Assuming that the diodes are ideal in figure the current in diode D_1, I_{D_1} is:

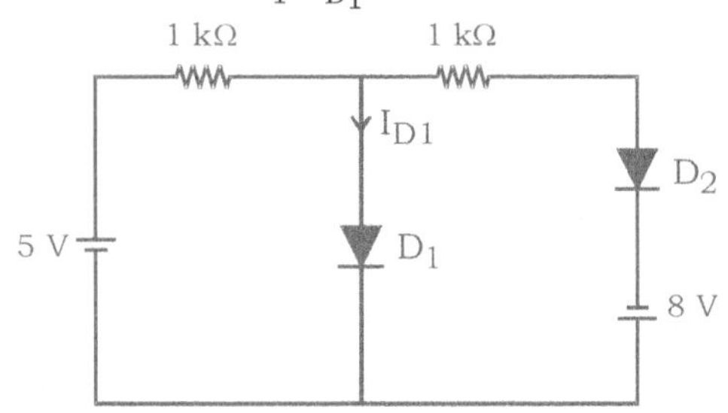

(*a*) 0 mA (*b*) 2 mA
(*c*) 5 mA (*d*) 7 mA

70. Assumning that the diodes in the given circuit are ideal, the voltage V_o is

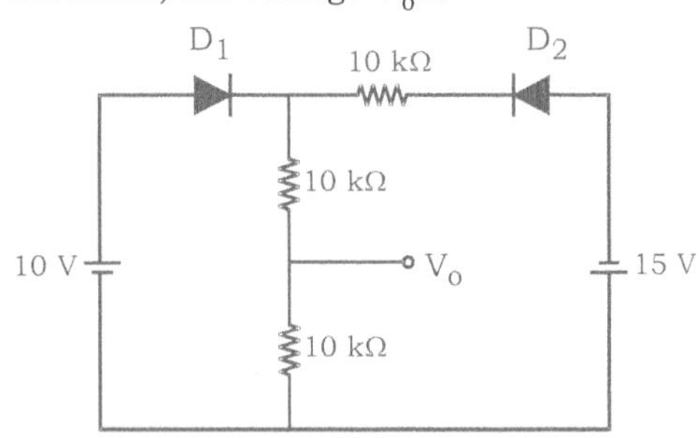

(*a*) 2 V (*b*) 3 V
(*c*) 5 V (*d*) 7 V

71. A voltage signal 10 sin ωt is applied to the circuit with ideal diodes as shown in figure. The maximum and minimum values of the waveform V_{out} of the circuit are respectively.

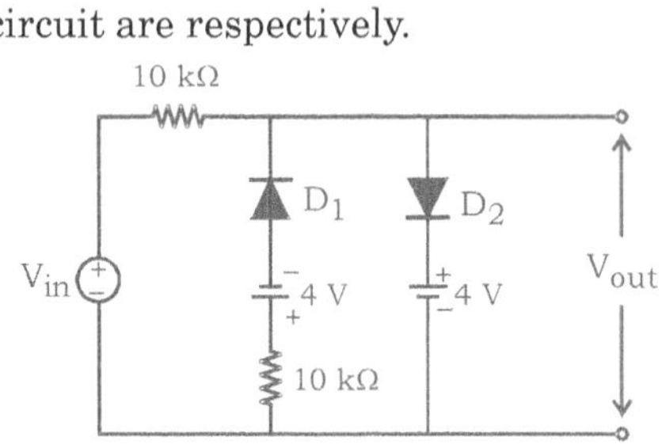

(*a*) – 5 V
(*b*) – 7 V
(*c*) – 3 V
(*d*) – 10 V

72. In the following limiter circuit, an input voltage $V_i = 10\sin 100\pi t$ is applied. Assume that the diode drop is 0.7 V, when it is forward biased. The zener breakdown voltage is 6.8 V. The maximum and minimum value of the output voltage respectively are

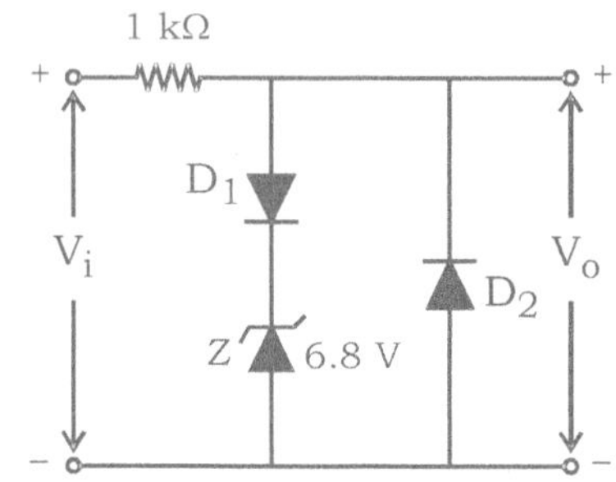

(*a*) – 0.7 V (*b*) – 2 V
(*c*) – 5 V (*d*) – 7 V

73. For an input of $V_s = 5\sin\omega t$, (assuming ideal diode), circuit shown in the figure will behave as a

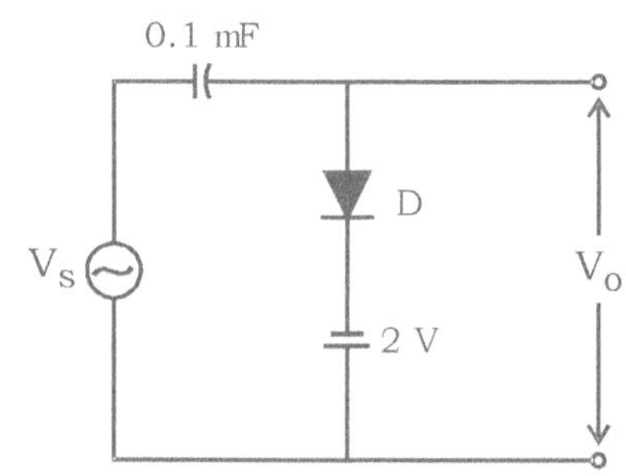

(*a*) – 3 V (*b*) – 5 V
(*c*) – 4 V (*d*) – 2 V

74. Consider the following limiter circuit an input voltage $V_i = 10.5\sin 100\pi t$ is applied. Assume that the diode drop is 0.7 V forward bias. The Zener breakdown voltage is 6.3 V

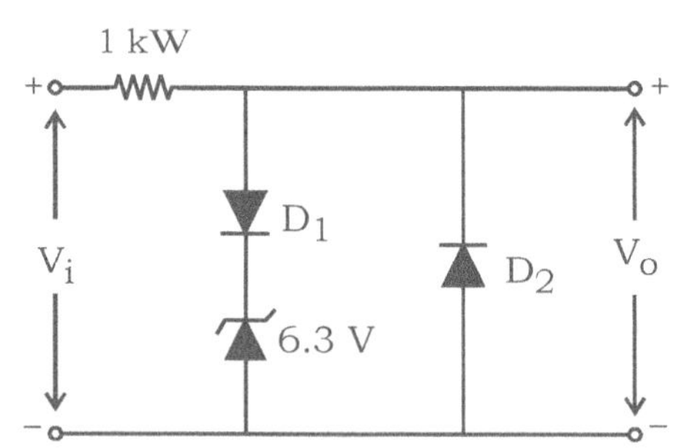

The maximum and minimum values of the output voltage respectively are

(*a*) – 1 V
(*b*) – 7 V
(*c*) –0.7 V
(*d*) – 10 V

75. The cut in voltage of both Zener diode D_z and diode D shown in figure is 0.7 V, while breakdown voltage of D_z is 3.3 V and reverse break down voltage of D is 50 V. The other parameters can be assumed to be the same as those of an ideal diode. The values of the peak output voltage (V_0) are

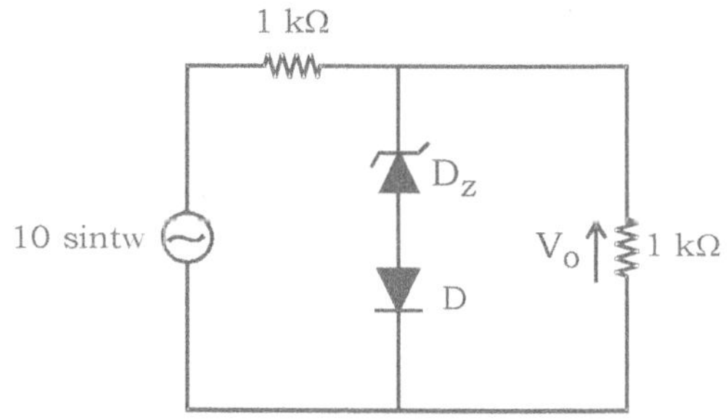

(*a*) 2 V and 5 V (*b*) 2 V and 4 V
(*c*) 8 V and 7 V (*d*) 4 V and 5 V

GENERAL INTELLIGENCE AND REASONING

Directions: In question nos. **76** and **77**, select the related letter/word from the given alternatives.

76. F-L : I-O :: P-V : ?
(*a*) R-Y (*b*) S-X
(*c*) S-Y (*d*) R-W

77. Baboon : Infant :: Beaver : ?
(*a*) Kitten (*b*) Kid
(*c*) Joey (*d*) Fawn

78. Which one set of letters when sequentially placed at the gaps in the given letter series will complete it?

a_bab_aab_bcaa_abc

(*a*) bbab (*b*) bcab
(*c*) abac (*d*) acab

Directions: In question nos. **79** to **87**, a series is given, with one term missing. Choose the correct alternative from the given ones that will complete the series.

79. BGL, FKP, JOT, __?__ .
(*a*) MSY (*b*) NSX
(*c*) NRX (*d*) MRY

80. 7, 13, 25, 49, 97, __?__ .
(*a*) 194 (*b*) 184
(*c*) 188 (*d*) 193

81. BX, EU, HR, KO, __?__ , QI, __?__ .
(*a*) NL and TF (*b*) ML and TE
(*c*) NK and TF (*d*) MK and TE

82. X T P L H D Z V R N J ? ? ? .
(*a*) E B Y (*b*) F B X
(*c*) F B Y (*d*) E C X

83. 72, 46, 521, 612, 343, __?__ .
(*a*) 678 (*b*) 545
(*c*) 215 (*d*) 154

84. BSG, HQJ, NOM, __?__ .
(*a*) TMP (*b*) SMO
(*c*) TPN (*d*) TMO

85. W 5 K, T 12 M, Q 20 O, N 29 Q, __?__ .
(*a*) L 45 R (*b*) K 39 S
(*c*) K 37 S (*d*) K 43 R

86. GJ XU MP RO __?__
(*a*) VS (*b*) IM
(*c*) LI (*d*) SV

87. 17, 150, __?__ , 1500, 1667, 15000, 16667.
(*a*) 1607 (*b*) 1067
(*c*) 167 (*d*) 177

88. Certain numbers have symbols as given below.

0 1 2 3 4 5 6 7 8 9
+ * ? $] # [& ^ @

What is the number indicated by these symbols?

&] ^ * [

(*a*) 3 4 8 9 1 (*b*) 9 6 8 0 5
(*c*) 7 6 8 1 4 (*d*) 7 4 8 1 6

Directions: In question nos. **89** to **90**, find the odd word/number from the given alternatives.

89. (*a*) French Beans (*b*) Gourd
(*c*) Pumpkin (*d*) Jackfruit

90. (*a*) 325 (*b*) 118
(*c*) 272 (*d*) 253

GENERAL AWARENESS

91. The disadvantages of liquid propellants are
(*a*) Difficult to store and handle
(*b*) Extreme toxicity
(*c*) Moderately cryogenic
(*d*) All of these

92. The concept 'Tragedy of Commons' is often cited in connection with
(*a*) Organic farming
(*b*) Sustainable development
(*c*) Cold war
(*d*) Nuclear disarmament

93. National Youth Day is celebrated on the birthday of
(*a*) Mahatma Gandhi
(*b*) Swami Vivekananda
(*c*) Jawaharlal Nehru
(*d*) Subhash Chandra Bose

94. The Hindi word 'rupiya' (Indian currency) is derived from Sanskrit word 'rupya' which literally means
(a) wrought iron (b) wrought silver
(c) wrought aluminium (d) wrought copper

95. 'ABO' blood group is known to be a 'Universal Recipient' because
(a) antibody A and B both are present in AB blood group
(b) no antigen is present in AB blood group
(c) antigen A and B both are present in AB blood group
(d) no antibody is present in AB blood group

96. Cracking in brass is caused by
(a) Ammonia (b) Nitric acid
(c) Hydrogen peroxide (d) Sulphuric acid

97. Which logic gate is a 'universal gate'?
(a) NAND gate (b) AND gate
(c) XOR gate (d) OR gate

98. Article 2 of the Indian constitution states
(a) Admission or establishment of new States
(b) Formation of new States and alteration of areas, boundaries or names of existing States
(c) a law that grants special autonomous status to Jammu and Kashmir
(d) deals with the failure of the Constitutional machinery of an Indian state

99. Strip cropping helps to stop
(a) Air pollution (b) Soil erosion
(c) Water pollution (d) Marine pollution

100. Which part of the human eye acts as an aperture?
(a) Retina (b) Pupil
(c) Sclera (d) Cornea

101. The manufacture of chlorofluorocarbon compounds has been phased out under the
(a) Kyoto protocol
(b) Montreal protocol
(c) Stockholm Convention on Persistent Organic Pollutants
(d) Protocol to the Convention on Long-range Transboundary Air Pollution on the Control of Emissions of Volatile Organic Compounds

102. Which personality is mentioned in the Buddhist text 'Milinda Panha'?
(a) Nagasena (b) Chanakya
(c) Kalidas (d) Hemachandra

103. Which battle resulted in a complete Macedonian victory and the annexation of the Punjab?
(a) Battle of Trafalgar (b) Battle of Hydaspas
(c) Battle of Wagram (d) Battle of Buxar

104. Which coating of mucins gives them considerable water-holding capacity and makes them resistant to proteolysis?
(a) Fat coating (b) Sugar coating
(c) Protein coating (d) Glycoprotein coating

105. Who developed the synthetic polymer DOPA (dihydroxyphenylalanine)?
(a) Messersmith (b) Henri Braconnot
(c) Friedrich Ludersdorf (d) Nathaniel Hayward

QUANTITATIVE APTITUDE

106. Ratio of the time taken by A alone to complete a certain piece of work while that by B and C to complete the same work together is 4 : 3. If the time taken by C to complete the work alone is 8 days and the ratio of the efficiencies of B and C is 1 : 3, then find the time in which A and B can complete the work while working together.
(a) 4.5 days (b) 6 days
(c) 7.5 days (d) 8 days

107. The length of each side of a rhombus is 29 cm while that of one of the diagonals is 40 cm. What will be the area of the rhombus?
(a) 840 cm^2 (b) 880 cm^2
(c) 820 cm^2 (d) 800 cm^2

108. Three successive discounts of 15%, 12.5%, and 20% are equivalent to a single discount of
(a) 38.5% (b) 41.5%
(c) 39.8% (d) 40.5%

109. A and B together have Rs.2,448. If 25% of A's amount is equal to 35% of B's amount, then find the amount with A.
(a) Rs. 1,020 (b) Rs. 1,428
(c) Rs. 1,462 (d) Rs. 1,468

110. If the cost price of 50 articles is equal to the selling price of 40 articles, then find the profit or loss percent.
(a) 20% (b) 16.67%
(c) 25% (d) 33.33%

111. A's income is 25% more than B's income while C's income is 20% more than B's income. By what percent is A's income more or less than C's income?
(a) 5.26% (b) 4.16%
(c) 4.28% (d) 4.46%

112. In what time will a 143 m long train running at the rate of 45 km/hr cross a man coming towards it at the rate of 4.5 km/hr?

(*a*) 11.4 sec (*b*) 10.4 sec

(*c*) 11.64 sec (*d*) 10.64 sec

113. If $3x + \frac{1}{2x} = 3$, then find the value of $32x^5 + \frac{1}{243x^5} + \frac{4}{27x} + \frac{8x}{9}$.

(*a*) $\frac{32}{3}$ (*b*) – 1

(*c*) 1 (*d*) 2

114. If $B : A + \sqrt{(A^2 + B^2)} = 2 : 3$ then find the value of B : A.

(*a*) 8 : 15 (*b*) 24 : 7

(*c*) 12 : 5 (*d*) 6 : 8

115. If $(\cos^2A - \sin A)(\sin A + \cos^2A) = -1$, then find the value of $2 + \cos^2A$.

(*a*) 0 (*b*) $\sec^2A$

(*c*) 1 (*d*) 2

116. If four interior angles of a pentagon are 140°, 90°, 70° and 80°, then find the value of the fifth interior angle.

(*a*) 20° (*b*) 160°

(*c*) 140° (*d*) 40°

117. Find the value of $\sec A(\sec A + \tan A + \frac{\cos A}{1+\sin A}) - 2\tan^2A$.

(*a*) 0 (*b*) 1

(*c*) 2 (*d*) 3

118. Find the value of $\frac{(2 \times 8^{n+1} + 4 \times 2^{3n-1})}{(4 \times 2^{2n+1} - 2 \times 4^{2n+1})}$ for n = 2.

(*a*) $\frac{3}{4}$ (*b*) $-\frac{3}{4}$

(*c*) $\frac{3}{5}$ (*d*) $-\frac{3}{5}$

119. If the average age of sixteen students is 18.5 years and the average age of these students along with their teacher is 22 years, then find the age of the teacher?

(*a*) 58 years (*b*) 78 years

(*c*) 68 years (*d*) 76 years

120. If a = b + 3, then the value of

$a^4 + b^4 + 6a^2b^2 - 4ab(a^2 + b^2) - 70$.

(*a*) 21 (*b*) 19

(*c*) 11 (*d*) 10

GENERAL ENGLISH

121. In the following question, out of the four alternatives, choose the word which best expresses the meaning of the word in **bold**.

TOLERATE

(*a*) Bear on (*b*) Bear at

(*c*) Bear of (*d*) Bear with

122. In the following question, out of the four alternatives, choose the word which is opposite in meaning to the word in **bold**.

ARABLE

(*a*) Cultivable (*b*) Parable

(*c*) Arid (*d*) Fable

123. Four words are given, out of which only one word is spelt correctly. Choose the correctly spelt word.

(*a*) Narcissist (*b*) Narcisist

(*c*) Narsisist (*d*) Narssicist

124. In the following questions, one part of the sentence may have an error. Find out which part of the sentence has an error and mark the appropriate option. If the sentence is free from error, choose the "No error" option.

She brought down (A) / the orphan (B) / as her own child (C) / No Error (D)

(*a*) A (*b*) B

(*c*) C (*d*) D

125. In the following questions, one part of the sentence may have an error. Find out which part of the sentence has an error and mark the appropriate option. If the sentence is free from error, choose the "No error" option.

Yesterday an European (A)/ came (B)/ to my office. (C)/ No error (D)

(*a*) A (*b*) B

(*c*) C (*d*) D

126. In the following questions, one part of the sentence may have an error. Find out which part of the sentence has an error and mark the appropriate option. If the sentence is free from error, choose the "No error" option.

The Prime Minister, (A)/ with his ministers, (B)/ are present. (C) No error (D)

(*a*) A (*b*) B

(*c*) C (*d*) D

127. The sentences given with blanks are to be filled with an appropriate word(s). Four alternatives are suggested. For each question, choose the correct alternative.

The robbers ______ the mansion.

(*a*) Broke at (*b*) Broke of

(*c*) Broke off (*d*) Broke into

128. The sentences given with blanks are to be filled with an appropriate word(s). Four alternatives are suggested. For each question, choose the correct alternative.

He will come, ______?

(*a*) Won't it (*b*) Won't

(*c*) Won't they (*d*) Won't he

129. The sentences given with blanks are to be filled with an appropriate word(s). Four alternatives are suggested. For each question, choose the correct alternative.

His speech on economic reforms _______.

(*a*) fell about (*b*) fell round

(*c*) fell flat (*d*) fell at

130. In each of the questions, four alternatives are given for the Idiom/Phrase. Choose the alternative which best expresses the meaning of the Idiom/Phrase.

Right hand man

(*a*) A one armed person

(*b*) Most efficient assistant

(*c*) Ambidextrous

(*d*) A person who rarely uses his left hand.

131. In each of the questions, four alternatives are given for the Idiom/Phrase. Choose the alternative which best expresses the meaning of the Idiom/Phrase.

All the rage

(*a*) Extremely popular (*b*) Very angry

(*c*) Out of fashion (*d*) Out of vogue

132. In each of the questions, four alternatives are given for the Idiom/Phrase. Choose the alternative which best expresses the meaning of the Idiom/Phrase.

Come to a head

(*a*) Strike a conversation

(*b*) Hit someone on the head

(*c*) Strike a rapport

(*d*) Reached a crisis

133. Out of the four alternatives, choose the one which can be substituted for the given words/sentences.

A statement which is accepted as true without proof

(*a*) Anthology (*b*) Alimony

(*c*) Axiom (*d*) Agenda

134. Out of the four alternatives, choose the one which can be substituted for the given words/sentences.

One who acts against religion

(*a*) Hedonist (*b*) Heretic

(*c*) Hermit (*d*) Hearse

135. Out of the four alternatives, choose the one which can be substituted for the given words/sentences.

One who compiles a dictionary

(*a*) Polyglot (*b*) Linguist

(*c*) Lexicographer (*d*) Cartographer

136. A sentence/a part of the sentence is underlined. Four alternatives are given to the underlined part which sentence. Choose the correct alternative. In case no improvement is needed, "No improvement" is to be marked.

He is rather blunt, but his heart <u>is in the right place</u>.

(*a*) is in the fourth place

(*b*) is in the fit place

(*c*) is in the true place

(*d*) No improvement

137. A sentence/a part of the sentence is underlined. Four alternatives are given to the underlined part which sentence. Choose the correct alternative. In case no improvement is needed, "No improvement" is to be marked.

The Arabian Nights <u>are still a great favourite</u>.

(*a*) were still a great favourite

(*b*) have been a great favourite

(*c*) is still a great favourite

(*d*) No improvement

138. A sentence/a part of the sentence is underlined. Four alternatives are given to the underlined part which sentence. Choose the correct alternative. In case no improvement is needed, "No improvement" is to be marked.

The rebels <u>laid at</u> their arms.

(*a*) laid down (*b*) laid up

(*c*) laid about (*d*) No improvement

139. A sentence/a part of the sentence is underlined. Four alternatives are given to the underlined part which sentence. Choose the correct alternative. In case no improvement is needed, "No improvement" is to be marked.

<u>At that time, she had been writing a novel for two months</u>.

(*a*) At that time, she wrote a novel for two months.

(*b*) At that time, she has written a novel for two months.

(*c*) At time time, she had written a novel for two months.

(*d*) No improvement

140. A sentence/a part of the sentence is underlined. Four alternatives are given to the underlined part which sentence. Choose the correct alternative. In case no improvement is needed, "No improvement" is to be marked.

He takes <u>at</u> his father.

(*a*) about

(*b*) after

(*c*) behind

(*d*) No improvement

Directions for questions 141 to 145: A passage is given with 5 questions following it. Read the passage carefully and choose the best answer out of the four alternatives.

The northern portion of the Caucasus is known as the Ciscaucasus and the southern portion as the Transcaucasus. The Ciscaucasus contains the larger majority of the Greater Caucasus Mountain range, also known as the Major Caucasus mountains. It includes Southwestern Russia and northern parts of Georgia and Azerbaijan. The Transcaucasus is bordered on the north by Russia, on the west by the Black Sea and Turkey, on the east by the Caspian Sea, and on the south by Iran. It includes the Caucasus Mountains and surrounding lowlands. All of Armenia, Azerbaijan (excluding the northern parts) and Georgia (excluding the northern parts) are in South Caucasus. The main Greater Caucasus range is generally **perceived** to be the dividing line between Asia and Europe. The highest peak in the Caucasus is Mount Elbrus (5,642 m) in the western Ciscaucasus in Russia, and is generally considered as the highest point in Europe.

The Caucasus is one of the most linguistically and culturally diverse regions on Earth. The nation states that comprise the Caucasus today are the post-Soviet states Georgia (including Adjara), Armenia, and Azerbaijan (including Nakhchivan). Three territories in the region claim independence but are recognized as such by only a handful or by no independent states: Nagorno-Karabakh, Abkhazia and South Ossetia. Abkhazia and South Ossetia are recognised by the majority of independent states as part of Georgia, and Nagorno-Karabakh is recognised as part of Azerbaijan. The Russian divisions include Krasnodar Krai, Stavropol Krai, and the autonomous republics of Adygea, Karachay-Cherkessia, Kabardino-Balkaria, North Ossetia, Ingushetia, Chechnya, and Dagestan.

141. As per the passage, the height of Mount Elbrus is

(*a*) More than 6000 metres.

(*b*) More than 6500 metres

(*c*) More than 5500 metres

(*d*) Can't be said

142. As per the passage, what divides Asia from Europe?

(*a*) The Durand line

(*b*) The Greater Caucasus range

(*c*) The Lesser Caucasus range

(*d*) The Caspian Sea

143. According to the passage, what can be said about the Caucasian region?

(*a*) The region is culturally homogeneous.

(*b*) The region has seen massive political turmoil.

(*c*) The region is culturally varied.

(*d*) None of the above

144. Out of the following options, which one is a synonym of 'perceive'?

(*a*) Understand (*b*) Return

(*c*) Rant (*d*) Taunt

145. Out of the following options, which one can be a possible title of the passage?

(*a*) The geographical account of post-Soviet states

(*b*) The mountains of Armenia

(*c*) The politically volatile Caucasian region

(*d*) An account of the Caucasian region

146. In the following question, out of the four alternatives, choose the word which best expresses the meaning of the word in bold.

ANNIHILATION

(*a*) Formation (*b*) Renovation

(*c*) Recuperation (*d*) Dissolution

147. In the following question, out of the four alternatives, choose the word which is opposite to the word given in bold.

FISSURE

(*a*) Fetish (*b*) Unity

(*c*) Puncture (*d*) Facile

148. Four words are given, out of which only one word is spelt correctly. Choose the correctly spelt word.

(*a*) Halucinate (*b*) Helucinate

(*c*) Hallucinete (*d*) Hallucinate

149. In the following questions, one part of the sentence may have an error. Find out which part of the sentence has an error and choose the appropriate option. If the sentence is free from error, choose the "No error" option.

The cistern (A)/ is running over (B)/ with water. (C)/ No error (D)

(*a*) A (*b*) B

(*c*) C (*d*) D

150. In the following questions, one part of the sentence may have an error. Find out which part of the sentence has an error and choose the appropriate option. If the sentence is free from error, choose the "No error" option.

Each of (A)/ these minerals (B)/ are found in India. (C)/ No error (D)

(*a*) A (*b*) B

(*c*) C (*d*) D

151. In the following questions, one part of the sentence may have an error. Find out which part of the sentence has an error and choose the appropriate option. If the sentence is free from error, choose the "No error" option.

Bread and butter (A)/are my (B)/ favourite breakfast. (C)/ No error (D)

(*a*) A (*b*) B

(*c*) C (*d*) D

152. The sentences given with blanks are to be filled with an appropriate word(s). Four alternatives are suggested. Choose the correct alternative.

The soldier dies fighting ______ his country.

(*a*) in behalf of (*b*) behalf of

(*c*) at behalf of (*d*) on behalf of

153. The sentences given with blanks are to be filled with an appropriate word(s). Four alternatives are suggested. Choose the correct alternative.

I have not slept ____ yesterday.

(*a*) until (*b*) to

(*c*) since (*d*) till

154. The sentences given with blanks are to be filled with an appropriate word(s). Four alternatives are suggested. Choose the correct alternative.

Whatever she does, she does ______ to the major prospects.

(*a*) about an eye (*b*) with an eye

(*c*) in an eye (*d*) for an eye

155. In each of the questions, four alternatives are given for the Idiom/Phrase. Choose the alternative which corresponds to the meaning of the Idiom/Phrase given in bold.

Jump on the bandwagon

(*a*) Board a bus

(*b*) Get involved in a scuffle

(*c*) A raucous gathering

(*d*) Join a popular trend or activity

156. In each of the questions, four alternatives are given for the Idiom/Phrase. Choose the alternative which corresponds to the meaning of the Idiom/Phrase given in bold.

Caught between two stools.

(*a*) When someone finds it difficult to adapt to his in-laws' place.

(*b*) When someone finds it difficult to choose between two alternatives.

(*c*) Difference between the developed and under-developed nations

(*d*) Inter-generational differences

157. In each of the questions, four alternatives are given for the Idiom/Phrase. Choose the alternative which corresponds to the meaning of the Idiom/Phrase given in bold.

Beat around the bush

(*a*) Not speaking directly about the issue

(*b*) Trying one's hand in gardening

(*c*) Assaulting a person severely

(*d*) Hitting the target accurately

158. Out of the four alternatives, choose the one which can be substituted for the given words/sentences.

A place for housing aeroplanes

(*a*) Hangar (*b*) Hanger

(*c*) Runway (*d*) Airport

159. Out of the four alternatives, choose the one which can be substituted for the given words/sentences.

A person, who walks in sleep

(*a*) Patient (*b*) Sleeplessness

(*c*) Somnambulist (*d*) Insomnia

160. Out of the four alternatives, choose the one which can be substituted for the given words/sentences.

Murder of a wife by her husband

(*a*) Suicide (*b*) Uxoricide

(*c*) Matricide (*d*) Genocide

161. A sentence/a part of the sentence is underlined. Alternatives are given which can replace the underlined part. Choose the appropriate option. In case no improvement is needed, choose "No improvement".

Neither the Chairman nor the directors <u>is</u> present.

(*a*) are (*b*) has

(*c*) was (*d*) No improvement

162. A sentence/a part of the sentence is underlined. Alternatives are given which can replace the underlined part. Choose the appropriate option. In case no improvement is needed, choose "No improvement".

Every boy and every girl <u>were</u> given a box of chocolates.

(*a*) have (*b*) have been

(*c*) was (*d*) No improvement

163. A sentence/a part of the sentence is underlined. Alternatives are given which can replace the underlined part. Choose the appropriate option. In case no improvement is needed, choose "No improvement".

Recently, she has <u>taken of</u> gambling.

(*a*) taken about (*b*) taken to

(*c*) taken in (*d*) No improvement

164. A sentence/a part of the sentence is underlined. Alternatives are given which can replace the underlined part. Choose the appropriate option. In case no improvement is needed, choose "No improvement".

I <u>saw through</u> the trick.

(*a*) saw in (*b*) saw to

(*c*) saw on (*d*) No improvement

165. A sentence/a part of the sentence is underlined. Alternatives are given which can replace the underlined part. Choose the appropriate option. In case no improvement is needed, choose "No improvement".

He seems to be <u>well of</u>.

(*a*) well off (*b*) well about

(*c*) well at (*d*) No improvement

Directions for questions 166 to 170: A passage is given with 5 questions following it. Read the passage carefully and choose the best answer out of the four alternatives.

Indian Peace Keeping Force (IPKF) was the Indian military contingent performing a peacekeeping operation in Sri Lanka between 1987 and 1990. It was formed under the mandate of the 1987 Indo-Sri Lankan Accord that aimed to end the Sri Lankan Civil War between militant Sri Lankan Tamil nationalists such as the Liberation Tigers of Tamil Eelam (LTTE) and the Sri Lankan military.

The main task of the IPKF was to disarm the different militant groups, not just the LTTE. It was to be quickly followed by the formation of an Interim Administrative Council. These were the tasks as per the terms of the Indo-Sri Lankan Accord, signed at the **behest** of Indian Prime Minister Rajiv Gandhi. Given the escalation of the conflict in Sri Lanka, and with the pouring of refugees into India, Gandhi took the decisive step to push this accord through. The IPKF was inducted into Sri Lanka on the request of Sri Lankan President J. R. Jayewardene under the terms of the Indo-Sri Lanka Accord.

The force was initially not expected to be involved in any significant combat by the Indian High Command. However, within a few months, the IPKF became embroiled in battle with the LTTE to enforce peace. The differences started with LTTE trying to dominate the Interim Administrative Council, and also refusing to disarm, which was a pre-condition to enforce peace in the island. Soon, these differences led to the LTTE attacking the IPKF, at which point the IPKF decided to disarm the LTTE militants, by force if required. In the two years it was in northern Sri Lanka, the IPKF launched a number of combat operations aimed at destroying the LTTE-led insurgency. Given LTTE's tactics in guerrilla warfare and using women and child soldiers to fight battles, it soon escalated into repeated skirmishes between the IPKF and LTTE.

The IPKF began withdrawing from Sri Lanka in 1989, following the election of the V. P. Singh government in India and on the request of the newly elected Sri Lankan President Ranasinghe Premadasa. The last IPKF contingents left Sri Lanka in March 1990.

166. The IPKF operations in Sri Lanka took place in the

(*a*) last quarter of the previous century.

(*b*) towards the middle of the previous century.

(*c*) 21st century.

(*d*) towards the beginning of the 20th century.

167. Out of the four alternatives, which one is closest in meaning to the word 'behest'?

(*a*) Remand (*b*) Request

(*c*) Resuscitation (*d*) Renovation

168. According to the passage, who requested the IPKF contingent to be removed?

(*a*) Former Prime Minister Rajiv Gandhi

(*b*) The UNO

(*c*) President Ranasinghe Premadasa.

(*d*) Former Prime Minister V.P. Singh

169. According to the passage, the main task of the IPKF was to

(*a*) disarm only the LTTE.

(*b*) disarm the different militant groups, including the LTTE.

(*c*) train the Sri Lankan forces

(*d*) be a part of the UN special forces.

170. According to the passage, the IPKF was formed

(*a*) under the mandate of the 1985 Indo-Sri Lankan Accord.

(*b*) under the mandate of the 1986 Indo-Sri Lankan Accord.

(*c*) owing to hostilities between India and Sri Lanka.

(*d*) under the mandate of the 1987 Indo-Sri Lankan Accord.

Directions (Question no. 171): In the following question, out of the four alternatives, choose the word which best expresses the meaning of the word given in bold.

Placid

(*a*) Reckless (*b*) Raucous

(*c*) Cacophony (*d*) Tranquil

Directions (Question no. 172): In the following question, out of the four alternatives, choose the word which is opposite in meaning to the word given in bold.

Pensive

(*a*) Blasé (*b*) Faux pas

(*c*) Couture (*d*) Crony

Directions (Question no. 173): Four words are given, out of which only one word is spelt correctly. Choose the correctly spelt word out of the given options.

(*a*) Osification (*b*) Osyfication

(*c*) Ossifikation (*d*) Ossification

Directions (Question nos. 174 to 176): In the following questions, one part of the sentence may have an error. Find out which part of the sentence has an error and choose the appropriate option. If the sentence is free from error, choose the "No error" option.

174. The patient (*a*)/ could not (*b*)/ call up past events. (*c*)/No error (*d*)

175. I said to her, (*a*)/ "those books (*b*)/ are your's." (*c*)/ No error (*d*)

176. Copper (*a*)/ is an (*b*)/ useful metal. (*c*)/ No error (*d*)

Directions (Question nos. 177 to 179): The sentences given with blanks are to be filled with an appropriate word(s). Four alternatives are suggested. Choose the correct option.

177. Horses prefer ________ dark stables.

(*a*) living on (*b*) living in

(*c*) living about (*d*) living for

178. Treasure Island is __________ pirate stories in English literature.

(*a*) one of the better

(*b*) a best

(*c*) one of the best

(*d*) best

179. I ______ him as my son.

(*a*) look on

(*b*) look around

(*c*) look about

(*d*) look for

Directions (Question 180). In each of the questions, four alternatives are given for the Idiom/Phrase. Choose the alternative which corresponds to the meaning of the given Idiom/Phrase.

180. On the spur of the moment

(*a*) Bravely

(*b*) Leniently

(*c*) Impulsively

(*d*) Sportively

OMR SHEET

1. (A) (B) (C) (D)
2. (A) (B) (C) (D)
3. (A) (B) (C) (D)
4. (A) (B) (C) (D)
5. (A) (B) (C) (D)
6. (A) (B) (C) (D)
7. (A) (B) (C) (D)
8. (A) (B) (C) (D)
9. (A) (B) (C) (D)
10. (A) (B) (C) (D)
11. (A) (B) (C) (D)
12. (A) (B) (C) (D)
13. (A) (B) (C) (D)
14. (A) (B) (C) (D)
15. (A) (B) (C) (D)
16. (A) (B) (C) (D)
17. (A) (B) (C) (D)
18. (A) (B) (C) (D)
19. (A) (B) (C) (D)
20. (A) (B) (C) (D)
21. (A) (B) (C) (D)
22. (A) (B) (C) (D)
23. (A) (B) (C) (D)
24. (A) (B) (C) (D)
25. (A) (B) (C) (D)
26. (A) (B) (C) (D)
27. (A) (B) (C) (D)
28. (A) (B) (C) (D)
29. (A) (B) (C) (D)
30. (A) (B) (C) (D)
31. (A) (B) (C) (D)
32. (A) (B) (C) (D)
33. (A) (B) (C) (D)
34. (A) (B) (C) (D)
35. (A) (B) (C) (D)
36. (A) (B) (C) (D)
37. (A) (B) (C) (D)
38. (A) (B) (C) (D)
39. (A) (B) (C) (D)
40. (A) (B) (C) (D)
41. (A) (B) (C) (D)
42. (A) (B) (C) (D)
43. (A) (B) (C) (D)
44. (A) (B) (C) (D)
45. (A) (B) (C) (D)
46. (A) (B) (C) (D)
47. (A) (B) (C) (D)
48. (A) (B) (C) (D)
49. (A) (B) (C) (D)
50. (A) (B) (C) (D)
51. (A) (B) (C) (D)
52. (A) (B) (C) (D)
53. (A) (B) (C) (D)
54. (A) (B) (C) (D)
55. (A) (B) (C) (D)
56. (A) (B) (C) (D)
57. (A) (B) (C) (D)
58. (A) (B) (C) (D)
59. (A) (B) (C) (D)
60. (A) (B) (C) (D)
61. (A) (B) (C) (D)
62. (A) (B) (C) (D)
63. (A) (B) (C) (D)
64. (A) (B) (C) (D)
65. (A) (B) (C) (D)
66. (A) (B) (C) (D)
67. (A) (B) (C) (D)
68. (A) (B) (C) (D)
69. (A) (B) (C) (D)
70. (A) (B) (C) (D)
71. (A) (B) (C) (D)
72. (A) (B) (C) (D)
73. (A) (B) (C) (D)
74. (A) (B) (C) (D)
75. (A) (B) (C) (D)
76. (A) (B) (C) (D)
77. (A) (B) (C) (D)
78. (A) (B) (C) (D)
79. (A) (B) (C) (D)
80. (A) (B) (C) (D)
81. (A) (B) (C) (D)
82. (A) (B) (C) (D)
83. (A) (B) (C) (D)
84. (A) (B) (C) (D)
85. (A) (B) (C) (D)
86. (A) (B) (C) (D)
87. (A) (B) (C) (D)
88. (A) (B) (C) (D)
89. (A) (B) (C) (D)
90. (A) (B) (C) (D)
91. (A) (B) (C) (D)
92. (A) (B) (C) (D)
93. (A) (B) (C) (D)
94. (A) (B) (C) (D)
95. (A) (B) (C) (D)
96. (A) (B) (C) (D)
97. (A) (B) (C) (D)
98. (A) (B) (C) (D)
99. (A) (B) (C) (D)
100. (A) (B) (C) (D)

OMR SHEET

101. Ⓐ Ⓑ Ⓒ Ⓓ	121. Ⓐ Ⓑ Ⓒ Ⓓ	141. Ⓐ Ⓑ Ⓒ Ⓓ	161. Ⓐ Ⓑ Ⓒ Ⓓ
102. Ⓐ Ⓑ Ⓒ Ⓓ	122. Ⓐ Ⓑ Ⓒ Ⓓ	142. Ⓐ Ⓑ Ⓒ Ⓓ	162. Ⓐ Ⓑ Ⓒ Ⓓ
103. Ⓐ Ⓑ Ⓒ Ⓓ	123. Ⓐ Ⓑ Ⓒ Ⓓ	143. Ⓐ Ⓑ Ⓒ Ⓓ	163. Ⓐ Ⓑ Ⓒ Ⓓ
104. Ⓐ Ⓑ Ⓒ Ⓓ	124. Ⓐ Ⓑ Ⓒ Ⓓ	144. Ⓐ Ⓑ Ⓒ Ⓓ	164. Ⓐ Ⓑ Ⓒ Ⓓ
105. Ⓐ Ⓑ Ⓒ Ⓓ	125. Ⓐ Ⓑ Ⓒ Ⓓ	145. Ⓐ Ⓑ Ⓒ Ⓓ	165. Ⓐ Ⓑ Ⓒ Ⓓ
106. Ⓐ Ⓑ Ⓒ Ⓓ	126. Ⓐ Ⓑ Ⓒ Ⓓ	146. Ⓐ Ⓑ Ⓒ Ⓓ	166. Ⓐ Ⓑ Ⓒ Ⓓ
107. Ⓐ Ⓑ Ⓒ Ⓓ	127. Ⓐ Ⓑ Ⓒ Ⓓ	147. Ⓐ Ⓑ Ⓒ Ⓓ	167. Ⓐ Ⓑ Ⓒ Ⓓ
108. Ⓐ Ⓑ Ⓒ Ⓓ	128. Ⓐ Ⓑ Ⓒ Ⓓ	148. Ⓐ Ⓑ Ⓒ Ⓓ	168. Ⓐ Ⓑ Ⓒ Ⓓ
109. Ⓐ Ⓑ Ⓒ Ⓓ	129. Ⓐ Ⓑ Ⓒ Ⓓ	149. Ⓐ Ⓑ Ⓒ Ⓓ	169. Ⓐ Ⓑ Ⓒ Ⓓ
110. Ⓐ Ⓑ Ⓒ Ⓓ	130. Ⓐ Ⓑ Ⓒ Ⓓ	150. Ⓐ Ⓑ Ⓒ Ⓓ	170. Ⓐ Ⓑ Ⓒ Ⓓ
111. Ⓐ Ⓑ Ⓒ Ⓓ	131. Ⓐ Ⓑ Ⓒ Ⓓ	151. Ⓐ Ⓑ Ⓒ Ⓓ	171. Ⓐ Ⓑ Ⓒ Ⓓ
112. Ⓐ Ⓑ Ⓒ Ⓓ	132. Ⓐ Ⓑ Ⓒ Ⓓ	152. Ⓐ Ⓑ Ⓒ Ⓓ	172. Ⓐ Ⓑ Ⓒ Ⓓ
113. Ⓐ Ⓑ Ⓒ Ⓓ	133. Ⓐ Ⓑ Ⓒ Ⓓ	153. Ⓐ Ⓑ Ⓒ Ⓓ	173. Ⓐ Ⓑ Ⓒ Ⓓ
114. Ⓐ Ⓑ Ⓒ Ⓓ	134. Ⓐ Ⓑ Ⓒ Ⓓ	154. Ⓐ Ⓑ Ⓒ Ⓓ	174. Ⓐ Ⓑ Ⓒ Ⓓ
115. Ⓐ Ⓑ Ⓒ Ⓓ	135. Ⓐ Ⓑ Ⓒ Ⓓ	155. Ⓐ Ⓑ Ⓒ Ⓓ	175. Ⓐ Ⓑ Ⓒ Ⓓ
116. Ⓐ Ⓑ Ⓒ Ⓓ	136. Ⓐ Ⓑ Ⓒ Ⓓ	156. Ⓐ Ⓑ Ⓒ Ⓓ	176. Ⓐ Ⓑ Ⓒ Ⓓ
117. Ⓐ Ⓑ Ⓒ Ⓓ	137. Ⓐ Ⓑ Ⓒ Ⓓ	157. Ⓐ Ⓑ Ⓒ Ⓓ	177. Ⓐ Ⓑ Ⓒ Ⓓ
118. Ⓐ Ⓑ Ⓒ Ⓓ	138. Ⓐ Ⓑ Ⓒ Ⓓ	158. Ⓐ Ⓑ Ⓒ Ⓓ	178. Ⓐ Ⓑ Ⓒ Ⓓ
119. Ⓐ Ⓑ Ⓒ Ⓓ	139. Ⓐ Ⓑ Ⓒ Ⓓ	159. Ⓐ Ⓑ Ⓒ Ⓓ	179. Ⓐ Ⓑ Ⓒ Ⓓ
120. Ⓐ Ⓑ Ⓒ Ⓓ	140. Ⓐ Ⓑ Ⓒ Ⓓ	160. Ⓐ Ⓑ Ⓒ Ⓓ	180. Ⓐ Ⓑ Ⓒ Ⓓ

ANSWERS

1. (d)	2. (b)	3. (b)	4. (a)	5. (c)	6. (b)	7. (b)	8. (a)	9. (b)	10. (b)
11. (a)	12. (b)	13. (d)	14. (c)	15. (a)	16. (c)	17. (a)	18. (c)	19. (a)	20. (d)
21. (b)	22. (b)	23. (d)	24. (b)	25. (c)	26. (d)	27. (b)	28. (b)	29. (d)	30. (b)
31. (d)	32. (b)	33. (d)	34. (c)	35. (b)	36. (c)	37. (d)	38. (d)	39. (a)	40. (d)
41. (b)	42. (d)	43. (d)	44. (b)	45. (a)	46. (a)	47. (b)	48. (a)	49. (c)	50. (b)
51. (a)	52. (b)	53. (b)	54. (b)	55. (c)	56. (b)	57. (b)	58. (b)	59. (a)	60. (c)
61. (a)	62. (b)	63. (c)	64. (a)	65. (b)	66. (a)	67. (c)	68. (b)	69. (a)	70. (c)
71. (b)	72. (a)	73. (d)	74. (c)	75. (d)	76. (c)	77. (a)	78. (d)	79. (b)	80. (d)
81. (a)	82. (b)	83. (c)	84. (a)	85. (b)	86. (d)	87. (c)	88. (d)	89. (d)	90. (c)
91. (d)	92. (b)	93. (b)	94. (b)	95. (d)	96. (a)	97. (a)	98. (a)	99. (b)	100. (b)
101. (b)	102. (a)	103. (b)	104. (b)	105. (a)	106. (b)	107. (a)	108. (d)	109. (b)	110. (c)
111. (b)	112. (b)	113. (a)	114. (c)	115. (d)	116. (b)	117. (c)	118. (d)	119. (b)	120. (c)
121. (d)	122. (c)	123. (a)	124. (a)	125. (a)	126. (c)	127. (d)	128. (d)	129. (c)	130. (b)
131. (a)	132. (d)	133. (c)	134. (b)	135. (c)	136. (d)	137. (c)	138. (a)	139. (d)	140. (b)
141. (c)	142. (b)	143. (c)	144. (a)	145. (d)	146. (d)	147. (b)	148. (d)	149. (d)	150. (c)
151. (b)	152. (d)	153. (c)	154. (b)	155. (d)	156. (b)	157. (a)	158. (a)	159. (c)	160. (b)
161. (a)	162. (c)	163. (b)	164. (d)	165. (a)	166. (a)	167. (b)	168. (c)	169. (b)	170. (d)
171. (d)	172. (a)	173. (d)	174. (d)	175. (c)	176. (b)	177. (b)	178. (c)	179. (a)	180. (c)

EXPLANATIONS

2. MOSFET is used for this frequency range.
3. The circuit should not turn off before making off the thyristor.
4. Let I = device current

   R_{ON} = ON state resistance of power MOSFET

   Conduction loss = P = $I^2 R_{ON}$

   Therefore, conduction loss versus device current characteristic can be best approximated by a parabola.
5. During one half cycle, SCR can be in forward blocking mode and applying any gate pulse, the SCR operates in forward conduction mode (ON state).

   But SCR can not be turned off by applying gate pulse.
6. $\because$ x and y sensitivities are same.

   $\therefore$ deflection by both will be same.
7. For finding R_{th} voltage source is short circuited and current source is open circuited.

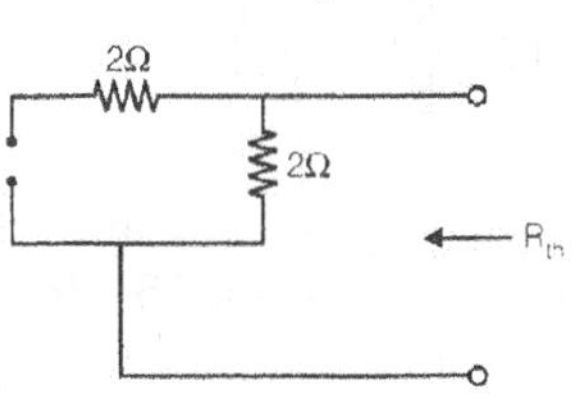

$R_{th} = 2\Omega$

8. For triangular wave

   $$\text{Avg value} = \frac{V_m}{2}$$

   $$\text{RMS value} = \frac{V_m}{\sqrt{3}}$$

   $$\therefore \quad \frac{V_m}{2} = 10\text{ V} \Rightarrow V_m = 20$$

   $$\text{RMS value} = \frac{V_m}{\sqrt{3}} = \frac{20}{\sqrt{3}}\text{ V}$$
9. A CRO uses a horizontal input voltage which is an internally generated ramp voltage (i.e. linear sweep voltage) called "Time Base of CRO". An oscilloscope is used to display a waveform which varies as a function of time. To reproduce the waveform accurately, the beam must have a constant horizontal velocity i.e. the deflecting voltage must increase linearly with time. This voltage is called a ramp voltage. Thus, a sawtooth waveform also called linear sweep voltage is applied to the horizontal deflecting system (or horizontal axis) which provides a time base to CRO.
10. $$r_D = \frac{V_T}{I_D} = \frac{25\text{ mV}}{2\text{ mA}} = 12.5\ \Omega$$
13. $$T = \frac{1}{f} = \frac{1}{20} = 0.05$$

14. Push pull is a power amplifier.

15. $p_1 = p_c\left(1+\frac{m_a^2}{2}\right)$

if $m_a = 0$

$p_{t_1} = p_c$

if $m_a = 1$

$p_{t_2} = \frac{3}{2}p_c$

$\Delta p_t = p_{t_2} - p_{t_1} = \frac{1}{2}p_c$

Increases by 50%

16. As wavelength decreases, frequency increases and maximum velocity of electron increases.

18. $VSWR = 10^{0.6}$

$3.98\rho = \frac{VSWR-1}{VSWR+1} = 0.6$

20. The unbypassed emitter resistor gives negative feedback which reduces the voltage gain.

21. $Z_i = Z_0\left[\frac{Z_L + jZ_0 \tan\beta d}{Z_0 + jZ_L \tan\beta d}\right], Z_L = 0$

if Z_i = $-jkZ_0$ then capacitive = jkZ_0 the inductive.

30. $\rho = \frac{Z_L - Z_0}{Z_L + Z_0}, P = V^2 / Z_L$

31. If line is OC or SC or terminated in pure reactance, SWR is 1.

33. $10 \log_{10}\frac{P}{5} = 20$ or P = 500W.

35. Television transmission is governed by line of sight distance which is about 100 km or less.

54. We know that I_{CEO} $(\beta + 1)I_{CO}$ and I_{CBO} is slightly greater than I_{CO} due to some current induced and flow at the surface and it also doubles for every ten degree rise in temperature.

55. Given that

$P = 2.25\times10^{15}$/CC

and $n_i = 1.5\times10^{10}$/CC

According to mass action low

$np = n_i^2$

$n = \frac{n_i^2}{p} = \frac{(1.5\times10^{10})2}{2.25\times10^{15}} = 10^5 / cc$

56. Voltage follower circuit is common collector configuration

57. At equilibrium,

$n_1p_1 = n_2p_2$

58. $\rho = \frac{RA}{l} = \frac{100\times0.0001}{1} = 0.01\,\Omega - m$

59. In parallel combination

$C = C_1 + C_2 = 1 + 1 = 2\ \mu F$

60. $Q = C_1V_1 = C_2V_2$

61. Voltage at B = $\frac{V_o\times10}{110} = \frac{V_o}{11}$

Voltage at A = voltage at B = $\frac{V_o}{11}$

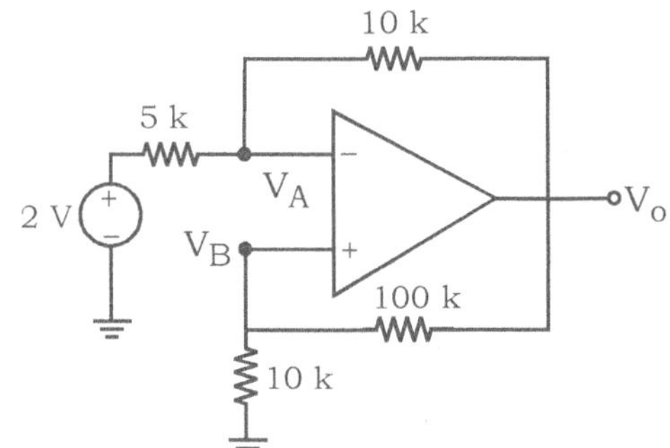

Using KCL at A

$$\frac{\frac{Vo}{11}-2}{5} + \frac{\frac{V_o}{11}-V_o}{10} = 0$$

$V_o = -5.5$ V

62.

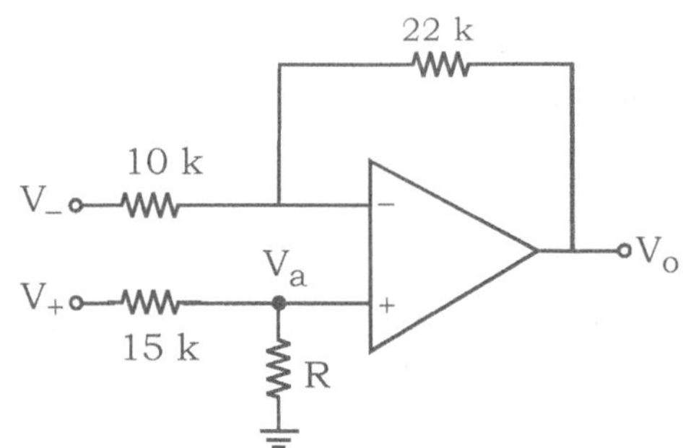

$V_a = \frac{V_+R}{15+R}$,

$V_o = -\left(\frac{R_f}{R}\right)V_i + \left(1+\frac{R_f}{R_1}\right)V_a$

$= -\left(\frac{22\,k\Omega}{10\,k\Omega}\right)V_- + \left(1+\frac{22\,k\Omega}{10\,k\Omega}\right)\left(\frac{V_+R}{15+R}\right)$

$\left(-\frac{R_f}{R}\right) = \left(1+\frac{R_f}{R_1}\right)\left(\frac{R}{15+R}\right)$

$\left|-\frac{22\,k\Omega}{10\,k\Omega}\right| = \left|1+\frac{22\,k\Omega}{10\,k\Omega}\right|\left|\frac{R}{15+R}\right|$

R = 33 kΩ

63. It is non inverting amplifier

$$V_o = \left(1+\frac{R_f}{R_1}\right)V_i = \left(1+\frac{2R}{R}\right)2$$

$$= 3 \times 2 = 6 \text{ V}$$

64. If a constant current is made to flow in a capacitor. The output voltage integration of input current and that is sawtooth waveform

$$V_c = \frac{1}{C}\int_0^t i dt$$

$$T = \frac{1}{f} = \frac{1}{500} = 2 \text{ms}$$

$$3 = \frac{1}{2\times10^{-6}}\int_0^{2\times10^{-3}} i dt$$

$i(2 \times 10^{-3} - 0) = 6 \times 10^{-6}$

$i = 3$ mA

Thus charging requires 3 mA current source for 2 msec.

65. The circuit is using ideal op-amp. The non inverting terminal of op-amp is at ground. Thus inverting terminal is also at virtual ground.

Thus current will flow from –ve terminal (0V) to –1 V source. The current is

$$I = \frac{0-(-1)}{100 \text{ k}\Omega} = \frac{1}{100 \text{ k}\Omega}$$

The current through diode is

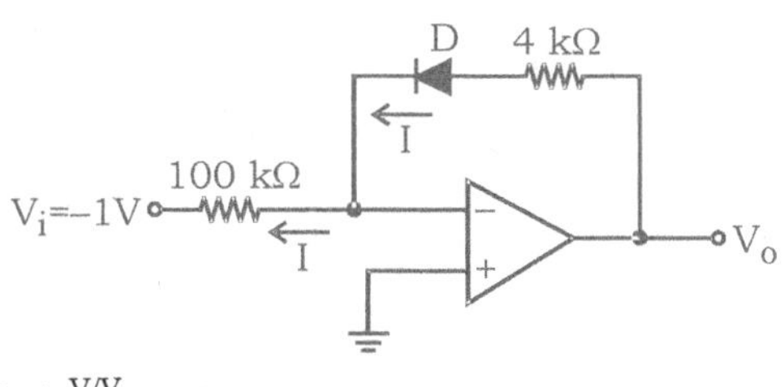

$I = I_o(e^{V/V_t} - 1)$

Now, $V_T = 25$ mV and $I_o = 1$ μA

$$\text{Thus } I = 10^{-6}[e^{V/25\times10^{-3}} - 1] = \frac{1}{10^5}$$

$V = 0.06$ V

$V_o = I \times 4 \text{ k}\Omega + V$

$$= \frac{1}{100 \text{ k}\Omega} \times 4\text{k}\Omega + 0.06 = 0.1 \text{ V}$$

66. $V_o = 5 \text{ mV} \times 10000 = 50$ V

But $V_o = \pm 15$ V in saturation

So it can never be exceeds ± 15 V

$V_o = \pm V_{sat} = \pm 15$ V

67. $I_f = 2$ mA

$0 - V_o = 2 \text{ mA} \times 2 \text{ k}\Omega$

$V_o = -4$V

$$I_L = \frac{V_o}{2 \text{ k}\Omega} = \frac{-4}{2 \text{ k}\Omega} = -2 \text{ mA}$$

$I_f + I = I_L$

$I = -2 \text{ mA} - 2 \text{ mA} = -4$ mA

68. $V_+ = 5$ V, $V_- = V_E$

The input current to op-amp is zero.

$i_{15} = i_z + i_c = i_z + \alpha I_E$

$$= \frac{15.5}{47 \text{ k}\Omega} + \frac{60}{61}\left(\frac{5}{100}\right) = 49.4 \text{ mA}$$

69. Assume D_1 is RB, D_2 is FB

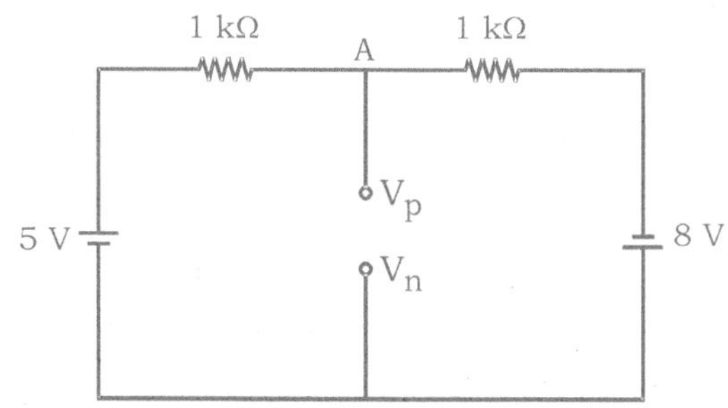

Apply node equation at A

$$\frac{V_P - 5}{1} + \frac{V_P + 8}{1} = 0$$

$2V_p = -3$

$V_P = -1.5$

$V_n = 0$; $V_p < V_n$

$I_{D_1} = 0$ mA

70. Given circuit is

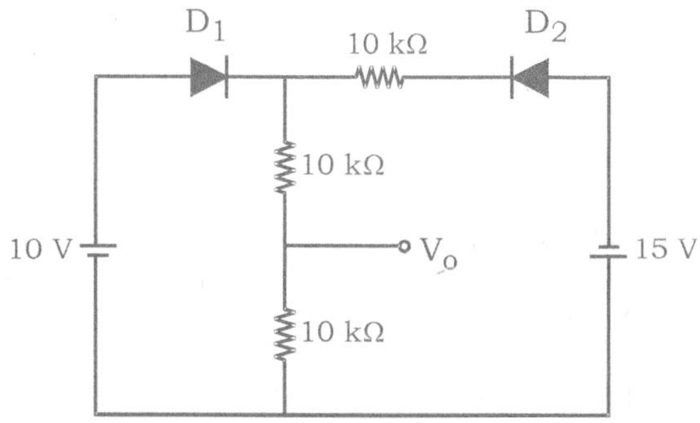

We can observe that diode D_2 is always OFF so equivalent circuit is

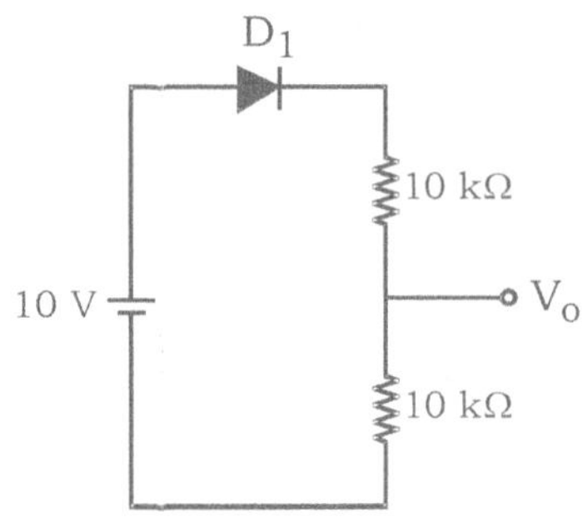

D_1 is ON in this condition

$$V_o = \frac{10}{10+10} \times 10 = 5\text{ V}$$

71. In positive half cycle when (V_{in} > 4 V) diodes D_2 conducts and D_1 OFF so the equivalent circuit is **ml**

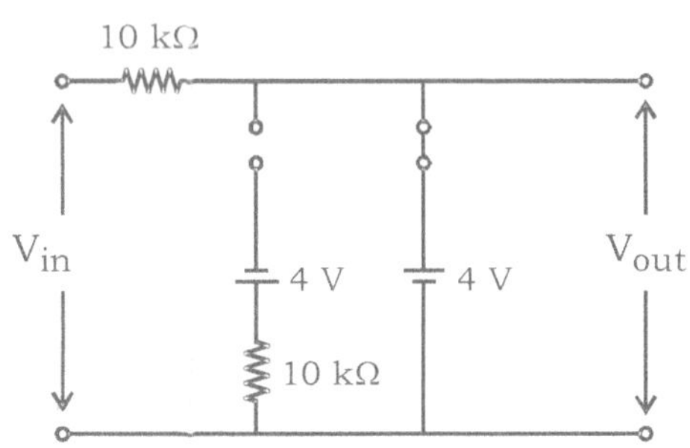

$\therefore$ $V_o = +4$ V

In negative half cycle diode D_1 conducts and D_2 will be OFF so that Apply kVL

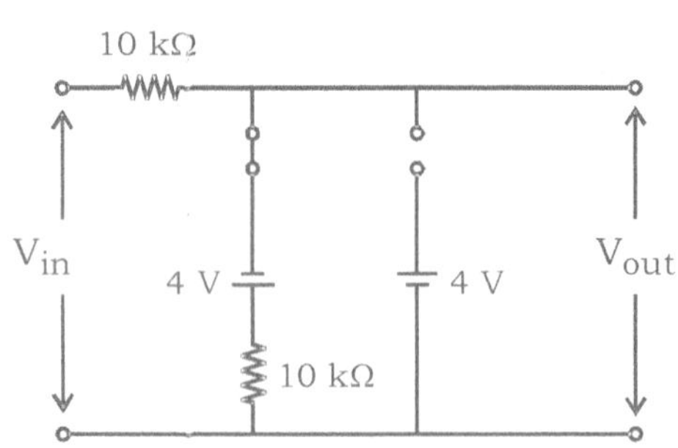

$$V_{in} - 10I + 4 - 10I = 0$$

$$\frac{V_{in}+4}{20} = I$$

$$V_{in} = -10\text{ V}$$

$$I = \frac{-10+4}{20}$$

$$= \frac{-3}{10}\text{ mA}$$

$$\frac{V_{in} - V_{out}}{10} = I$$

$$\frac{-10 - V_{out}}{10} = \frac{-3}{10}$$

$$V_{out} = -(10-3) = -7\text{ V}$$

72. For the positive half of V_i, the diode D_1 is forward bias, D_2 is reverse bias and the zener diode is in breakdown stage $V_i > 6.8$ V.

The output voltage is

$V_o = 0.7 + 6.8 = 7.5$ V

For the negative half of V_i, D_2 is forward bias thus,

$V_o = -0.7$ V.

73. Clamper, sinewave clamped at – 2 V.

74. For positive value of V_i, D_1 is FB and Zener diode regulates the voltage at 6.3 V across it.

$V_o = 6.3 + 0.7 = 7$ V

and for negative half of V_i,

D_2 is FB $V_o = -0.7$ V

75. In the positive half cycle zener diode D_z will be in reverse bias (behaves or a constant voltage source) and diode D is forward bias.

$V_o = 0.7 + 3.3 = 4$ V

In the negative half cycle, Zener (D_z) is in forward bias and diode (*d*) is in revese bias mode so

$$V_o = \frac{10}{(1+1)} \times 1$$

$V_o = 5$ V

$\therefore$ 4 V in the positive half cycle and 5 V in the negative half cycle.

76.

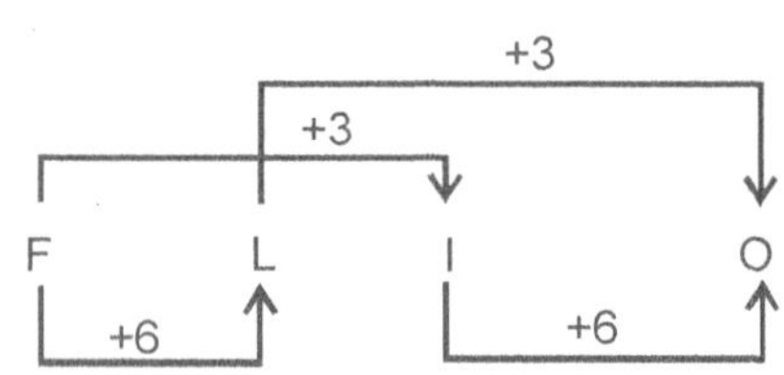

Similarly,

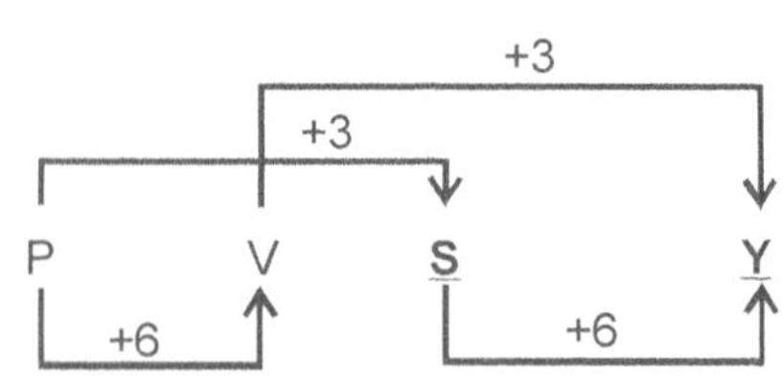

77. The young one of a baboon is called an infant while the young one of a beaver is called a kitten. Hence, option (A) is the correct answer.

78. The series is a,ab,abc,a,ab,abc,...

79.
B G L
↓+4 ↓+4 ↓+4
F K P
↓+4 ↓+4 ↓+4
J O T
↓+4 ↓+4 ↓+4
<u>N</u> <u>S</u> <u>X</u>

80.
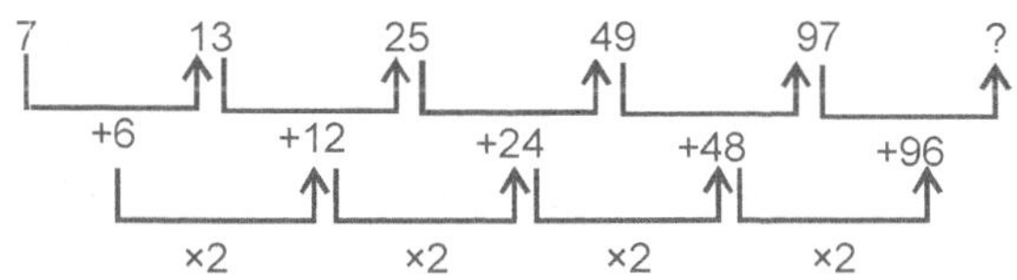

∴ ? = 97 + 96 = 193.

81.
B X
↓+3 ↓–3
E U
↓+3 ↓–3
H R
↓+3 ↓–3
K O
↓+3 ↓–3
<u>N</u> <u>L</u>
↓+3 ↓–3
Q I
↓+3 ↓–3
<u>T</u> <u>F</u>

82. The letters in the series are moving four places back in the alphabetical series. So, the next letters in the series will beF B X.

83. By reversing the digits of each term, the series will be:

27, 64, 125, 216, 343.

The series is formed by the cubes of consecutive numbers starting from 3. Hence, the next term in the series will be $8^3 = 512$ and by reversing the digits of the same, we get 215 as the next term in the series.

84.
B S G
↓+6 ↓–2 ↓+3
H Q J
↓+6 ↓–2 ↓+3
N O M
↓+6 ↓–2 ↓+3
<u>T</u> <u>M</u> <u>P</u>

85.
W 5 K
↓–3 ↓+7 ↓+2
T 12 M
↓–3 ↓+8 ↓+2
Q 20 O
↓–3 ↓+9 ↓+2
N 29 Q
↓–3 ↓+10 ↓+2
<u>K</u> <u>39</u> <u>S</u>

86.
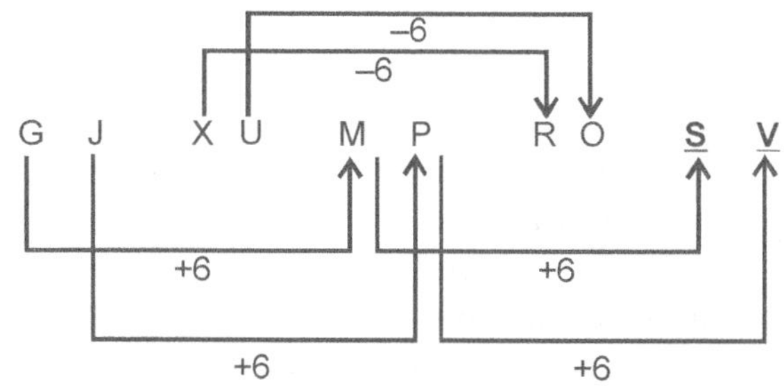

87.
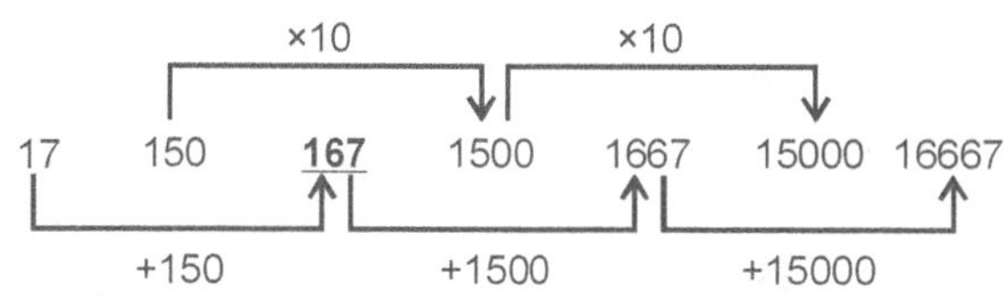

88. The number indicated by these symbols is 7 4 8 1 6.

89. Gourd, Pumpkin and French Beans grow on vines while Jackfruit grows on trees. Hence, option (D) is the answer.

90. Sum of the digits of 272 is 11. While in rest of the three options, sum of the digits of the numbers is 10.

91. The main difficulties with liquid propellants are with the oxidizers. These are generally at least moderately difficult to store and handle due to their low reactivity with common materials, may have extreme toxicity (nitric acids), moderately cryogenic (liquid oxygen), or both.

92. The tragedy of the commons is an economics theory by Garrett Hardin. The concept is often cited in connection with sustainable development, meshing economic growth and environmental protection, as well as in the debate over global warming. "Commons" can include the atmosphere, oceans, rivers, fish stocks, national parks, advertising, and even parking meters.

93. National Youth Day is celebrated in India on 12 January on the birthday of Swami Vivekananda. In 1984, the Government of India declared the day as the National Youth Day and since 1985, the event is celebrated in India every year.

94. The first currency called rupiya was introduced in the 16th century by Sher Shah Suri. The term originated from rûpya, a Sanskrit term for silver coin.

96. Brass is an alloy made of copper and zinc; the proportions of zinc and copper can be varied to create a range of brasses with varying properties. Cracking in brass caused by ammonia attack.

97. The NAND gate is a logic gate used to build digital logic circuits. Given two inputs, A and B, A NAND B will be true if at most one of A and B is true. In other words, A NAND B is false if both A and B are true, and true otherwise. The NAND gate is a "universal gate."

98. Admission or establishment of new States: Parliament may by law admit into the Union, or establish, new States on such terms and conditions.

99. Strip cropping is a method of farming used when a slope is too steep or too long, or otherwise, when one does not have an alternative method of preventing soil erosion.

100. The eye includes a lens not dissimilar to lenses found in optical instruments such as cameras and the same principles can be applied. The pupil of the human eye is its aperture; the iris is the diaphragm that serves as the aperture stop.

101. A chlorofluorocarbon (CFC) is an organic compound that contains only carbon, chlorine, and fluorine, produced as a volatile derivative of methane, ethane, and propane. They are also commonly known by the DuPont brand name Freon. The manufacture of such compounds has been phased out under the Montreal Protocol.

102. The Milinda Panha is a Buddhist text which dates from approximately 100 BCE. It purports to record a dialogue in which the Indo-Greek king Menander I (Milinda in Pali) of Bactria, who reigned in the 2nd century BCE, poses questions on Buddhism to the sage Nâgasena.

103. Alexander the Great in 326 BC against King Porus of the Hindu Paurava kingdom on the banks of the Hydaspas River (Jhelum River).

104. The dense "sugar coating" of mucins gives them considerable water-holding capacity and makes them resistant to proteolysis.

105. Messersmith first developed the polymer DOPA in 2002.

106. Let the total work be $8 \times 3 = 24$ units.

$\Rightarrow$ Work done by C in 1 day = 3 units and that by B = 1 unit.

Thus, time taken by B and C together = 6 days while that by A alone = 8 days.

$\Rightarrow$ Work done by A in 1 day = 3 units

Hence, required time $= \frac{24}{4} = 6$ days.

107.

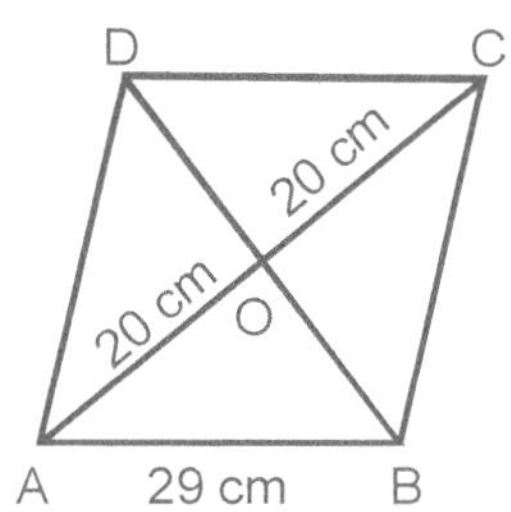

$AB^2 = AO^2 + OB^2$

$\Rightarrow 29^2 = 20^2 + OB^2$

$\Rightarrow OB = 21$ cm

Hence, required area

$= 4 \times \frac{1}{2} \times 21 \times 20 = 840\ cm^2$.

108. $x \times \frac{17}{20} \times \frac{7}{8} \times \frac{4}{5} = \frac{119}{200} = 59.5\%$ of x

Hence, required discount = 40.5%.

109. 25% of A = 35% of B

$\Rightarrow A : B = 7 : 5$

Hence, required amount $= \frac{7}{12} \times 2448 =$ Rs. 1,428.

110. Let the selling price of 1 article be Re.1

$\Rightarrow$ Cost price of 50 articles = Rs. 40 and their selling price = Rs. 50

Hence, there would be profit equivalent to

$\frac{50 - 40}{40} \times 100 = 25\%$.

111. Let the income of B = Rs. 100

$\Rightarrow$ A's income = Rs. 125 and C's income = Rs. 120

Hence, required percentage

$$= \frac{125 - 120}{120} \times 100 = 4.16\%.$$

112. Relative speed = 45 + 4.5 = 49.5 km/hr

$$= 49.5 \times \frac{5}{18}$$

$$= 13.75 \text{ m/s}$$

Hence, required time $= \frac{143}{13.75}$

= 10.4 seconds.

113. $3x + \frac{1}{2x} = 3$

$\Rightarrow 2x + \frac{1}{3x} = 2$ ... (i)

Squaring equation (i) we get,

$$4x^2 + \frac{1}{9x^2} + \frac{4}{3} = 4$$

$\Rightarrow 4x^2 + \frac{1}{9x^2} = \frac{8}{3}$ ... (ii)

Similarly, on cubing equation (i) we get,

$$8x^3 + \frac{1}{27x^3} + 2 \times 2 = 8$$

$\Rightarrow 8x^3 + \frac{1}{27x^3} = 4$ ... (iii)

Multiplying equations (ii) and (iii) we get,

$$32x^5 + \frac{4}{27x} + \frac{8x}{9} + \frac{1}{243x^5} = \frac{32}{3}$$

Hence, the required value is $\frac{32}{3}$.

114. Going through the options we get the required answer as 12 : 5.

115. $(\cos^2A - \sin A)(\sin A + \cos^2A) = -1$

$\Rightarrow \cos^4A - \sin^2A = -1$

$\Rightarrow \cos^4A + \cos^2A = 0$

$\Rightarrow \cos^2A(\cos^2A + 1) = 0$

Since $\cos^2A$ can never be negative in the given range thus A = 90° and $(\cos^2A + 2) = 2$.

116. Let the fifth angle be x°.

Sum of all the exterior angles of any polygon is always 360°.

$$\Rightarrow (180 - 140)° + (180 - 90)° + (180 - 70)° + (180 - 80)° + (180 - x)° = 360°$$

Hence, x = 160°.

117. Putting A = 45° in $\sec A(\sec A + \tan A + \frac{\cos A}{1 + \sin A}) - 2\tan^2A$, we get the required value as 2.

118. $\frac{(2 \times 8^{n+1} + 4 \times 2^{3n-1})}{(4 \times 2^{2n+1} - 2 \times 4^{2n+1})}$

$$= \frac{(2^{3n+4} + 2^{3n+1})}{(2^{2n+3} - 2^{4n+3})}$$

$$= \frac{(2^{10} + 2^7)}{(2^7 - 2^{11})} = -\frac{3}{5}.$$

119. Age of the teacher = 17 × 22 – 16 × 18.5 = 78 years.

120. Putting a = 3 and b = 0 in the equation

$$a^4 + b^4 + 6a^2b^2 - 4ab(a^2 + b^2) - 70$$

we get, $3^4 - 70 = 11$.

124. The correct expression should be 'brought up' which means 'to rear.'

125. The article should be 'a.'

126. The verb 'are' is wrong. 'Is' should have been used.

127. The phrase 'broke into' means to 'enter by force.'

129. To 'fall flat' refers to 'producing no effect.'

136. Refers to a good and kind person.

137. 'Arabian Nights' is the name of a book. Hence, singular verb is to be used.

138. 'Lay down' implies 'surrender.'

139. In the context of the sentence, the past perfect continuous tense is to be used.

140. 'Take after' implies resembling somebody.

141. The height of Mount Elbrus is 5642 metres.

142. Refer to the first paragraph for the answer.

143. Refer to the first sentence of the last paragraph for the answer.

144. The other options are narrow in scope.

149. 'Running over' means overflowing.

150. The correct verb should be 'is'.

151. The verb used should be 'is' since 'bread and butter' is considered to be a single entity.

161. When one of the subjects joined by 'or' or 'nor' is plural, the verb must be plural, and the plural subject should be placed nearest the verb.

162. Two nouns qualified by 'every', even though connected by 'and', require a singular verb.

163. 'Taken to' refers to 'being addicted to something.'

164. 'Saw through' implies to detect something.

165. 'Well off' means to be in comfortable circumstances.

166. The IPKF operations took place in the years between 1987 to 1990.

168. Refer to the last paragraph for the answer.

169. Refer to the first sentence of the second paragraph for the answer.

170. Refer to the first paragraph for the answer.

174. The sentence is correct. "Call up" means "to recollect".

175. "Your's" is wrong. "Yours" should be used.

176. The article should be "a".

PRACTICE PAPER-2

ELECTRONICS ENGINEERING

Time Allowed : 2:15 Hours **Maximum Marks : 180**

Read the following instructions carefully before you begin to answer the questions. This Booklet contains questions in English.

INSTRUCTIONS

1. This Booklet contains 180 questions in total comprising of the following five parts:

Part – I:	Domain Knowledge	[1 – 75 Questions]
Part – II:	General Intelligence and Reasoning	[76 – 90 Questions]
Part – III:	General Awareness	[91 – 105 Questions]
Part – IV:	Quantitative Aptitude	[106 – 120 Questions]
Part – V:	General English	[121 – 180 Questions]

2. All questions are compulsory and carry 1 mark each.

3. Before you start to answer the questions you must check this Booklet to ensure that it contains all the pages and see that no page is missing or repeated.

4. For every wrong answer 1/3 marks will be deducted.

5. You are supplied with the OMR-Sheet separately.

6. Answers must be indicated by completely blackening one of the corresponding ovals on OMR-Sheet against the relevant question number, by black/blue ball pen only. Answers which are not indicated by black/blue ball pen will not be awarded any mark.

7. The OMR-Sheet must be handed over to the Invigilator before you leave the Examination Hall.

8. Failure to comply with any of the above instructions will render a candidate liable to such action/penalty as may be deemed fit.

PRACTICE PAPER – 2

DOMAIN KNOWLEDGE

1. At room temperature the current in an intrinsic semiconductor is due to

(*a*) holes (*b*) electrons

(*c*) ions (*d*) holes and electrons

2. In which of these is reverse recovery time nearly zero?

(*a*) Zener diode (*b*) Tunnel diode

(*c*) Schottky diode (*d*) PIN diode

3. A transistor has a current gain of 0.99 in the CB mode. Its current gain in the CC mode is

(*a*) 100 (*b*) 99

(*c*) 1.01 (*d*) 0.99

4. In an n-channel JFET, the gate is

(*a*) n-type (*b*) p-type

(*c*) either n or p (*d*) partially n & partially p

5. In a p type material the Fermi level is 0.3 eV above valence band. The concentration of acceptor atoms is increased. The new position of Fermi level is likely to be

(*a*) 0.5 eV above valence band

(*b*) 0.28 eV above valence band

(*c*) 0.1 eV above valence band

(*d*) below the valence band

6. A silicon (PN) junction at a temperature of 20°C has a reverse saturation current of 10 pico Ampere. The reverse saturation current at 40°C for the same bias is approximately.

(*a*) 30 pA (*b*) 40 pA

(*c*) 50 pA (*d*) 60 pA

7. In a bipolar transistor the barrier potential is

(*a*) 0

(*b*) a total of 0.7 V

(*c*) 0.7 V across each depletion layer

(*d*) 0.35 V

8. An amplifier without feedback has a voltage gain of 50, input resistance of 1 kΩ and output resistance of 2.5 kΩ. The input resistance of the current shunt -ve feedback amplifier using the above amplifier with a feedback factor of 0.2 is

(*a*) 1/11 kΩ (*b*) 1/5 kΩ

(*c*) 5 kW (*d*) 11 kW

9. As compared to an ordinary semiconductor diode, a Schottky diode

(*a*) has lower cut in voltage

(*b*) has higher cut in voltage

(*c*) lower reverse saturation current

(*d*) both (b) and (c)

10. If α_{ac} for transistor is 0.98 then β_{ac} is equal to

(*a*) 51 (*b*) 49

(*c*) 47 (*d*) 45

11. In an n-p-n transistor biased for operation in forward active region

(*a*) emitter is positive with respect to base

(*b*) collector is positive with respect to base

(*c*) base is positive with respect to emitter and collector is positive with respect to base

(*d*) none of the above

12. A zener diode is used in

(*a*) voltage regulator circuits

(*b*) amplifier circuits

(*c*) both voltage regulator and amplifier circuits

(*d*) none of the above

13. In a degenerate n type semiconductor material, the Fermi level,

(*a*) is in valence band

(*b*) is in conduction band

(*c*) is at the centre in between valence and conduction bands

(*d*) is very near valence band

14. A potential of 7 V is applied to a silicon diode. A resistance of 1 kilo ohm is also in series with the diode. The current is

(*a*) 7 mA

(*b*) 6.3 mA

(*c*) 0.7 mA

(*d*) 0

15. In which region of a CE bipolar transistor is collector current almost constant?

(*a*) Saturation region

(*b*) Active region

(*c*) Breakdown region

(*d*) Both saturation and active region

16. Ideally the stability factor should be

(*a*) 0 (*b*) ∞

(*c*) 1 (*d*) -∞

17. If J_E junction is forward biased and J_C junction is reverse biased than the transistor will operate

(*a*) cut-off region

(*b*) active region

(*c*) saturation region

(*d*) reverse cut-off region

18. The type of noise reduced by limiters in FM receivers is

(*a*) Avalanche noise

(*b*) Burst noise

(*c*) Narrow band-pass noise

(*d*) Impulse noise

19. A sinusoidal voltage has peak to peak value of 100 V. The rms value is

(*a*) 50 V (*b*) 70.7 V

(*c*) 35.35 V (*d*) 141.41 V

20. A capacitor is needed for an ac circuit of 230 V, 50 Hz the peak voltage rating of the capacitor should be

(*a*) 230 V (*b*) 0.5 × 230 V

(*c*) $\sqrt{2}$ × 230V (*d*) 230/$\sqrt{2}$ V

21. RMS value of i(t) = 10[1 + sin (- t)] is

(*a*) 10 (*b*) 10/$\sqrt{2}$

(*c*) $\sqrt{150}$ (*d*) 10$\sqrt{2}$

22. The time constant of the circuit in figure shown.

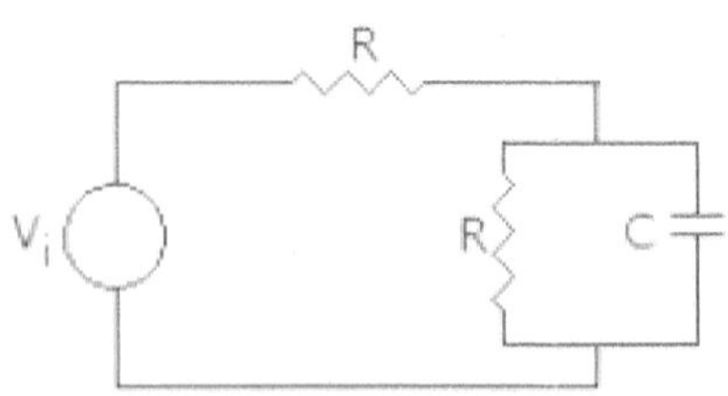

(*a*) RC (*b*) RC/2

(*c*) 1/2RC (*d*) 2/RC

23. In an ac circuit, the maximum and minimum values of power factor can be

(*a*) 2 and 0 (*b*) 1 and 0

(*c*) 0 and -1 (*d*) 1 and -1

24. The impedance of an RC series circuit is 12Ω at f= 50 Hz. At f= 200 Hz, the impedance will be

(*a*) zero (*b*) unchange

(*c*) decreases (*d*) increases

25. A capacitor stores 0.15C at 5 V. Its capacitance is

(*a*) 0.75 F (*b*) 0.75   F

(*c*) 0.03 F (*d*) 0.03   F

26. A 0.5   F capacitor is connected across a 10 V battery. After a long time, the circuit current and voltage across capacitor will be

(*a*) 0.5 A and 0 V (*b*) 0 A and 10 V

(*c*) 20 A and 5 V (*d*) 0.05 A and 5 V

27. In the circuit of figure, the switch is closed at t = 0, the current through C is

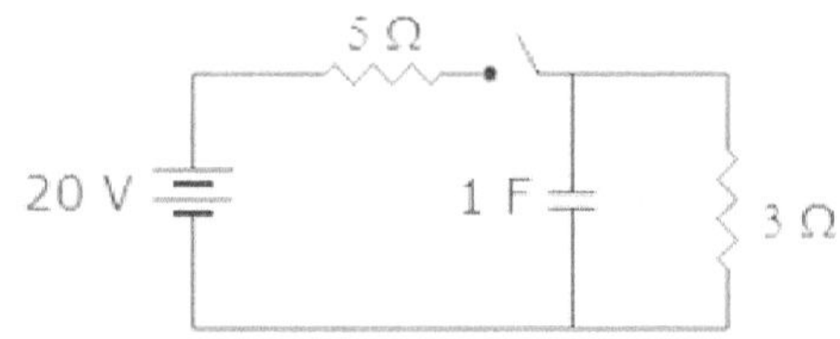

(*a*) 4 A (*b*) 2.5 A

(*c*) 3.1 A (*d*) 0

28. The resistance of a 230 V, 100 W lamp is

(*a*) 529 Ω (*b*) 2300 Ω

(*c*) 5290 Ω (*d*) 23 Ω

29. In the given circuit, viewed from AB, the circuit can be reduced to an equivalent circuit as

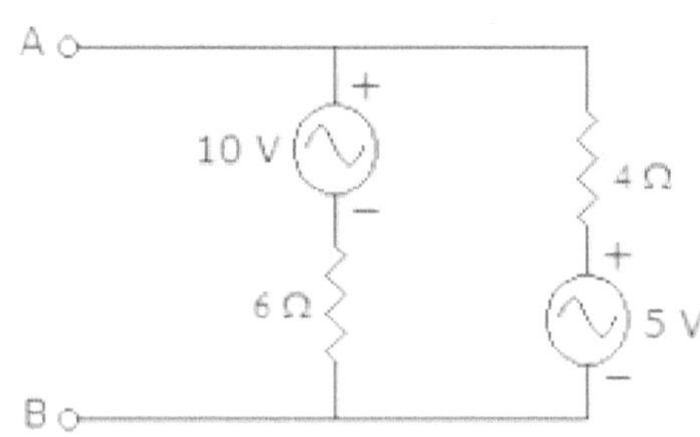

(*a*) 5 volts source in series with 10Ω resistor

(*b*) 7 volts source in series with 2.4 Ω resistor

(*c*) 15 volts source in series with 2.4 Ω resistor

(*d*) 1 volt source in series with 10 Ω resistor

30. Two coils having self inductances of 10 mH and 40 mH are mutually coupled. The mutual inductance is

(*a*) 40 mH

(*b*) 20 mH

(*c*) 10 mH

(*d*) 5 mH

31. In the circuit shown in figure the current I of sinusoidal source is

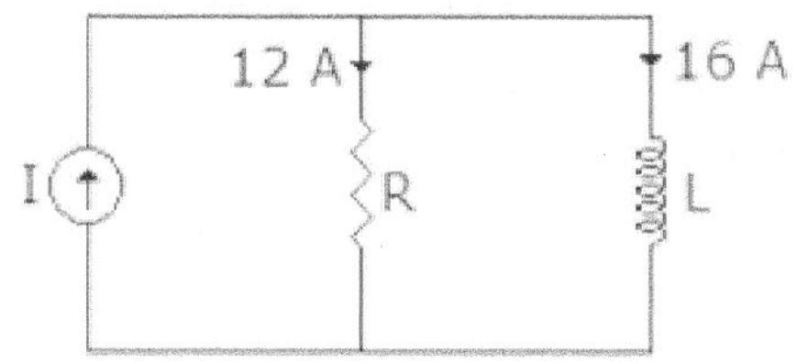

(*a*) 25 A

(*b*) 4 A

(*c*) 20 A

(*d*) Not determinable from given data

32. Which of these has a layer of intrinsic semiconductor?

(*a*) Zener diode (*b*) PIN diode

(*c*) Photo diode (*d*) Schottky diode

33. In a varactor diode the increase in width of depletion layer results in

(*a*) decrease in capacitance

(*b*) increase in capacitance

(*c*) no change in capacitance

(*d*) either (a) or (b)

34. When a p-n junction is reverse biased

(*a*) holes and electrons move away from the junction

(*b*) holes and electrons move towards the junction

(*c*) holes move towards junction and electrons move away from junction

(*d*) holes move away from junction and electrons move towards junction

35. At room temperature the barrier potential in a silicon diode is

(*a*) 0.1 V (*b*) 0.3 V

(*c*) 0.7 V (*d*) 1 V

36. Two sinusoidal signals of same amplitude and frequencies 10 kHz and 10.1 kHz are added together. The combined signal is given to an ideal frequency detector. The output of the detector is

(*a*) 0.1 kHz sinusoidal

(*b*) 10.1 kHz sinusoidal

(*c*) a linear function

(*d*) a constant

37. The percentage saving in power of 100% modulated suppressed carrier AM signal is

(*a*) 100 (*b*) 75

(*c*) 66.6 (*d*) 50

38. A carrier is phase modulated (PM) with frequency deviation of 10 kHz by a single tone frequency of 1 kHz. If the single tone frequency is increased to 2 kHz, assuming that phase deviation remains unchanged, the bandwidth of the PM signal is

(*a*) 21 kHz (*b*) 22 kHz

(*c*) 44 kHz (*d*) 42 kHz

39. In PCM, the information lies

(*a*) in one or two properties of the pulse

(*b*) in absence of pulse

(*c*) in presence of pulse

(*d*) in presence or absence of pulse

40. A carrier is simultaneously modulated by two sine waves having modulation indices of 0.4 and 0.3. The total modulation index will be

(*a*) 0.1 (*b*) 0.7

(*c*) 0.5 (*d*) 0.35

41. The biasing shown in the below circuit is.

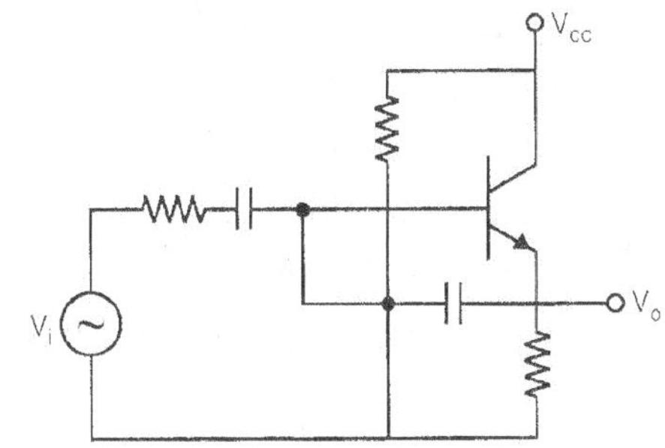

(*a*) Emitter bias

(*b*) Self bias

(*c*) Potential divider bias

(*d*) bootstrap bias

42. In PCM the amplitude levels are transmitted in a 7 unit code. Sampling frequency is 10 kHz. The bandwidth should be

(*a*) 5 kHz (*b*) 35 kHz

(*c*) 70 kHz (*d*) 5 MHz

43. Which of the following has highest power gain?

(*a*) CE (*b*) CC

(*c*) CB (*d*) CB or CC

44. Two messages m_1, and m_2 have a probability of 0.5 each. The entropy is

(*a*) 0.25 (*b*) 0.5

(*c*) 0.75 (*d*) 1

45. In figure, the steady state current through 1H inductance is

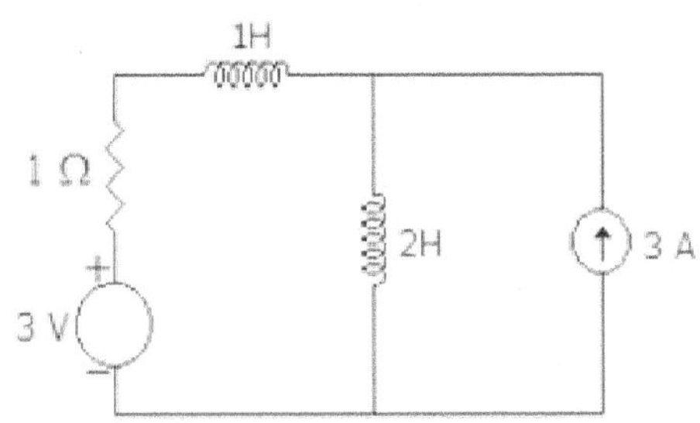

(*a*) Zero (*b*) 3 A

(*c*) 5 A (*d*) 6 A

46. A 400 W carrier is amplitude modulated with m=0.75. The total power in AM is

(*a*) 400 W (*b*) 512 W

(*c*) 588 W (*d*) 650 W

47. Non-coherently detection is not possible for

(*a*) PSK (*b*) ASK

(*c*) FSK (*d*) both (*a*) & (*c*)

48. In radar systems PRF stands for

(*a*) Power Return Factor

(*b*) Pulse Return Factor

(*c*) Pulse Repetition Frequency

(*d*) Pulse Response Factor

49. The velocity factor of a transmission line depends on

(*a*) temperature

(*b*) skin effect

(*c*) relative permittivity of dielectric

(*d*) none of the above

50. A loss less line of characteristic impedance Z_0 is terminated in pure reactance of $-jZ_0$ value. VSWR is

(*a*) 10 (*b*) 2

(*c*) 1 (*d*) infinity

51. In a klystron amplifier the input cavity is called

(*a*) buncher

(*b*) catcher

(*c*) Pierce gun

(*d*) collector

52. Microwave resonators are used in

(*a*) microwave oscillators

(*b*) microwave narrow band amplifier

(*c*) microwave frequency metres

(*d*) all of the above

53. Capacitive susceptance is a measure of

(*a*) a purely capacitive circuit's ability to pass current

(*b*) a purely capacitive circuit's ability to resist the flow of current

(*c*) the extent of neutralization of reactive power in a circuit

(*d*) reactive power in a circuit

54. The main advantage of a crystal oscillator is that its output is

(*a*) A constant frequency

(*b*) dc

(*c*) 50 Hz to 60 Hz

(*d*) Variable frequency

55. The gain of amplifier in RC phase shift oscillator should be

(*a*) 1 (*b*) Greater than 29

(*c*) 29 (*d*) Less than 29

56. Wien-bridge oscillator is basically a

(*a*) Pulse generator

(*b*) sine wave generator

(*c*) Square wave generator

(*d*) Triangular wave generator

57. An amplifier with voltage gain of 60 dB uses 1/2 of its output in negative feedback. Determine the gain with feedback.

(*a*) 30.8 dB (*b*) 25.85 dB

(*c*) 40.86 dB (*d*) 75.5 dB

58. The voltage gain of an amplifier without feedback is 60 dB. It decreases to 40 dB with feedback. Determine the feedback factor.

(*a*) 12 (*b*) 15

(*c*) 10 (*d*) 9

59. A single stage transistor amplifier has a voltage gain of 600 without feedback and 50 with feedback. Find the percentage of output which is feedback to the input side.

(*a*) 2.0 % (*b*) 8.5 %

(*c*) 1.83 % (*d*) 6.0 %

60. The value of V_{GS} and V_G for the circuit shown in figure is

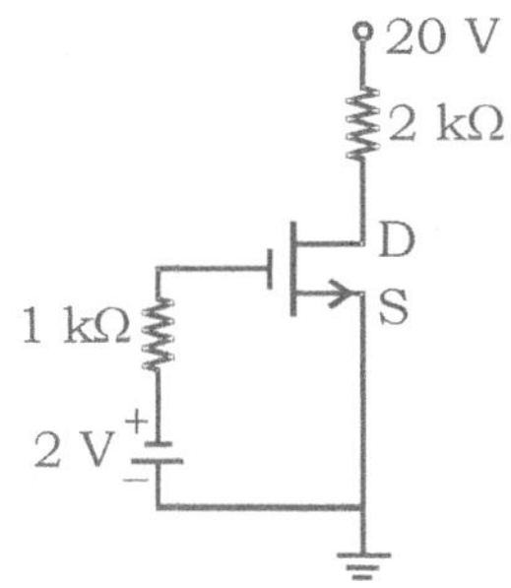

(*a*) -2 V (*b*) 2 V
(*c*) 3 V (*d*) - 3 V

61. Consider an N channel MOSFET with parameters k_n = 0.25 mA/V^2, V_{Th} = 1 V, γ = 0, C_{gd} = 0.04 pF and C_{gs} = 0.2 pF. The transistor is biased at V_{GN} = 3 V. The unity gain band width of an FET is.

(*a*) 663 MHz (*b*) 750 MHz
(*c*) 500 MHz (*d*) 400 MHz

62. Given for an FET, g_m = 95 mA/V, total capacitance = 500 pF. For a voltage gain of –30. The bandwidth with be

(*a*) 6.3 MHz (*b*) 6.9 MHz
(*c*) 7.5 MHz (*d*) 8 MHz

63. In a JFET

(*a*) drain current is very nearly equal to source current
(*b*) drain current is much less than source current
(*c*) drain current may be equal to or less than source current
(*d*) drain current may be even more than source current

64. The early effect in a BJT is caused by

(*a*) fast turn on
(*b*) fast turn off
(*c*) large collector base reverse bias
(*d*) large emitter base forward bias

65. Barkhausen criterion of oscillation is

(*a*) $A\beta > 1$ (*b*) $A\beta = 1$
(*c*) $A\beta \geq 1$ (*d*) $A\beta \leq 1$

66. Which of the following is used for generating time varying wave forms?

(*a*) MOSFET
(*b*) PIN diode
(*c*) Tunnel diode
(*d*) UJT

67. The voltage across a zener diode

(*a*) is constant in forward direction
(*b*) is constant in reverse direction
(*c*) is constant in both forward and reverse directions
(*d*) none of the above

68. A 10 GHz signal propagates in a waveguide with a phase velocity of 4×10^8 m/s. If the signal produces standing wave in the guide, then the distance between successive minimal will be

(*a*) 1 cm (*b*) 2 cm
(*c*) 3 cm (*d*) 4 cm

69. Bandwidth occupied by 100 MHz carrier, AM modulated by signal frequency of 10 kHz is

(*a*) 100 MHz (*b*) 20 kHz
(*c*) 10 kHz (*d*) 110 MHz

70. The length of half-wave antenna is at 30 MHz

(*a*) 2.5 meter (*b*) 5 meter
(*c*) 10 meter (*d*) 100 meter

71. All types of linear modulation can be detected by

(*a*) Product demodulator
(*b*) Envelop detector
(*c*) Filtering
(*d*) Linear detector

72. Ideally, for linear operation, a transistor should be biased so that the Q-point is

(*a*) near saturation.
(*b*) near cutoff.
(*c*) where IC is maximum.
(*d*) halfway between cutoff and saturation.

73. The most stable biasing technique used is the

(*a*) voltage-divider bias.
(*b*) base bias.
(*c*) emitter bias.
(*d*) collector bias.

74. For a BJT α = 0.98, I_B = 26 μA, then β and I_C will be

(*a*) 49, 0.02 mA (*b*) 49, 2 mA
(*c*) 49, 1.27 mA (*d*) 48, 3 mA

75. In a circuit using a full wave converter (M- 2 connection) the PIV of each thyristor is 400 V. For the same output voltage and fully controlled bridge converter, PIV will be

(*a*) 400 V (*b*) 800 V
(*c*) 200 V (*d*) $200\sqrt{2}$ V

GENERAL INTELLIGENCE AND REASONING

Directions: In question nos. **76** to **79**, select the related letters /word /number from the given alternatives.

76. TALENT : VYNCPR :: NORMAL : ?

(a) PMYKBK (b) PMTKCJ
(c) PMTLBJ (d) PNTKDK

77. Vitamin A : Night blindness :: Vitamin C : ?

(a) Nerve Contraction (b) Rickets
(c) Scurvy (d) Goitre

78. Sick : Dog :: Busy : ?

(a) Toad (b) Crocodile
(c) Cuckoo (d) Bee

79. H : 512 : : M : ?

(a) 2197 (b) 1728
(c) 2744 (d) 1566

80. Which one of the given responses would be a meaningful order of the following?

(1) Idea Generation and Screening
(2) Market Testing
(3) Concept Development/Testing
(4) Product Development
(5) Commercialization
(6) Analysis

(a) 1, 2, 4, 6, 5, 3 (b) 1, 2, 3, 4, 5, 6
(c) 1, 3, 6, 4, 2, 5 (d) 1, 4, 3, 2, 6, 5

81. If 462 * 3 = 18 and 564 * 2 = 12, then 617 * 4 =?

(a) 13 (b) 14
(c) 20 (d) 16

Directions: In question nos. **82** and **83**, which one set of letters when sequentially placed at the gaps in the given letter series shall complete it?

82. c_ba_bba_bbac_ba

(a) bbab (b) bccb
(c) abac (d) acab

83. a_bcabb_abc_aa_c

(a) baab (b) bccb
(c) abac (d) accb

Directions: In question nos. **84** to **88**, find the odd word/number / number-pair from the given alternatives.

84. (a) Black gram (b) Groundnut
(c) Wheat (d) Fennel

85. 1, 9, 52, 94, 18, 57, 961

(a) 18 (b) 961
(c) 57 (d) 52

86. (a) 253 – 351 (b) 478 – 692
(c) 123 – 22 (d) 971 – 816

87. (a) Alfalfa (b) Aloe Vera
(c) Bonsai (d) Lavender

88. (a) Primitive (b) Protest
(c) Problem (d) Project

Directions: In question nos. **89** to **90**, a series is given, with one term missing. Choose the correct alternative from the given ones that will complete the series.

89. ac, abd, eg, efh, ik, _?_ .

(a) ikl (b) ijl
(c) jkm (d) jkn

90. ACE, GIK, MOQ, SUW, _?_, _?_.

(a) XZB, DFH (b) YZB, DFH
(c) XZA, CEG (d) YAC, EGI

GENERAL AWARENESS

91. Methane gas is produced from

(a) Ruminants (b) Rice fields
(c) Marshy fields (d) All of the above

92. Anandwan was founded by

(a) Baba Amte (b) Rajiv Gandhi
(c) Mahatma Gandhi (d) Anna Hazare

93. Where is Kansai International Airport located?

(a) South Korea (b) North Korea
(c) Japan (d) China

94. Noori, a cloned pashmina goat was created by which of the following countries?

(a) India (b) Pakistan
(c) Israel (d) Bangladesh

95. Which of the following prize/award is also known as Asia's Nobel Prize?

(a) Ramon Magsaysay Award
(b) Simon Bolivar Prize
(c) Jnanpith Award
(d) Chameli Devi Award

96. BCCI was formed in

(a) 1928 (b) 1938
(c) 1948 (d) 1958

97. Smog is a type of

(a) Air Pollution (b) Water Pollution
(c) Soil Pollution (d) Marine pollution

98. Which of the following article of Indian constitution is related to right to information?

(*a*) Art 19 (*b*) Art 21

(*c*) Art 14 (*d*) Art 32

99. Which of the following is not true about Indian constitution?

(*a*) It is the longest written constitution of any sovereign nation.

(*b*) Jawaharlal Nehru is considered as the 'Father of Indian Constitution'.

(*c*) The constitution of India is federal in structure.

(*d*) Besides the English version, there is an official Hindi translation.

100. Which of the following odorant is added to LPG to detect its leakage?

(*a*) Ethane (*b*) Methane

(*c*) Ethanethiol (*d*) Ethanol

101. Barabati Stadium is situated at

(*a*) Kolkata (*b*) Cuttack

(*c*) Chennai (*d*) Karnataka

102. Ilbert Bill was introduced during the tenure of

(*a*) Lord Ripon (*b*) Lord Macaulay

(*c*) Lord Dalhousie (*d*) Lord Willington

103. 74th amendment of Indian Constitution is associated with the

(*a*) Panchayati Raj Institutions

(*b*) Urban local Bodies

(*c*) Both (*a*) & (*b*)

(*d*) Reservation of women in Panchayat elections

104. Origin of which of the following musical instrument was not India?

(*a*) Conga

(*b*) Pakhavaj

(*c*) Talking drum

(*d*) Mridanga

105. Giffen goods are

(*a*) A product that people consume more often as the price rises

(*b*) A product that people consume more often as the price falls

(*c*) A product that people consume more or less in the same quantity irrespective of price changes

(*d*) None of the above

QUANTITATIVE APTITUDE

106. The sum and the difference of two positive integers are respectively 166 and 8. What is the sum of square of these numbers?

(*a*) 13810 (*b*) 24960

(*c*) 30270 (*d*) 27556

107. The difference between the selling prices of two articles, with same cost price, sold at 4% and 8% profits is Rs. 3.75. The selling price of the dearer article is

(*a*) Rs. 97. 50 (*b*) Rs. 93.75

(*c*) Rs. 101.25 (*d*) Rs. 97.25

108. In a school there are 1200 students, out of whom 55 percent are girls. Monthly fee of each boy is Rs . 750 and monthly fee of each girl is 20 percent less than each boy. What is the monthly revenue generated by the school from all the students?

(*a*) Rs. 7,91,000 (*b*) Rs. 8,11,000

(*c*) Rs. 8,01,000 (*d*) Rs. 8,21,000

109. A train of length 105 m takes 45 sec to cross a bridge of length 210 m. What is the speed of the train in km/hr?

(*a*) 24.8 (*b*) 25.2

(*c*) 32.2 (*d*) 25.4

110. Simplify the question:

$$\frac{(0.87+0.13)^2-(0.87-0.13)^2}{0.2262}$$

(*a*) 4 (*b*) 1

(*c*) 8 (*d*) 2

111. If $x+\frac{1}{x}=-1$, then the value of $x^7+\frac{1}{x^7}$ is

(*a*) 4 (*b*) –2

(*c*) –1 (*d*) –4

112. In a circle, two arcs of unequal lengths subtend 60° and 105° at the center. If the length of the smaller arc is 7 cm, then the length of larger arc is

(*a*) 12.25 cm (*b*) 9 cm

(*c*) 11.15 cm (*d*) 10.25 cm

113. The length of three sides of a right angled triangle are x, (x – 7) and (x + 2) cm respectively. What is the value of x?

(*a*) 3 (*b*) 17

(*c*) 13 (*d*) 15

114. The value of $\frac{3\cot 30^\circ - 5\cos 30^\circ + 3\sin 30^\circ}{\sin 60^\circ + \tan 60^\circ - 3\cos 60^\circ}$ is

(a) $\sqrt{3}\left(2+\sqrt{3}\right)$ (b) $\frac{2}{\sqrt{3}}+1$

(c) $1-\frac{2}{\sqrt{3}}$ (d) $\sqrt{3}\left(2-\sqrt{3}\right)$

115. P, Q and R together can complete a work in $9\frac{3}{4}$ days. If P and R together can complete the same work in $15\frac{3}{4}$ days, then in how many days can Q alone finish the work?

(a) $24\frac{17}{32}$ (b) $25\frac{19}{32}$

(c) $24\frac{19}{32}$ (d) $25\frac{17}{32}$

116. The diameters of two cones whose volumes are equal, is in the ratio 2 : 3. The ratio of their height is

(a) 2 : 3 (b) 4 : 9

(c) $\sqrt{2}:\sqrt{3}$ (d) 3 : 2

117. A trader marks his goods 37.5% above the cost price and allows a discount of 20 percent. He

(a) earns a profit of 10%.

(b) suffers a loss of 6%.

(c) earns a profit of 8%.

(d) suffers a loss of 10%.

118. A bicycle covers a distance of 278.6 metres in 3 minutes whereas a motorcycle covers 40% of the distance covered by bicycle in 1 minute 12 sec. The ratio of their speeds is

(a) 2 : 3 (b) 1 : 3

(c) 1 : 1 (d) 3 : 5

119. Three years hence, the average age of A, B and C will be 24 years. If A is five years older than B and C is two years older than A, then the sum of the ages of A six years ago and that of C five years hence will be:

(a) 45 years (b) 46 years

(c) 41 years (d) 43 years

120. If a – b = c and $a^2 + b^2 + c^2 = 4$, then the value of (ab – bc + ac) is

(a) –1 (b) 0

(c) 2 (d) –2

GENERAL ENGLISH

Directions (Question no. 121): In the following question, out of the four alternatives, choose the word which best expresses the meaning of the word given in bold.

121. Laceration

(a) Machination (b) Mutilation

(c) Alliteration (d) Assimilation

Directions (Question no. 122): In the following question, out of the four alternatives, choose the word which is opposite in meaning to the word given in bold.

122. Benevolence

(a) Animosity (b) Altruism

(c) Benediction (d) Generosity

Directions (Question no. 123): Four words are given, out of which only one word is spelt correctly. Choose the correctly spelt word out of the given options.

123. (a) Acquesce (b) Acquiesce

(c) Acquisce (d) Acqueisce

Directions (Question nos. 124 to 126): In the following questions, one part of the sentence may have an error. Find out which part of the sentence has an error and choose the appropriate letter. If the sentence is free from error, choose the "No error" option.

124. I think (a)/ your's (b)/ is a better plan. (c)/ No error (d)

125. Neither of (a)/ the two dogs (b)/ were black. (c)/ No error (d)

126. It is impossible (a)/ to trust (b)/ the word of a habitual liar. (c)/ No error (d)

Directions (Question nos. 127 to 129): The sentences given with blanks are to be filled with an appropriate word(s). Four alternatives are suggested. Choose the correct option.

127. He is still in vigorous health, although he is on the _______ side of sixty.

(a) left (b) top

(c) full (d) wrong

128. He says he has _______ the examination.

(a) passing (b) past

(c) passed (d) pass

129. There are black ________ in every department.

(a) horses

(b) sheep

(c) ship

(d) goats

Directions (Question nos. 130 to 132): In each of the questions, four alternatives are given for the Idiom/Phrase. Choose the alternative which corresponds to the meaning of the given Idiom/Phrase.

130. Cock and bull story

(*a*) A factual tale (*b*) An unbelievable tale

(*c*) A mythological story (*d*) A folktale

131. Hit the sack

(*a*) Get into bad

(*b*) Hit someone angrily

(*c*) Meet with an accident

(*d*) Shout at someone

132. Hit the books

(*a*) To sell the books (*b*) To study hard

(*c*) Become a bestseller (*d*) Publish a book

Directions (Question nos. 133 to 135): Out of the four alternatives, choose the one which can be substituted for the given words/sentences. Give answer.

133. Fear of public places

(*a*) Acrophobia (*b*) Agoraphobia

(*c*) Arachnophobia (*d*) Xenophobia

134. One who has two wives

(*a*) Bilingual (*b*) Bipolar

(*c*) Bigamist (*d*) Polyandry

135. An intense craving for alcohol

(*a*) Kleptomania (*b*) Andromania

(*c*) Anglomania (*d*) Dipsomania

Directions: (Question nos. 136 to 140): A sentence/a part of the sentence is underlined. Four alternatives are given which can replace the underlined part. Choose the correct alternative. In case no improvement is needed, choose the option corresponding to "No improvement".

136. They were playing since six o' clock.

(*a*) play (*b*) have been playing

(*c*) have playing (*d*) No improvement

137. I met him in Mumbai in 1997.

(*a*) have met him

(*b*) have meet him

(*c*) meet him

(*d*) No improvement

138. I have been teaching for fifteen years next December.

(*a*) have taught

(*b*) will teach

(*c*) will have been teaching

(*d*) No improvement

139. I have written the letter before he arrived.

(*a*) had written (*b*) will write

(*c*) shall write (*d*) No improvement

140. The war being ended, the soldiers returned.

(*a*) ends (*b*) have ended

(*c*) can end (*d*) No improvement

Directions (Question nos. 141 to 145): A passage is given with 5 questions following it. Read the passage carefully and choose the best answer out of the four alternatives.

Rhino poaching in Assam is one of the major environmental issues in India which continues in the region of Kaziranga National Park, Manas National Park and some other grasslands of Assam. The one horn rhino or Indian rhino is surviving in the north-east corner of India, Assam. Kaziranga National Park, Pobitora in Marigaon district and Orang National Park in Darrang district of Assam account almost 95% of the total wild one horned rhino in the world. These rhinos inhabit most of the floodplains of the Indogangetic and Brahmaputra riverine tracts and the neighboring foothills.

Sport hunting became common in the late 1800s and early 1900s.Indian rhinos were hunted relentlessly and **persistently**. Reports from the middle of the 19th century claim that some military officers in Assam individually shot more than 200 rhinos. By 1908, the population in Kaziranga had decreased to around 12 individuals.In the early 1900s, the species had declined to near extinction.

Poaching for rhinoceros horn became the single most important reason for the decline of the Indian rhino after conservation measures were put in place from the beginning of the 20th century, when legal hunting ended. From 1980 to 1993, 692 rhinos were poached in India. In India's Laokhowa Wildlife Sanctuary 41 rhinos were killed in 1983, virtually the entire population of the sanctuary. By the mid-1990s, poaching had rendered the species extinct there.

Illegal rhino horn trade has been the main problem facing managers of the rhino-protected areas of Assam. Some other parts like nails, skins have very high value in Asian traditional medicinal market. According to research by Traffic (conservation programme) and World Wide Fund for Nature, some Vietnamese buyers believe horn to be a cure for cancer when ground to a fine powder. According to a survey conducted by World Wide Fund for Nature in South Africa, it is kept by wealthy people in Vietnam as a "peace of mind" cure.

In 1993, rhino horn was removed from the official lists of Traditional Chinese Medicine. It is now only sold in Vietnam following an unsubstantiated rumour that horn cured a high-ranking official in Vietnam of cancer.

141. According to the passage, what is the main problem plaguing rhino conservation?

(*a*) Climatic change

(*b*) Illegal rhino horn trade

(*c*) Changing character of the grasslands

(*d*) Shrinking fodder

142. According to the passage, for what purpose are the nails of rhinos used?

(*a*) The nails of rhinos can be used as weapons.

(*b*) The nails of rhinos are mainly used for decorative purposes.

(*c*) People believe that the nails of rhinos bring good luck to families.

(*d*) People believe that the nails of rhinos have medicinal properties.

143. Out of the following options, which one comes closest to the word 'persistent'?

(*a*) Precedent (*b*) Resolute

(*c*) Residue (*d*) Reticent

144. By when the rhinos had become extinct in Laokhowa Wildlife sanctuary?

(*a*) Towards the beginning of this century.

(*b*) Towards the beginning of the previous century.

(*c*) Towards the middle of the last decade of the previous century.

(*d*) Towards the middle of the previous century.

145. How many rhinos were killed in India between the years 1980 and 1993?

(*a*) A little more than 700 rhinos

(*b*) A little less than 700 rhinos

(*c*) A little more than 1000 rhinos

(*d*) A little more than 1100 rhinos

Directions (Question no. 146): Out of the four alternatives, choose the one which best expresses the meaning of the given word and mark it.

146. Malicious

(*a*) Malapropism (*b*) Callous

(*c*) Benign (*d*) Pernicious

Directions (Question no. 147): Out of the four alternatives, choose the one which best expresses the opposite meaning of the given word and mark it.

147. Responsible

(*a*) Remnant (*b*) Ramification

(*c*) Baleful (*d*) Immature

148. In the following, four words are given, out of which only one word is correctly spelt. Find the correctly spelt word and mark your answer.

(*a*) Cynycal (*b*) Missanthrope

(*c*) Misoginist (*d*) Pessimistic

Directions (Question nos. 149 to 151): One part of the sentence has an error while the other parts are correct. Find which part of the sentence has an error. If no error is found, mark no error (option [D])

149. Despite promises to return him (*a*)/, the sailors were held (*b*)/ for 13 days. (*c*)/ No error (*d*)

150. On Mondays, Ramesh finds it difficult (*a*)/ to reach the school in time (*b*)/ because he feels lazy on the mornings after weekends. (*c*)/ No error (*d*)

151. Gabriel admires (*a*)/ his mother and brother (*b*)/ and all that they did for he and his sister. (*c*)/ No error (*d*)

Directions (Question nos. 152 to 154): A sentence is given with a blank space which is to be filled in with appropriate word(s). Four alternatives are suggested. Choose the correct alternative out of four.

152. She is quite confident ___ success.

(*a*) in

(*b*) for

(*c*) of

(*d*) on

153. These computers are ________ enough to be ________ to most people.

(*a*) easy, availability (*b*) difficult, given

(*c*) proper, bought (*d*) cheap, accessible

154. Your car differs from mine in several ______.

(*a*) respect

(*b*) respects

(*c*) respectability

(*d*) responses

Directions (Question nos. 155 to 157): Four alternatives are given for the idiom/phrase underlined in the sentence. Choose the idiom/phrase which best expresses the meaning.

155. He <u>burnt his fingers</u>by interfering in his colleague's personal life.

(*a*) hurt himself

(*b*) got himself into trouble

(*c*) committed an unpardonable act.

(*d*) earned a bad reputation

156. The retired manager was murdered in cold blood.

(*a*) stabbed in the back

(*b*) killed accidentally

(*c*) murdered a long time back

(*d*) murdered deliberately

157. The actor was born with a silver spoon in his mouth.

(*a*) born in wealth and luxury

(*b*) born amidst hardships

(*c*) born in poverty

(*d*) born with a spoon

Directions (Question nos. 158 to 160): Out of the four alternatives, choose the one which can be substituted for the given words/sentence.

158. The ability to produce a large profit with ease

(*a*) Touchstone (*b*) Magic wand

(*c*) Midas touch (d) Black magic

159. Fear of Strangers

(*a*) Vertigo (*b*) Xenophobia

(*c*) Arachnophobia (*d*) Cynophobia

160. Someone who refuses to play by the rules

(*a*) Poltroon (*b*) Maverick

(*c*) Tyrant (*d*) Dogmatist

Directions (Question nos. 161 to 165): A part of the sentence/whole sentence is underlined. Below are gfiven alternatives that imporve the sentence. Choose the correct option. In case no improvement is desired, your answer is D.

161. He lives adjacent the railway station.

(*a*) adjacent of the (*b*) adjacent from the

(*c*) adjacent to the (*d*) No improvement

162. I made in my mind to retire from the business.

(*a*) made on my mind (*b*) made up my mind

(*c*) made my mind (*d*) No improvement

163. The attack can happen any time and so, you must be on alert.

(*a*) you must be at alert

(*b*) you must be in alert

(*c*) you might be remaining to alert

(*d*) No improvement

164. We shall fight tooth to nail for our rights.

(*a*) tooth and nail (*b*) tooth in nail

(*c*) tooth from nail (*d*) No improvement

165. Of the three, tea, coffee and cocoa, the latter is her favourite.

(*a*) the latter last (*b*) the last

(*c*) the later (*d*) No improvement

Directions (Question nos. 166 to 170): You have a passage with 5 questions following. Read the passage carefully and choose the best answer to each question out of the four alternatives and mark it.

Passage

Dharavi, isn't just Asia's largest slum, it is its own mini-industry comprising of many artisans and craftsmen, who make some of the finest products in the country. So instead of the dark lanes and open sewage, Megha Gupta saw a gold mine of potential. That's what led her to start DharaviMarket.com, a website that provides a listing of Dharavi's businesses, which is helping these small-scale workers to dream big. So how does an online business like DharaviMarket.com make any difference to them? What do they do with the extra money they earn from online sales - the money that doesn't go to a wholesaler or traditional retailer? Well, Abbas, one of the artisans, has a simple answer to this: instead of working every day, they now allow themselves to take a day off every now and again. "Now with this additional income, we can take our families out to enjoy", he says.

I like the idea that my buying of a wallet or bag could contribute to a parent taking some time off from a 15-hour-a-day, 7-day work week, to spend it with his children.

166. What made Megha feel that Dharavi had great potential?

(*a*) It had dark lanes and open sewage

(*b*) There was a gold mine in it

(*c*) Its artisans and craftsmen made some of the finest products in the country

(*d*) It inhabited small scale workers that dreamt big

167. What did Megha do for Dharavi?

(*a*) She got Dharavi cleaned

(*b*) She gave leaves to the Dharavi artists working for her

(*c*) She created an online business website for Dharavi artisans

(*d*) She gave the Dharavi artisans extra money for the work they did for her.

168. What is sold at Dharavimarket.com?

(*a*) Gold

(*b*) Products made by Dharavi artisans

(*c*) Clothes made by Dharavi artisans

(*d*) Dreams of Dharavi artisans

169. What makes Dharavi artisans earn extra money from online sale?

(*a*) They do not hire any people to run their business

(*b*) They sell their products at a higher rate

(*c*) They have become very popular among their customers.

(*d*) The money that goes to a wholesaler or traditional retailer is saved

170. What do the people do after earning extra money from online sales?

(*a*) They buy more raw materials.

(*b*) They are able to meet their basic needs easily.

(*c*) They give a part of it to Megha.

(*d*) They take leaves to spend time with family.

Directions: In question nos. **171** to **173**, out of the four alternatives, choose the one which can be substituted for the given words/sentences and indicate it.

171. A ghost or a spirit

(*a*) Elf

(*b*) Phantom

(*c*) Caricature

(*d*) Jester

172. Word to describe a river following a winding course is

(*a*) Uphill (*b*) Descending

(*c*) Meandering (*d*) Linear

173. A task requiring great strength and effort is

(*a*) Facile (*b*) Towering

(*c*) Herculean (*d*) Degrading

Directions: In question nos. **174** to **178**, a part of the sentence/whole sentence is underlined. Below are given alternatives that improve the sentence. Choose the correct option.

In case no improvement is desired, your answer is (D).

174. The militants, using the cover of thick, wintry fog, scaled the walls of the Bacha Khan University in Charsadda, northwestern Khyber Pakhtunkhwa province, before entering buildings and <u>had opened fire on students</u> and teachers in classrooms and hostels, police said.

(*a*) open fire on students

(*b*) opening fire on students

(*c*) open the fire on students

(*d*) No improvement

175. What is there to <u>prevent him of leaving</u> Mumbai?

(*a*) prevent him to leaving

(*b*) prevent from leaving

(*c*) prevent him from leaving

(*d*) No improvement

176. He is <u>averse to play</u> cards.

(*a*) averse to playing (*b*) averse playing

(*c*) averse to played (*d*) No improvement

177. His manager generally <u>passed by</u> the faults of his subordinates.

(*a*) passed of (*b*) passed off

(*c*) passed over (*d*) No improvement

178. They <u>kept in</u> a long conversation.

(*a*) kept to

(*b*) kept up

(*c*) kept of

(*d*) No improvement

Directions: In question nos. **179** to **180**, you have a brief passage with 2 questions. Read the passage and choose the best answer to each question.

<u>Passage</u>

Since these pujas and pandals are organized by various local committees, our donations could help them organize eco-friendly pujas. People need to be taught and encouraged about the reuse of rough idols of gods made of wood and hay post their immersion in water bodies. NASA's satellite caught the country shining bright, beautifully from space during Diwali. But the country would shine brighter if we succeed in celebrating Diwali with earthen pots and lights. Apart from environmental benefits, imagine how many low-income families that sell diyas and other earthenware can benefit with a single effort. While lighting our own house, we can light up theirs too. Moreover, giving alms to the poor instead of shelling out on crackers could earn you extra 'Karmic' brownie points. So let's not wreck our planet anymore. It's time to 'Go Eco'.

179. Which of the following is the most apt title for the passage?

(*a*) Burn diyas not crackers

(*b*) Save the Earth

(*c*) Diwali - A festival of lights

(*d*) Celebrate eco-friendly Diwali

180. What are the benefits of lighting diyas instead of crackers on diwali?

(*a*) It does not cause pollution

(*b*) It helps the poor families selling diyas

(*c*) It purifies the air in one's household

(*d*) It improves one's karmic account

OMR SHEET

1. Ⓐ Ⓑ Ⓒ Ⓓ	26. Ⓐ Ⓑ Ⓒ Ⓓ	51. Ⓐ Ⓑ Ⓒ Ⓓ	76. Ⓐ Ⓑ Ⓒ Ⓓ
2. Ⓐ Ⓑ Ⓒ Ⓓ	27. Ⓐ Ⓑ Ⓒ Ⓓ	52. Ⓐ Ⓑ Ⓒ Ⓓ	77. Ⓐ Ⓑ Ⓒ Ⓓ
3. Ⓐ Ⓑ Ⓒ Ⓓ	28. Ⓐ Ⓑ Ⓒ Ⓓ	53. Ⓐ Ⓑ Ⓒ Ⓓ	78. Ⓐ Ⓑ Ⓒ Ⓓ
4. Ⓐ Ⓑ Ⓒ Ⓓ	29. Ⓐ Ⓑ Ⓒ Ⓓ	54. Ⓐ Ⓑ Ⓒ Ⓓ	79. Ⓐ Ⓑ Ⓒ Ⓓ
5. Ⓐ Ⓑ Ⓒ Ⓓ	30. Ⓐ Ⓑ Ⓒ Ⓓ	55. Ⓐ Ⓑ Ⓒ Ⓓ	80. Ⓐ Ⓑ Ⓒ Ⓓ
6. Ⓐ Ⓑ Ⓒ Ⓓ	31. Ⓐ Ⓑ Ⓒ Ⓓ	56. Ⓐ Ⓑ Ⓒ Ⓓ	81. Ⓐ Ⓑ Ⓒ Ⓓ
7. Ⓐ Ⓑ Ⓒ Ⓓ	32. Ⓐ Ⓑ Ⓒ Ⓓ	57. Ⓐ Ⓑ Ⓒ Ⓓ	82. Ⓐ Ⓑ Ⓒ Ⓓ
8. Ⓐ Ⓑ Ⓒ Ⓓ	33. Ⓐ Ⓑ Ⓒ Ⓓ	58. Ⓐ Ⓑ Ⓒ Ⓓ	83. Ⓐ Ⓑ Ⓒ Ⓓ
9. Ⓐ Ⓑ Ⓒ Ⓓ	34. Ⓐ Ⓑ Ⓒ Ⓓ	59. Ⓐ Ⓑ Ⓒ Ⓓ	84. Ⓐ Ⓑ Ⓒ Ⓓ
10. Ⓐ Ⓑ Ⓒ Ⓓ	35. Ⓐ Ⓑ Ⓒ Ⓓ	60. Ⓐ Ⓑ Ⓒ Ⓓ	85. Ⓐ Ⓑ Ⓒ Ⓓ
11. Ⓐ Ⓑ Ⓒ Ⓓ	36. Ⓐ Ⓑ Ⓒ Ⓓ	61. Ⓐ Ⓑ Ⓒ Ⓓ	86. Ⓐ Ⓑ Ⓒ Ⓓ
12. Ⓐ Ⓑ Ⓒ Ⓓ	37. Ⓐ Ⓑ Ⓒ Ⓓ	62. Ⓐ Ⓑ Ⓒ Ⓓ	87. Ⓐ Ⓑ Ⓒ Ⓓ
13. Ⓐ Ⓑ Ⓒ Ⓓ	38. Ⓐ Ⓑ Ⓒ Ⓓ	63. Ⓐ Ⓑ Ⓒ Ⓓ	88. Ⓐ Ⓑ Ⓒ Ⓓ
14. Ⓐ Ⓑ Ⓒ Ⓓ	39. Ⓐ Ⓑ Ⓒ Ⓓ	64. Ⓐ Ⓑ Ⓒ Ⓓ	89. Ⓐ Ⓑ Ⓒ Ⓓ
15. Ⓐ Ⓑ Ⓒ Ⓓ	40. Ⓐ Ⓑ Ⓒ Ⓓ	65. Ⓐ Ⓑ Ⓒ Ⓓ	90. Ⓐ Ⓑ Ⓒ Ⓓ
16. Ⓐ Ⓑ Ⓒ Ⓓ	41. Ⓐ Ⓑ Ⓒ Ⓓ	66. Ⓐ Ⓑ Ⓒ Ⓓ	91. Ⓐ Ⓑ Ⓒ Ⓓ
17. Ⓐ Ⓑ Ⓒ Ⓓ	42. Ⓐ Ⓑ Ⓒ Ⓓ	67. Ⓐ Ⓑ Ⓒ Ⓓ	92. Ⓐ Ⓑ Ⓒ Ⓓ
18. Ⓐ Ⓑ Ⓒ Ⓓ	43. Ⓐ Ⓑ Ⓒ Ⓓ	68. Ⓐ Ⓑ Ⓒ Ⓓ	93. Ⓐ Ⓑ Ⓒ Ⓓ
19. Ⓐ Ⓑ Ⓒ Ⓓ	44. Ⓐ Ⓑ Ⓒ Ⓓ	69. Ⓐ Ⓑ Ⓒ Ⓓ	94. Ⓐ Ⓑ Ⓒ Ⓓ
20. Ⓐ Ⓑ Ⓒ Ⓓ	45. Ⓐ Ⓑ Ⓒ Ⓓ	70. Ⓐ Ⓑ Ⓒ Ⓓ	95. Ⓐ Ⓑ Ⓒ Ⓓ
21. Ⓐ Ⓑ Ⓒ Ⓓ	46. Ⓐ Ⓑ Ⓒ Ⓓ	71. Ⓐ Ⓑ Ⓒ Ⓓ	96. Ⓐ Ⓑ Ⓒ Ⓓ
22. Ⓐ Ⓑ Ⓒ Ⓓ	47. Ⓐ Ⓑ Ⓒ Ⓓ	72. Ⓐ Ⓑ Ⓒ Ⓓ	97. Ⓐ Ⓑ Ⓒ Ⓓ
23. Ⓐ Ⓑ Ⓒ Ⓓ	48. Ⓐ Ⓑ Ⓒ Ⓓ	73. Ⓐ Ⓑ Ⓒ Ⓓ	98. Ⓐ Ⓑ Ⓒ Ⓓ
24. Ⓐ Ⓑ Ⓒ Ⓓ	49. Ⓐ Ⓑ Ⓒ Ⓓ	74. Ⓐ Ⓑ Ⓒ Ⓓ	99. Ⓐ Ⓑ Ⓒ Ⓓ
25. Ⓐ Ⓑ Ⓒ Ⓓ	50. Ⓐ Ⓑ Ⓒ Ⓓ	75. Ⓐ Ⓑ Ⓒ Ⓓ	100. Ⓐ Ⓑ Ⓒ Ⓓ

OMR SHEET

101. (A) (B) (C) (D)	121. (A) (B) (C) (D)	141. (A) (B) (C) (D)	161. (A) (B) (C) (D)
102. (A) (B) (C) (D)	122. (A) (B) (C) (D)	142. (A) (B) (C) (D)	162. (A) (B) (C) (D)
103. (A) (B) (C) (D)	123. (A) (B) (C) (D)	143. (A) (B) (C) (D)	163. (A) (B) (C) (D)
104. (A) (B) (C) (D)	124. (A) (B) (C) (D)	144. (A) (B) (C) (D)	164. (A) (B) (C) (D)
105. (A) (B) (C) (D)	125. (A) (B) (C) (D)	145. (A) (B) (C) (D)	165. (A) (B) (C) (D)
106. (A) (B) (C) (D)	126. (A) (B) (C) (D)	146. (A) (B) (C) (D)	166. (A) (B) (C) (D)
107. (A) (B) (C) (D)	127. (A) (B) (C) (D)	147. (A) (B) (C) (D)	167. (A) (B) (C) (D)
108. (A) (B) (C) (D)	128. (A) (B) (C) (D)	148. (A) (B) (C) (D)	168. (A) (B) (C) (D)
109. (A) (B) (C) (D)	129. (A) (B) (C) (D)	149. (A) (B) (C) (D)	169. (A) (B) (C) (D)
110. (A) (B) (C) (D)	130. (A) (B) (C) (D)	150. (A) (B) (C) (D)	170. (A) (B) (C) (D)
111. (A) (B) (C) (D)	131. (A) (B) (C) (D)	151. (A) (B) (C) (D)	171. (A) (B) (C) (D)
112. (A) (B) (C) (D)	132. (A) (B) (C) (D)	152. (A) (B) (C) (D)	172. (A) (B) (C) (D)
113. (A) (B) (C) (D)	133. (A) (B) (C) (D)	153. (A) (B) (C) (D)	173. (A) (B) (C) (D)
114. (A) (B) (C) (D)	134. (A) (B) (C) (D)	154. (A) (B) (C) (D)	174. (A) (B) (C) (D)
115. (A) (B) (C) (D)	135. (A) (B) (C) (D)	155. (A) (B) (C) (D)	175. (A) (B) (C) (D)
116. (A) (B) (C) (D)	136. (A) (B) (C) (D)	156. (A) (B) (C) (D)	176. (A) (B) (C) (D)
117. (A) (B) (C) (D)	137. (A) (B) (C) (D)	157. (A) (B) (C) (D)	177. (A) (B) (C) (D)
118. (A) (B) (C) (D)	138. (A) (B) (C) (D)	158. (A) (B) (C) (D)	178. (A) (B) (C) (D)
119. (A) (B) (C) (D)	139. (A) (B) (C) (D)	159. (A) (B) (C) (D)	179. (A) (B) (C) (D)
120. (A) (B) (C) (D)	140. (A) (B) (C) (D)	160. (A) (B) (C) (D)	180. (A) (B) (C) (D)

ANSWERS

1. (*d*)	2. (*c*)	3. (*a*)	4. (*b*)	5. (*a*)	6. (*b*)	7. (*c*)	8. (*a*)	9. (*a*)	10. (*b*)
11. (*c*)	12. (*a*)	13. (*b*)	14. (*b*)	15. (*b*)	16. (*a*)	17. (*b*)	18. (*d*)	19. (*c*)	20. (*c*)
21. (*c*)	22. (*b*)	23. (*b*)	24. (*c*)	25. (*c*)	26. (*b*)	27. (*a*)	28. (*a*)	29. (*b*)	30. (*b*)
31. (*c*)	32. (*b*)	33. (*a*)	34. (*a*)	35. (*c*)	36. (*b*)	37. (*c*)	38. (*c*)	39. (*d*)	40. (*c*)
41. (*d*)	42. (*b*)	43. (*a*)	44. (*d*)	45. (*b*)	46. (*b*)	47. (*a*)	48. (*c*)	49. (*c*)	50. (*d*)
51. (*a*)	52. (*d*)	53. (*a*)	54. (*a*)	55. (*b*)	56. (*b*)	57. (*b*)	58. (*d*)	59. (*c*)	60. (*a*)
61. (*a*)	62. (*a*)	63. (*a*)	64. (*c*)	65. (*b*)	66. (*d*)	67. (*b*)	68. (*b*)	69. (*b*)	70. (*b*)
71. (*a*)	72. (*d*)	73. (*a*)	74. (*d*)	75. (*c*)	76. (*b*)	77. (*c*)	78. (*d*)	79. (*a*)	80. (*c*)
81. (*c*)	82. (*b*)	83. (*d*)	84. (*c*)	85. (*c*)	86. (*a*)	87. (*c*)	88. (*a*)	89. (*b*)	90. (*d*)
91. (*d*)	92. (*a*)	93. (*c*)	94. (*a*)	95. (*a*)	96. (*a*)	97. (*a*)	98. (*a*)	99. (*b*)	100. (*c*)
101. (*b*)	102. (*a*)	103. (*b*)	104. (*a*)	105. (*a*)	106. (*a*)	107. (*c*)	108. (*c*)	109. (*b*)	110. (*d*)
111. (*c*)	112. (*a*)	113. (*d*)	114. (*b*)	115. (*b*)	116. (*b*)	117. (*a*)	118. (*c*)	119. (*a*)	120. (*c*)
121. (*b*)	122. (*a*)	123. (*b*)	124. (*b*)	125. (*c*)	126. (*d*)	127. (*d*)	128. (*c*)	129. (*b*)	130. (*b*)
131. (*a*)	132. (*b*)	133. (*b*)	134. (*c*)	135. (*d*)	136. (*b*)	137. (*d*)	138. (*c*)	139. (*a*)	140. (*d*)
141. (*b*)	142. (*d*)	143. (*b*)	144. (*c*)	145. (*b*)	146. (*d*)	147. (*d*)	148. (*d*)	149. (*a*)	150. (*b*)
151. (*c*)	152. (*c*)	153. (*d*)	154. (*b*)	155. (*b*)	156. (*d*)	157. (*a*)	158. (*c*)	159. (*b*)	160. (*b*)
161. (*c*)	162. (*b*)	163. (*d*)	164. (*a*)	165. (*b*)	166. (*c*)	167. (*c*)	168. (*b*)	169. (*d*)	170. (*d*)
171. (*b*)	172. (*c*)	173. (*c*)	174. (*b*)	175. (*c*)	176. (*a*)	177. (*d*)	178. (*b*)	179. (*d*)	180. (*d*)

EXPLANATIONS

1. Intrinsic material has equal number of holes and electrons.

2. In schottky diode there is no charge storage and hence almost zero reverse recovery time.

3. $\beta = \frac{\alpha}{1-\alpha} \Rightarrow \frac{0.99}{1-0.99} = 99$ = Current gain

$= 1 + \beta = 100.$

4. Since channel is n type gate must be p type.

6. By increasing of temperature by 10°C, I_0 become double so by increasing temperature 20°C, I_0 become 4 time than initial value... and it is 40 PA

7. Since there are two p-n junctions, there are two depletion layers and 0.7 V across each layer.

8. Input Resistance with feedback for current shunt,

$$R_j = \frac{R_j}{1+\beta A}$$

9. Cut in voltage in Schottky diode is about 0.3 V as compared to 0.7 V in ordinary semiconductor diode.

10. $\beta_{ac} = \frac{\alpha_{ac}}{1-\alpha_{ac}} = \frac{0.98}{0.02} = 49$

11. In forward active mode emitter base junction is forward biased and base collector junction is reverse biased.

15. It is used as amplifier when it operates in this region.

16. Ideally the stability factor should be 0. Practically it should be minimum value.

19. Peak value = 100 V, rms value = $\frac{100}{\sqrt{2}}$ = 70.71 V.

20. Peak voltage rating = $\sqrt{2}$ (rms voltage rating).

21. RMS value = $\sqrt{10^2 + \left(\frac{10}{\sqrt{2}}\right)^2} = \sqrt{150}$

22. Time constant $= R_{eq}.C$

when V is shorted and R is seen from C then $R_{eq} = R \,||\, R = R/2$

$$T = \frac{CR}{2}$$

23. Power factor is maximum and equal to 1 for purely resistive load. Power factor is minimum and equal to zero for purely reactive load.

24. $Z = \sqrt{R^2 + X_C^{\ 2}}$. When f is made four times, X_C becomes one-fourth but R remains the same. So the total impedance will decreases.

25. Q = CV or 0.15 = C(5) or C = 0.03 F.

26. When the capacitor is fully charged, i = 0 and voltage across capacitor is equal to battery voltage.

27. Capacitor behaves as a short-circuit at t = 0.

28. $P = \frac{V^2}{R}$ or $R = \frac{V^2}{P}$ or

$R = \frac{230 \times 230}{100} = 529$ ohm

29. For R_{Th}, Voltage source will be short circuited

$R_{Th} = 2.4\Omega$, $I = \frac{10-5}{10} = 0.5$Amp

$V_{AB} = 10 - 6I = 7$ V.

30. Maximum mutual inductance = $\sqrt{L_1 L_2} = 20$ mH.

31. Currents in resistance and inductance are out of phase by 90°.

Hence $I = \sqrt{12^2 + 16^2} = 20$A

32. The intrinsic layer is in between p and n layers.

33. Capacitance is inversely proportional to distance between plates.

34. Holes and electrons move away from junction and therefore resistance increases to a high value.

35. This 0.7 V is also called cut in voltage. When forward voltage is 0.7 V or less there is no current.

36. $Y(t) = A\cos(2\pi \times 10^4 t) + A\cos(2\pi \times 10.1 \times 10^3 t)$

when passed through high pass filter,

Output will be $A\cos(2\pi \times 10.1 \times 10^3 t)$ only.

37. $P_c = 400$ W, $m = 0.75$

$P_t = P_c\left(1 + \frac{m^2}{2}\right)$

38. Frequency deviation = 10 kHz, $f_m = 1$ kHz

$2(f_m + Df)$

Bandwidth = 22 kHz

$2(10 + 1) = 22$ kHz

f_m is doubled

Hence $22 \times 2 = 44$ kHz bandwidth.

40. $m_1 = 0.4$, $m_2 = 0.3$.

$m_t = \sqrt{m_1^2 + m_2^2}$

41. This arrangement is bootstrap bias.

42. Bandwidth = $\frac{1}{2}(10 \times 7) = 35$ kHz.

44. Entropy = $0.5 \log_2 2 + 0.5 \log_2 2 = 1$.

45. In steady state inductance behaves as short-circuit.

57. Given that

Open-loop voltage gain

A = 60 dB or A = 1000

and feedback ratio,

$\beta = \frac{1}{20} = 0.05$

Now, gain with feedback is given by

$A_f = \frac{A}{1 + \beta A}$

Substituting given values, we get

$A_f = \frac{1,000}{1 + 0.05 \times 1,000}$

Solving we get

$A_f = 19.6$ or $20 \log_{10} 19.6$ dB = 25.85 dB

58. Given that

Voltage gain of amplifier without feedback

A = 60 dB or A = 1,000

Voltage gain with feedback will be

$A_f = 40$ dB or 100

Therefore, gain with feedback is given by

$A_f = \frac{A}{1 + \beta A}$

Now, feedback factor, $\beta A = 9$

59. Given that

Voltage gain without feedback, A= 600

Voltage gain without feedback, $A_f = 50$

We know that $A_f = \frac{A}{1 + \beta A}$

Substituting values, we get

$50 = \frac{600}{1 + 600\beta}$

Solving, we get

$\beta = 0.01833$

Now, the percentage of output voltage which is feedback to the input side, is give as

$= \frac{V_f}{V_{out}} \times 100 = \beta \times 100$

$= 0.01833 \times 100 = 1.833\%$

60. It is a fixed bias circuit

$(V_{GS})_Q = -V_{GS} = -2V$

$V_G = V_{GS} + V_S$(where $V_S = 0$)

$V_{GS} = V_G = -2V$

61. The transconductance is

$g_m = 2k_n(V_{GN} - V_{TN}) = 1$ mA/V

Unity gain bandwdith is

$f_T = \frac{g_m}{2\pi(C_{gs} + C_{gd})} = 663$ MHz

62. $\frac{g_m}{C} = A_V \times BW$

$\frac{95 \times 10^{-3}}{500 \times 10^{-12}} = 30 \times BW$

BW= 6.3 MHz

64. Since gate current is very small. I_D and I_s are almost equal.

66. Its output is used to trigger SCR.

67. Zener diode is always reverse biased.

68. f = 10 GHz, v = 4×10^8 m/sec

$$\lambda = \frac{v}{f} \Rightarrow \frac{4\times10^8}{10\times10^9} \Rightarrow 0.04\text{m} = 4\text{cm}$$

$$l = \frac{\lambda}{2} \Rightarrow \frac{4}{2} \Rightarrow 2\text{cm}$$

69. The bandwidth required for amplitude modulation is twice the frequency of the modulating signal.

$BW = 2f_m = 2 \times 10 = 20\text{kHz}$

70. $l = \frac{\lambda}{2} \Rightarrow \frac{C}{f.2} \Rightarrow \frac{3\times10^8}{2\times30\times10^6}$

$$\Rightarrow \frac{30}{6} \Rightarrow 5 \text{ metre}$$

74. $\beta = \frac{\alpha}{1-\alpha} = \frac{0.98}{1-0.98} = 49, I_C = I_B\,\beta$

75. In bridge connection PIV is half of that in M-2 connection.

76.

(20)	(1)	(12)	(5)	(14)	(20)
T	A	L	E	N	T
+2↓	−2↓	+2↓	−2↓	+2↓	−2↓
V	Y	N	C	P	R
(22)	(25)	(14)	(3)	(16)	(18)

Similarly,

(14)	(15)	(18)	(13)	(1)	(12)
N	O	R	M	A	L
+2↓	−2↓	+2↓	−2↓	+2↓	−2↓
P	M	T	K	C	J
(16)	(13)	(20)	(11)	(3)	(10)

77. Night blindness is caused due to the deficiency of vitamin A. Similarly, scurvy is caused due to the deficiency of vitamin C. Hence, option (C) is the correct answer. Nerve contraction is caused due to the deficiency of calcium, rickets is caused due to the deficiency of vitamin D and goiter is caused due to the deficiency of iodine.

78. This analogy is about the idioms which are used to draw comparisons. The idioms are as sick as a dog and as busy as a bee. Hence, option (D) is the correct answer.

79. H(8) $\rightarrow 8^3 = 512$,

Similarly, M(13) $\rightarrow 13^3 = 2197$.

80. The sequence shows the steps involved in a new product development. The correct answer is option (C), i.e., 1, 3, 6, 4, 2, 5.

81. $462 * 3 \Rightarrow 462 \times 3 = 1386 \rightarrow 1 + 3 + 8 + 6 = 18$

$564 * 2 \Rightarrow 564 \times 2 = 1128 \rightarrow 1 + 1 + 2 + 8 = 12$

Similarly, $617 * 4 \Rightarrow 617 \times 4 = 2468$

$\rightarrow 2 + 4 + 6 + 8 = 20$.

82. The series is cbba, cbba, cbba,...

83. The series is aabc, abbc, abcc,...

84. Wheat is a rabi crop while black gram, groundnut and fennel are kharif crops. Hence, option (c) is the correct answer.

85. Reversing the digits of the numbers in series, we get 1, 9, 25, 49, 81, 75, 169

All numbers are squares of odd consecutive numbers except 75.

Hence, answer is 57.

86. $253 - 351 \Rightarrow 2 + 5 + 3 - (3 + 5 + 1) = 1$

$478 - 692 \Rightarrow 4 + 7 + 8 - (6 + 9 + 2) = 2$

$123 - 22 \Rightarrow 1 + 2 + 3 - (2 + 2) = 2$

$971 - 816 \Rightarrow 9 + 7 + 1 - (8 + 1 + 6) = 2$

Hence, 253 – 351 is an odd pair.

87. Out of the given options, only bonsai is not a plant. Bonsai is a Japanese art form using miniature trees grown in containers. Hence, option (c) is the correct answer.

88. Options (b), (c) and (d) have two vowels viz. 'o' and 'e'. Option (a) has 4 vowels, three 'i's' and an 'e'. Hence, option (a) is the correct answer.

89. The series in numerical form can be written as 1-3, 1-2-4, 5-7, 5-6-8, 9-11. So, the next term in the series will be 9-10-12, i.e., ijl.

90. The series is A(1) C(3) E(5), G(7) I(9) K(11), and so on. So, the required terms will be Y(25) A(1) C(3), E(5) G(7) I(9).

91. The various sources of methane gas are ruminants, rice fields, marshy fileds, biogas, Methane hydrates/clathrates (ice-like combinations of methane and water on the sea floor) etc.

92. Anandwan in Chandrapur district in the state of Maharashtra, is an ashram, and a community rehabilitation centre for leprosy patients and the disabled from downtrodden sections of society. It was founded in 1948 by noted social activist, Baba Amte.

94. Noori is a female pashmina goat, the first pashmina goat to be cloned using the process of nuclear transfer at the Faculty of Veterinary Sciences and Animal Husbandry, Sher-e-Kashmir University of Agricultural Sciences and Technology of Kashmir, Shuhama, Srinagar.

95. The Ramon Magsaysay Award is often considered to be Asia's Nobel Prize. The prize was established in April 1957 by the trustees of the Rockefeller Brothers Fund based in New York City with the concurrence of the Philippine government.

96. The Board of Control for Cricket in India (BCCI), headquartered at Mumbai, India, is the national governing body for all cricket in India. The board was formed in December 1928 as BCCI replaced Calcutta Cricket Club.

97. Smog is a type of air pollution.

98. Article 19(1) says that every citizen has freedom of speech and expression.Right to information is embedded in article 19.

99. Dr. BhimraoRamjiAmbedkar is widely regarded as the father of the Indian Constitution.

100. A powerful odorant, ethanethiol, is added so that leaks can be detected easily.

101. The Barabati Stadium is a cricket venue in the city of Cuttack, Odishaand is operated by Odisha Cricket Association.

102. The Ilbert Bill was a bill introduced in 1883 by Viceroy Lord Ripon that proposed an amendment for existing laws in the country at the time to allow Indian judges and magistrates the jurisdiction to try British offenders in criminal cases at the District level, something that was disallowed at the time.

103. The Nagarpalika Act, i.e. the Constitutional (74th Amendment) Act, 1992 had been enacted to accord constitutional recognition to the Urban Local Bodies as third tier of Government.

104. Conga is of Caribbean origin.

106. $x + y = 166$ ... (i)

$x - y = 8$ ... (ii)

On solving, $x = 87$ and $y = 79$

Thus, required sum = $87^2 + 79^2 = 13810$.

107. Difference in profit = 4%

Thus, 4% of CP = Rs. 3.75

Therefore, required SP = 108% of CP

$= 3.75 \times \frac{1.08}{0.04}$ = Rs. 101.25.

108. No. of boys = 45% of 1200 = 540

No. of girls = 1200 – 540 = 660

Total monthly fee of boys = 540 × 750

= Rs. 4,05,000

Total monthly fee of girls = 660 × 80% of 750

= 660 × 600 = Rs. 3,96,000

Therefore, the total monthly revenue

= 4,05,000 + 3,96,000

= Rs. 8,01,000.

109. (105 + 210) m = S × 45 sec

$$S = \frac{315}{45} \times \frac{18}{5} = 25.2 \text{ km/hr}$$

110. $$\frac{(a+b)^2 - (a-b)^2}{2ab} = \frac{4ab}{2ab} = 2$$

$$\therefore \frac{(0.87+0.13)^2 - (0.87-0.13)^2}{2 \times 0.87 \times 0.13} = 2$$

111. $x + \frac{1}{x} = -1$ ... (i)

Squaring (i) on both sides

$x^2 + \frac{1}{x^2} = -1$ ... (ii)

Again, squaring (ii) on both sides

$x^4 + \frac{1}{x^4} = -1$ ... (iii)

Cubing (i) on both sides

$x^3 + \frac{1}{x^3} = 2$ ... (iv)

Now, multiply (iii) and (iv), we get

$$\left(x^4 + \frac{1}{x^4}\right)\left(x^3 + \frac{1}{x^3}\right) = -2$$

$$\therefore x^7 + \frac{1}{x^7} = -1$$

112. $l = r\theta$

$$\therefore \frac{l_1}{l_2} = \frac{r\theta_1}{r\theta_2} \Rightarrow \frac{7}{l_2} = \frac{60°}{105°}$$

$\therefore l_2 = 12.25$ cm.

113. $(x+2)^2 = x^2 + (x-7)^2$

$\Rightarrow x^2 - 18x + 45 = 0$

$\therefore x = 3, 15$

x = 3 is not possible with given condition. Hence, x = 15.

114. $\dfrac{3\cot 30^\circ - 5\cos 30^\circ + 3\sin 30^\circ}{\sin 60^\circ + \tan 60^\circ - 3\cos 60^\circ}$

$= \dfrac{3\times\sqrt{3} - 5\dfrac{\sqrt{3}}{2} + 3\times\dfrac{1}{2}}{\dfrac{\sqrt{3}}{2} + \sqrt{3} - 3\times\dfrac{1}{2}}$

$= \dfrac{6\sqrt{3} - 5\sqrt{3} + 3}{\sqrt{3} + 2\sqrt{3} - 3} = \dfrac{\sqrt{3}+3}{3\sqrt{3}-3} = \dfrac{\sqrt{3}(1+\sqrt{3})}{3(\sqrt{3}-1)}$

$= \dfrac{3+1+2\sqrt{3}}{2\sqrt{3}} = \dfrac{2}{\sqrt{3}} + 1.$

115. Let total work (LCM of $9\frac{3}{4}$ and $15\frac{3}{4}$) be $\frac{3\times 13\times 21}{4}$ units.

Combined efficiency of P, Q and R = 21 units

Combined efficiency of P and R = 13 units

So, efficiency of Q = 8 units

Hence, time taken by Q

$= \dfrac{3\times 13\times 21}{4}\times\dfrac{1}{8} = 25\dfrac{19}{32}$ days.

116. Ratio of diameter = $\dfrac{d_1}{d_2} = \dfrac{2}{3}$

$\Rightarrow \dfrac{r_1}{r_2} = \dfrac{2}{3}$

Let their radii be 2x and 3x and respective heights be h_1 and h_2.

Thus,

$\frac{1}{3}\pi(r_1)^2 h_1 = \frac{1}{3}\pi(r_2)^2 h_2$

$\Rightarrow (2x)^2 h_1 = (3x)^2 h_2$

$\therefore \dfrac{h_1}{h_2} = \dfrac{9}{4}$

117. Let the CP of goods be Rs. 100.

Then, MP = Rs. 137.50

After allowing a discount of 20%, SP = 137.5 × 0.8 = Rs. 110

Hence, he earns a profit of 10%.

118. The required ratio would be independent of absolute distance.

Therefore, we can consider the distance in both the two cases to be 100 m and 40 m respectively.

Hence, ratio of speeds = $\dfrac{100}{3\times 60}\times\dfrac{72}{40} = \dfrac{1}{1}$.

119. A + B + C + 3 × 3 = 3 × 24 ... (i)

Let B be x years, then A = x + 5 and C = x + 7.

On substituting these values in (i), we get

x = 17.

Therefore, A = 22 years, B = 17 years and C = 24 years.

Hence, required answer = (22 – 6) + (24 + 5)

= 45 years.

120. a – b – c = 0

On squaring both sides, we get

$a^2 + b^2 + c^2 - 2ab + 2bc - 2ac = 0$

$\Rightarrow 4 - 2(ab - bc + ac) = 0$

$\Rightarrow (ab - bc + ac) = 2$

124. "Your's" should be replaced by "yours".

125. The correct verb should be singular. Hence, "was" can be used.

126. The sentence is grammatically correct.

127. "Wrong side of sixty" means more than sixty years of age.

128. "Passed" is grammatically correct.

129. "Black sheep" refers to bad or dishonest individuals.

136. The present perfect continuous tense is appropriate in the context of the sentence.

137. The sentence is grammatically correct.

138. The future perfect continuous tense should be used because an action is indicated which will be in progress over a period of time that will end in the future.

139. The past perfect tense is appropriate for the underlined part.

140. The sentence is grammatically correct.

141. Refer to the first sentence of the fourth paragraph for the answer.

142. Refer to the second sentence of the fourth paragraph for the answer.

144. Refer to the last sentence of the third paragraph.

145. Refer to the second sentence of the third paragraph for the answer.

149. Instead of "him", "them" should have been used.

150. 'On the mornings' should be replaced by 'in the mornings'.

151. Since Gabriel's mother and brother did something for Gabriel and his sister, we should use objective pronoun 'him' for Gabriel and not 'he'. Therefore, the correct sentence should be "Gabriel admires his mother and brother and all that they did for him and his sister."

156. The idiom 'in cold blood' refers to a crime committed deliberately and not in the heat of the moment.

157. The idiom refers to the state of being born amidst opulence.

158. If a person has Midas touch, that person is financially successful in everything he does. Touchstone is the standard or criterion by which something is judged.

159. Vertigo is the fear of heights. Arachnophobia is the fear of spiders. Cynophobia is the fear of dogs. 'Xeno', means alien and therefore, Xenophobia means the fear of foreigners or strangers.

160. Maverick means an unorthodox or independent-minded person, who does not follow conventional rules. A poltroon is a coward, a tyrant is a cruel and oppressive ruler and a dogmatist is a person who believes in his opinions very strongly.

161. 'Adjacent' is always followed by the preposition 'to.'

162. 'To make up one's mind' refers to taking a decision.

163. The sentence is grammatically correct.

164. The correct idiom is tooth and nail which means 'to fight with all possible means and power.'

165. The correct exzpression should be 'the last.'

166. The passage states that Megha saw gold mine of a potential in the slums of Dharavi as the slum consisted of "artisans and craftsmen, who make some of the finest products in the country." Hence, option (c) is the answer.

167. The passage states that Megha Gupta created an online website Dharavimarket.com that sold the products made by Dharavi artisans and craftsmen. Therefore, option (c) is the answer.

168. The passage states that Megha Gupta created an online website Dharavimarket.com that sold the products made by Dharavi artisans and craftsmen. Therefore, option (b) is the answer.

169. Refer to the following line from the passage -" What do they do with the extra money they earn from online sales - the money that doesn't go to a wholesaler or traditional retailer?" This renders option (d) correct.

170. Abbas, an artisan, in the passage states that "Now with this additional income, we can take our families out to enjoy". Hence, option (d) is correct.

171. A phantom is an apparition, a ghost or a spirit of a dead person. An elf is a supernatural creature from folktales, characterized by pointed ears, magical powers and capricious nature. A caricature is a picture of a person whose certain striking characteristics are exaggerated to create comic or grotesque effect. A jester is a professional joker or a fool at a medieval court.

172. Meandering refers to something following a winding or turning course. Linear means straight, uphill means upward and descending means downward.

173. Facile means simple, towering means extremely tall and degrading is humiliating. Herculean means arduous or laborious.

174. This is an instance of parallelism. The verb 'opening' should be in the same form as the other verb 'entering'.

175. The grammatically correct usage is given in option (c).The correct preposition should be 'from'.

176. The preposition 'to' always follows 'averse'.

177. The sentence is grammatically correct. 'Pass by' refers to the act of 'overlooking.'

178. 'Kept up' refers to the act of continuing the long conversation.

179. Since the passage is talking about various ways of celebrating an eco-friendly Diwali, option (d) is the answer. Option (a) can be eliminated because it talks about one of the ways of making one's Diwali eco-friendly. Option (b) does not refer to Diwali at all. Option (c) does not refer to making Diwali eco-friendly.

180. Refer to the following lines from the passage, "Apart from environmental benefits, imagine how many low-income families that sell diyas and other earthenware can benefit with a single effort" and "Moreover, giving alms to the poor instead of shelling out on crackers could earn you extra 'Karmic' brownie points." So, options (a), (b) and (c) are the benefits mentioned in the passage. Option (d), is the answer.

PRACTICE PAPER-3

ELECTRONICS ENGINEERING

Time Allowed : 2:15 Hours — **Maximum Marks : 180**

Read the following instructions carefully before you begin to answer the questions. This Booklet contains questions in English.

INSTRUCTIONS

1. This Booklet contains 180 questions in total comprising of the following five parts:

Part – I:	Domain Knowledge	[1 – 75 Questions]
Part – II:	General Intelligence and Reasoning	[76 – 90 Questions]
Part – III:	General Awareness	[91 – 105 Questions]
Part – IV:	Quantitative Aptitude	[106 – 120 Questions]
Part – V:	General English	[121 – 180 Questions]

2. All questions are compulsory and carry 1 mark each.
3. Before you start to answer the questions you must check this Booklet to ensure that it contains all the pages and see that no page is missing or repeated.
4. For every wrong answer 1/3 marks will be deducted.
5. You are supplied with the OMR-Sheet separately.
6. Answers must be indicated by completely blackening one of the corresponding ovals on OMR-Sheet against the relevant question number, by black/blue ball pen only. Answers which are not indicated by black/blue ball pen will not be awarded any mark.
7. The OMR-Sheet must be handed over to the Invigilator before you leave the Examination Hall.
8. Failure to comply with any of the above instructions will render a candidate liable to such action/penalty as may be deemed fit.

PRACTICE PAPER – 3

DOMAIN KNOWLEDGE

1. 1 cm^3 of pure Germanium at 20°C contains about 4.2×10^{22} atoms, 2.5×10^{13} free electrons and 2.5×10^{13} holes. 0.0001% of Arsenic doping donates an extra 10^{17} free electrons in the same volume. The approximate number of holes in one cm^3 in the doped semiconductor under equilibrium condition is

(*a*) 6.25×10^9 (*b*) 2.5×10^9
(*c*) 10.5×10^9 (*d*) 10^{17}

2. Measurement of Hall coefficient enables the determination of

(*a*) Temperature coefficient and thermal conductivity
(*b*) Mobility and concentration of charge carriers
(*c*) Fermi level and forbidden energy gap
(*d*) None of the above

3. Silicon is not suitable for fabrication of light emitting diodes because it is

(*a*) An indirect band gap semiconductor
(*b*) A direct band gap semiconductor
(*c*) A wide band gap semiconductor
(*d*) A narrow band gap semiconductor

4. The band gap of elements arranged in ascending order is

(*a*) Diamond, Ge, Si (*b*) Si, Ge, Diamond
(*c*) Ge, Si, Diamond (*d*) Diamond, Si, Ge

5. The following statements are made for FETs:

1. In n-channel depletion mode MOS in the active region, the control voltage V_{GS} is negative.
2. NMOS in depletion mode is cut off for $V_{GS} = 0$
3. NMOS in enhance mode is cut off for $V_{GS} = 0$
4. There is no path between source and drain in the enhancement mode

Of these, the true statements are

(*a*) 1 and 2 only
(*b*) 1 and 3 only
(*c*) 3 and 4 only
(*d*) 1, 3 and 4 only

6. If the input to the circuit of Figure is a sine wave, then the output will be

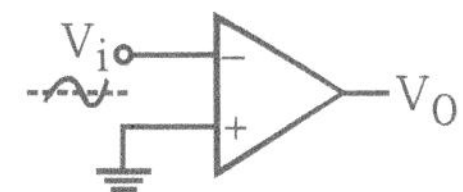

(*a*) Half wave rectified sine wave
(*b*) A full wave rectified sine wave
(*c*) A square wave
(*d*) A triangular wave

7. A high gain Op-Amp has $R_f = 10^5\ \Omega$ and $R_i = 10^3\ \Omega$. The correct statement is

(*a*) The non-inverting gain and inverting gain are nearly same
(*b*) The inverting gain is very much higher than non-inverting gain
(*c*) The non-inverting gain is very much higher than inverting gain
(*d*) The inverting and non-inverting gains are unrelated

8. For various types of oscillators, the correct statement is

(*a*) LC oscillators are more stable than crystal oscillators
(*b*) Crystal oscillators have highest Q
(*c*) Phase-shift oscillators have the widest range of frequency
(*d*) Wein bridge oscillator is used where a single frequency oscillator is required

9. For Butterworth and Chebyshev filters, the correct statement is

(*a*) Butterworth response has a sharp cut-off
(*b*) Chebyshev response has a flat response in the pass band
(*c*) Butterworth response has ripples in the pass band
(*d*) Chebyshev response has ripples in the pass band

10. In a feedback series regulator circuit, the output voltage is regulated by controlling the

(*a*) Magnitude of the input voltage
(*b*) Gain of the feedback transistor
(*c*) Voltage drop across the series pass transistor
(*d*) Reference voltage

11. The current through the Zener diode in figure is

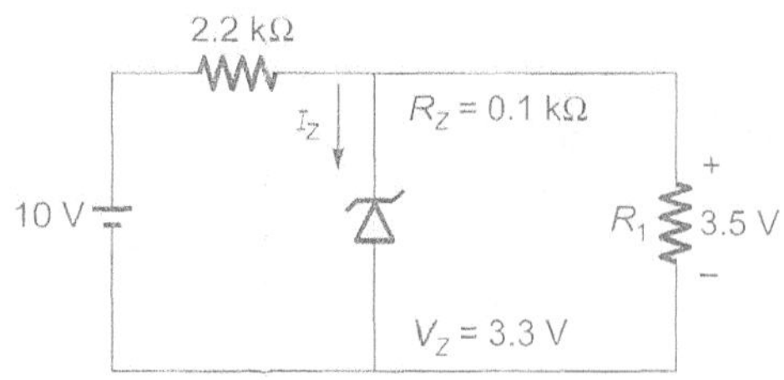

(*a*) 33 mA (*b*) 3.3 mA
(*c*) 2 mA (*d*) 0 mA

12. Assuming that the diodes are ideal in figure, the current in diode D_1 is

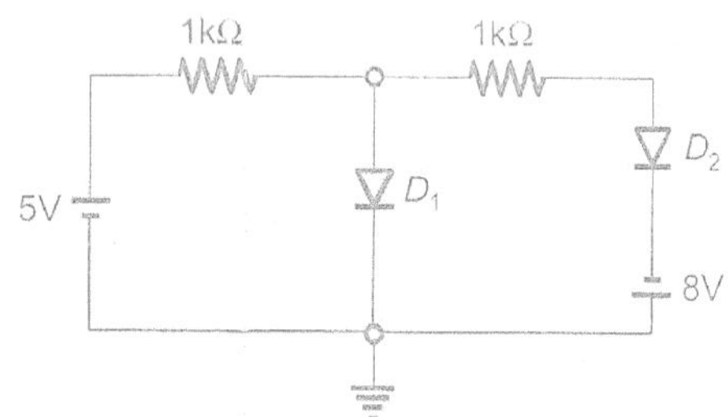

(*a*) 8 mA (*b*) 5 mA
(*c*) 0 mA (*d*) –3 mA

13. Assume that D_1 and D_2 in figure are ideal diodes. The value of current is

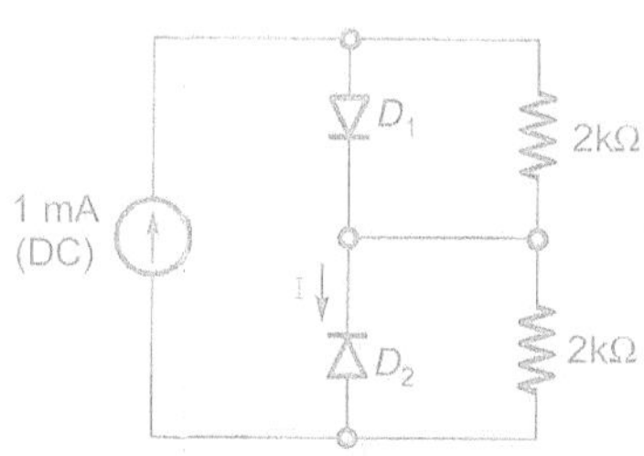

(*a*) 0 mA (*b*) 0.5 mA
(*c*) 1 mA (*d*) 2 mA

14. One of the applications of current mirror is

(*a*) output current limiting
(*b*) obtaining a very high current gain
(*c*) current feedback
(*d*) temperature stabilized biasing

15. Figure is a 24 V stabilized power supply. The zener is 24 V, 600 mW. The minimum zener current is 10 mA. Find the value of R and maximum load current are

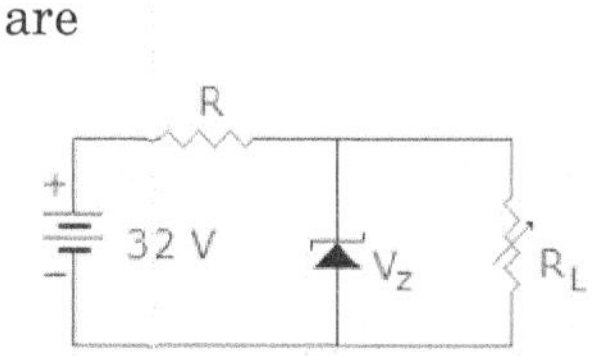

(*a*) 320 Ω, 10 mA (*b*) 400 Ω, 15 mA
(*c*) 400 Ω, 10 mA (*d*) 320 Ω, 15 mA

16. In the circuit of figure, the value of the base current I_B will be

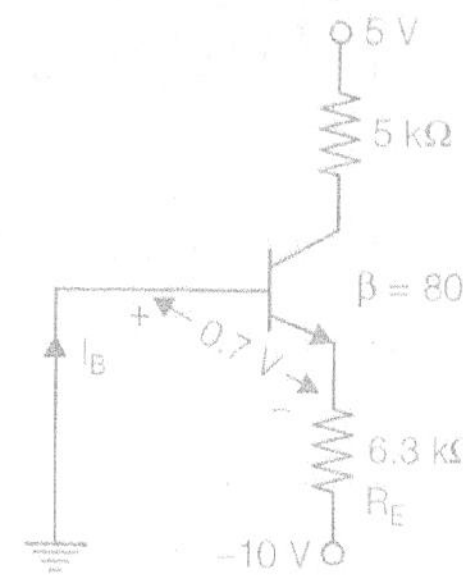

(*a*) 0 A (*c*) 18.2 A
(*b*) 26.7 A (*d*) 40 A

17. An n-channel JFET having a pinch off voltage (V_p) of –5 V shows a transcon- ductance (g_m) of 1 mA/V when the applied Gate to source voltage (V_{GS}) is –3 V. Its maximum transconduc- tance (in mA/V) is

(*a*) 1.5 (*b*) 2.0
(*c*) 2.5 (*d*) 3.0

18. An op-amp having a slew rate of 62.8 V/ sec is connected in a voltage follower configuration, if the maximum amplitude of the input sinusoidal is 10 Volts, then the minimum frequency at which the slew rate limited distortion would set in the output is

(*a*) 1 MHz (*b*) 6.8 MHz
(*c*) 10 MHz (*d*) 62.8 MHz

19. Transformer and emitter follower can both be used for impedance matching at the output of an audio amplifier. The basic relationship between the input power P_{in} and output power P_{out} in both the cases is

(*a*) $P_{in} = P_{out}$ for both transformer and emitter follower
(*b*) $P_{in} > P_{out}$ for both transformer and emitter follower
(*c*) $P_{in} < P_{out}$ for transformer and $P_{in} = P_{out}$ for emitter follower
(*d*) $P_{in} = P_{out}$ for transformer and $P_{in} < P_{out}$ for emitter follower

20. If $R_1 = R_2 = R_A$ and $R_3 = R_4 = R_B$, the circuit acts as a

(*a*) All pass filter
(*b*) Band pass filter
(*c*) High pass filter
(*d*) Low pass filter

21. A low-pass filter with a cut-off frequency of 30 Hz is cascaded with a high-pass filter with a cut-off frequency of 20 Hz. The resultant system of filters will function as (*a*) An all-pass filter

(*b*) An all-stop filter

(*c*) A band stop (band renect) filter

(*d*) A band-pass filter

22. Of the four characteristics given below, which are the major requirements for an instrumentation amplifier

P. High common mode rejection ratio

Q. High input impedance

R. High linearity

S. High output impedance

(*a*) P, Q and R only (*b*) P and R only

(*c*) P, Q and S only (*d*) Q, R and S only

23. The circuit shown below is an example of a

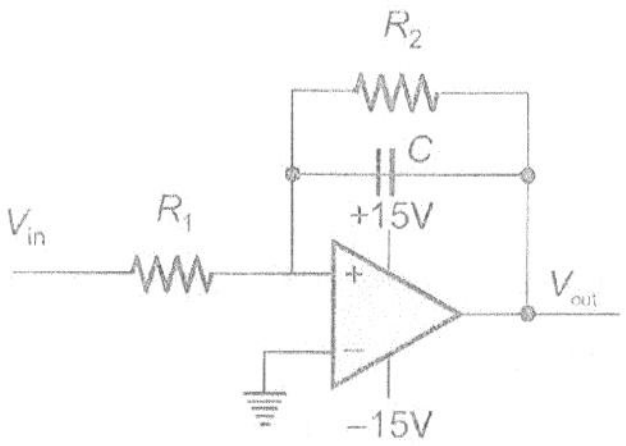

(*a*) low pass filter (*b*) band pass filter

(*c*) High pass filter (*d*) notch filter

24. In a common emitter amplifier, the unbypassed emitter resistance provides

(*a*) Voltage shunt feedback

(*b*) current sereis feedback

(*c*) negative voltage feedback

(*d*) Positive current feedback

25. The type and order of a system having transfer function

$T(s) = \dfrac{s+2}{s(s+1)(s+3)}$ are respectively

(*a*) 0, 2 (*b*) 0, 3

(*c*) 1, 2 (*d*) 1, 3

26. What is the purpose of impedance matching between the output of previous stage and input of next stage in a cascaded amplifier?

(*a*) To achieve high efficiency

(*b*) To achieve maximum power transfer

(*c*) To achieve reduced distortion

(*d*) To achieve reduced noise

27. As compared to the impedance of the signal source, the input impedance of a voltage amplifier should be

(*a*) small (*b*) large

(*c*) equal (*d*) none

28. Transformer coupling provides high efficiency because the

(*a*) collector voltage is stepped up

(*b*) dc resistance in the collected circuit is low

(*c*) collector voltage is stepped down

(*d*) flux linkages are incomplete

29. In a multistage RC coupled amplifier the coupling capacitor

(*a*) limits the low frequency response

(*b*) limits the high frequency response

(*c*) does not affect the frequency response

(*d*) none

30. A cascode amplifier stage is equivalent to

(*a*) CE-CB (*b*) CE-CE

(*c*) CE-CC (*d*) CC-CC

31. Which of the following amplifier has higher voltage gain and impedance matching?

(*a*) RC coupling

(*b*) Inductive coupling

(*c*) Transformer coupling

(*d*) Direct coupling

32. Effect of cascading several amplifier stages is to

(*a*) reduce the overall gain

(*b*) reduce the overall frequency response

(*c*) increases overall gain and reduce the frequency response

(*d*) double tuned circuits

33. The energy stored in capacitor is given by

(*a*) $\frac{1}{2}C\,V^2_{rms}$ (*b*) $C\,V^2_{rms}$

(*c*) $\frac{1}{2}C^2V_{rms}$ (*d*) $C^2\,V_{rms}$

34. Which of the following theorem is applicable to any circuit

(*a*) Superposition (*b*) Millman

(*c*) Tellegen (*d*) Thevenin

35. Which one of the following is a valid value of coefficient of coupling between two inductors

(*a*) 1.414 (*b*) 0.9

(*c*) 1.732 (*d*) 1.01

36. The principle of dynamically induced emf is utilized in

(*a*) choke (*b*) thermocouple

(*c*) transformer (*d*) generator

37. What is the ac otuput voltage and input impedance of single ended input and single ended output of an amplifier?

(*a*) 150, 2 (*b*) 151, 2

(*c*) 159, 2 (*d*) 154, 3

38. An op-amp with A = 200,000 and $CMRR_{dB}$ = 90 dB. What is the common-mode voltage gain, if the desired and common-mode signal each has a value of 1 μV, what is the output voltage?

(*a*) 6.32, 6.32 (*b*) 5, 6

(*c*) 7.89, 6.59 (*d*) 2.4, 3.8

39. An amplifier has an open-loop gain of 100, an input impedance of 1 kΩ, and an output impedance of 100 Ω. A feedback network with a feedabck factor of 0.99 is connected to the amplifier in a voltage series feedabck mode. The new input and output imedances, respectively, are

(*a*) 10 Ω & 1 Ω (*b*) 10 Ω & 10 Ω

(*c*) 100 kΩ & 1 Ω (*d*) 100 kΩ & 1 kΩ

40. An amplifier without feedback has a voltage gain of 50, input resistance of 1 kΩ and output resistance of 2.5 kΩ. The input resistance of the current shunt negative feedback amplifier using the above amplifier with a feedback factor of 0.2 is

(*a*) 1/11 kΩ (*b*) 1/5 kΩ

(*c*) 5 kΩ (d) 11 kΩ

41. The effect of current shunt feedback in an amplifier is to

(*a*) Increase the input resistance and decrease the output resistance

(*b*) Increase both input and output resistances

(*c*) Decrease both input and output resistances

(*d*) Decrease the input resistance and increase the output resistance

42. An amplifier has an open loop gain of 1000 ±10. Negative feedback is provided such that the gain variation remains within 0.1%. What is the amount of feedback β_f?

(*a*) 1/10 (*b*) 1/9

(*c*) 9/100 (*d*) 9/1000

43. A feedback amplifier has an open loop gain of –100. If 4% of the output is feedback in a degenerative loop. What is the closed loop gain of the amplifier?

(*a*) –33.3 (*b*) –25

(*c*) –20 (*d*) +25

44. The circuit in the figure employs positive feedback and is intended to generate sinusoidal oscillation. If at a frequency f_0,

$B(f) = \frac{V_f(f)}{V_0(f)} = \frac{1}{6} \angle 0°$ then to sustain oscillation at this frequency

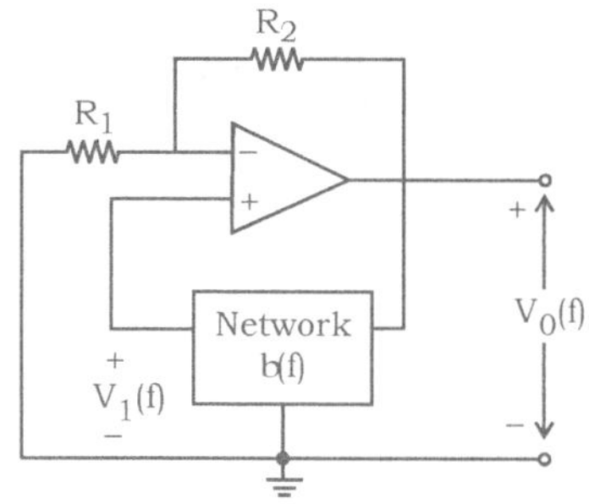

(*a*) $R_2 = 5R_1$ (*b*) $R_2 = 6R_1$

(*c*) $R_2 = \frac{R_1}{6}$ (*d*) $R_2 = \frac{R_1}{5}$

45. For a transistor colpitt's oscillator L = 100 μH, C_1 = 0.005 μF and C_2 = 0.01 μF. Calculate the freqeuency of oscillations generated.

(*a*) 277.05 kHz (*b*) 987.05 kHz

(*c*) 577.05 kHz (*d*) 497.05 kHz

46. Calculate the ferequency of oscillations of a Colpitt's oscillator with

$C_1 = C_2$ = 500 PF and L = 1 mH

(*a*) 318.3 kHz (*b*) 418.3 kHz

(*c*) 718.3 kHz (*d*) 818.3 kHz

47. Find the frequency of oscillations of a Hartley oscillator with component values L_1 = 1mH, L_2 = 100 μH, M = 50 μH and C = 100 pF.

(*a*) 459 kHz (*b*) 959 kHz

(*c*) 659 kHz (*d*) 859 kHz

48. The parameters of a crystal oscillator equivalent circuit are given as; L_S = 0.08 H, C_S = 0.08 pF, R_S = 5 kΩ and C_P = 1.0 pF.

Find the resonance fequency f_s

(*a*) 629 kHz (*b*) 700 kHz

(*c*) 750 kHz (*d*) 800 kHz

49. Given that L_S = 800 mH, C_S = 0.01 PF, R_S = 1000 Ω and C_P = 20 pF are the various values of an ac equivalent circuit of a piezoelectric crystal, calculate the value of f_s.

(*a*) 2000.2 kHz (*b*) 2500. 0 kHz

(*c*) 1779.40 kHz (*d*) 1000.5 kHz

50. Determine the frequency of oscillations of a clapp oscillator if the component values are as follows.

C_1 = 120 pF, C_2 = 1.5 pF, C_3 = 15 pF L_3 = 120μF

(*a*) 10.99 MHz (*b*) 12.99 MHz

(*c*) 15.5 MHz (*d*) 20.0 MHz

51. What is the frequency of oscillations of a Hartley oscillator is the component values are L_1= L_2 = 50 μH and C_3 = 100 pF.

(*a*) 1.59 MHz (*b*) 3.0 MHz

(*c*) 5.0 MHz (*d*) 6.5 MHz

52. In a phase shift oscillator, $R_1 = R_2 = R_3 = 800\ \Omega$ and $C_2 = C_2 = C_3$ = 100 pF. Find the frequency of oscillation.

(*a*) 812 Hz (*b*) 912 Hz

(*c*) 712 Hz (*d*) 1012 Hz

53. The cut in voltage of a diode is nearly equal to

(*a*) applied forward voltage

(*b*) applied reverse voltage

(*c*) barrier potential

(*d*) none of the above

54. In degenerate p-type semiconductor material, the Fermi level,

(*a*) is in the valence band

(*b*) is in conduction band

(*c*) is at the centre in between valence and conduction bands

(*d*) is very near conduction band

55. As comparated to an ordinary p-n diode, the extent of impurity atoms in a tunnel diode

(*a*) is more

(*b*) is less

(*c*) may be more or less

(*d*) is almost the same

56. What is the purpose of impedance matching between the output of previous stage and input of next stage in a cascaded amplifier?

(*a*) To achieve high efficiency

(*b*) To achieve maximum power transfer

(*c*) To achieve reduced distortion

(*d*) To achieve reduced noise

57. Intermediate (I) layer of PIN-diode imparts Which one of the following features to a pn junction diode?

(*a*) High reverse blocking capability

(*b*) High forward current rating

(*c*) Inverting capability

(*d*) Poor turn off performance

58. Negative feedback in an amplifier leads to which one of the following?

(*a*) Decrease in bandwidth

(*b*) Increase in current gain

(*c*) Increase in voltage gain

(*d*) Decrease in voltage gain

59. If A and B are Boolean variables, then what is $(A + B).(A+\overline{B})$ equal to?

(*a*) B (*b*) A

(*c*) Λ ı B (*d*) AB

60. What is the Boolean expression $A \oplus B$ equivalent to?

(*a*) $AB + \overline{A}\,\overline{B}$ (*b*) $\overline{A}B + A\overline{B}$

(*c*) B (*d*) $\overline{A}$

61. The minimum number of NAND gates required to implement the Boolean function $A + A\overline{B} + A\overline{B}C$ is equal to

(*a*) 0 (*b*) 1

(*c*) 4 (*d*) 7

62. Consider the Boolean expression X = ABCD + $ABCD + A\overline{B}CD + \overline{A}BCD + \overline{A}C\overline{B}D$ The simplified form of X is

(*a*) $\overline{C} + \overline{D}$ (*b*) BC

(*c*) CD (*d*) None of these

63. The minimized expression for the given K map (x:don't care) is

CD \ AB	00	01	11	10
00	0	0	1	1
01	0	x	x	1
11	x	x	1	x
10	1	0	1	1

(*a*) $A+\overline{B}C$ (*b*) B + AC

(*c*) C+AB (*d*) ABC

64. Which one of the following is equivalent to the Boolean expresson $Y = \overline{AB} + \overline{BC} + \overline{CA}$?

(*a*) $\overline{AB + BC + CA}$

(*b*) $(\overline{A} + \overline{B})(B + \overline{C})(\overline{A} + \overline{C})$

(*c*) $\overline{(A+B)(B+C)(C+A)}$

(*d*) $\overline{(A+B)}\,\overline{(B+C)}\,\overline{(C+A)}$

65. A half-wave rectifier has an input voltage of 240 V rms. If the step-down transformer has turns ratio of 8:1, what is the peak load voltage? Ignore diode drop.

(*a*) 27.5 V (*b*) 86.5 V

(*c*) 30.0 V (*d*) 42.5 V

66. The PIV rating of the diodes used in power supply circuits are chosen by which one of the following criteria? (V_m is the peak input supply voltage to the rectifier circuit used in the power supply)

(*a*) The diodes that are to be used in a full wave rectifier should be rated 2 V_m and in bridge rectifier equal to V_m

(*b*) The diodes that are to be used in a full wave rectifier should be rated V_m and in bridge rectifier equal to 2 V_m

(*c*) All diodes should be rated for V_m only

(*d*) All diodes should be rated for 2 V_m

67. An operational amplifier possesses

(*a*) very large input resistance and very large output resistance

(*b*) very small input resistance and very small output resistance

(*c*) very large input resistance and very small output resistance

(*d*) very small input resistance and very large output

68. The condition to be satisfied to prevent thermal runaway in a transistor amplifier where (P_c = Power dissipated at Collector. T_j = Junction temperature, T_A = Ambient temperature, Q = Thermal resistance) is

(*a*) $\frac{\delta P_c}{\delta T_j} > \frac{1}{Q}$

(*b*) $\frac{\delta P_c}{\delta T_A} < \frac{1}{Q}$

(*c*) $\frac{\delta P_c}{\delta T_j} < \frac{1}{Q}$

(*d*) $\frac{\delta P_c}{\delta T_A} > \frac{1}{Q}$

69. CE configuration is the most preferred transistor configuration when used as a switch because it

(*a*) requires only one power supply

(*b*) requires low voltage or current

(*c*) is easily understood by every one

(*d*) has small I_{CEO}

70. The given circuit has a feedback factor of

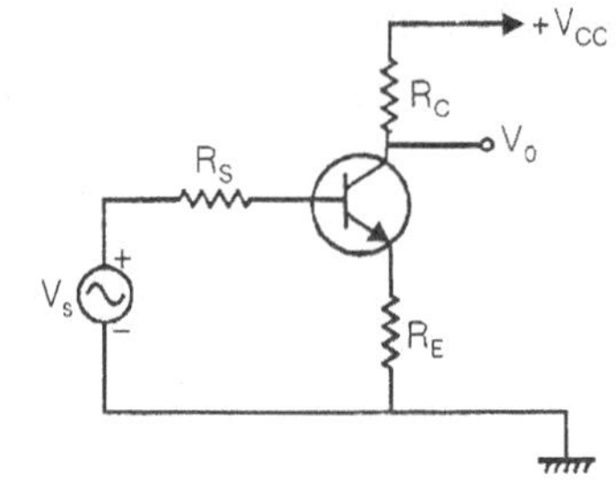

(*a*) $-R_C/R_S$ (*b*) $-R_E/R_C$

(*c*) $-R_E/R_S$ (*d*) $-R_C/R_E$

71. Thermal runaway will take place if the quiescent point is such that

(*a*) $V_{CE} > \frac{1}{2} V_{CC}$ (*b*) $V_{CE} < V_{CC}$

(*c*) $V_{CE} < 2V_{CC}$ (*d*) $V_{CE} < \frac{1}{2} V_{CC}$

72. The collector voltage V_C of the circuit shown in the given figure is approximately

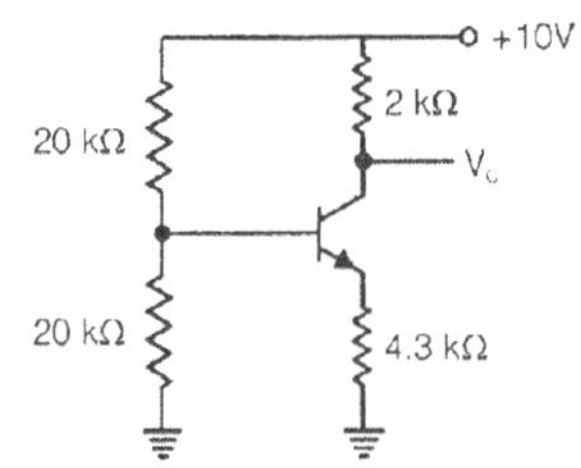

(*a*) 2V (*b*) 4.6V

(*c*) 8V (*d*) 8.6V

73. What is the simplified form of the boolean expression

$T = (X+Y)(X+\overline{Y})(\overline{X}+Y)$?

(*a*) $\overline{X}\overline{Y}$

(*b*) $\overline{X}Y$

(*c*) XY

(*d*) $X\overline{Y}$

74. The circuit shown is

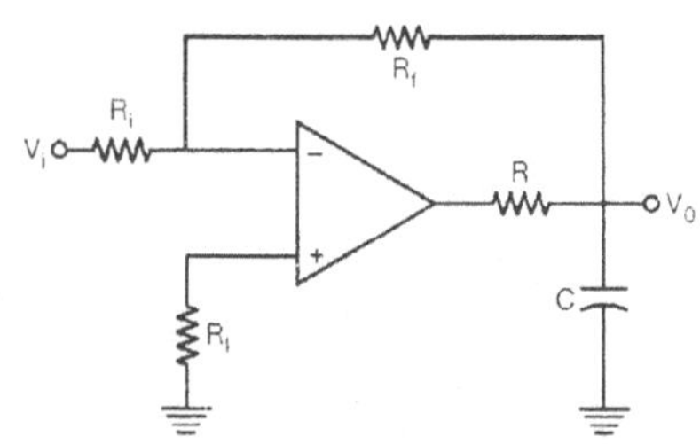

(*a*) A low pass filter

(*b*) A clamper

(*c*) A lag compensated inverting amplifier

(*d*) A narrow band video amplifier

75. For a Silicon n-p-n transistor, the base to emitter voltage (V_{BE}) is 0.7 V and the collector to base voltage (V_{CB}) is 0.2 V. Then the transistor is operating in the

(*a*) Normal active mode

(*b*) Saturation mode

(*c*) Inverse active mode

(*d*) Cut-off mode

GENERAL INTELLIGENCE AND REASONING

76. If SMART is coded as RTASM, then UNDER is coded as

(*a*) NUDRE (*b*) REDNU
(*c*) ERDUN (*d*) RDENU

77. If NOVELTY is coded as 5270813 and ESSAY is coded as 04463, then SEVEN will be coded as

(*a*) 48785 (*b*) 40085
(*c*) 08285 (*d*) 40705

78. If STYLE is written as ELSTY, then FLIRT is written as

(*a*) LFRIT (*b*) TRILF
(*c*) TRLFI (*d*) TRFLI

79. 35% of 1180 + 65% of 860 = ? % of 8100

(*a*) 8 (*b*) 12
(*c*) 15 (*d*) 21

80. Which of the following interchange of signs would make the given equation correct?

$30 + 5 \times 4 \div 3 - 2 = 25$

(*a*) and – (*b*) + and
(*c*) and – (*d*) + and

81. 'A' stands for multiplication, 'D' stands for equal to, 'B' stands for greater than, 'F' stands for subtraction, 'G' stands for less than, 'E' stands for addition and 'I' stands for division. In the following four alternatives, only one expression is correct according to the letter symbol. Identify that expression.

(*a*) 18 I 6 E 4 G 30 F 11 A 2

(*b*) 18 F 6 I 4 B 30 E 11 A 2

(*c*) 18 F 6 I 4 E 30 G 11 A 2

(*d*) 18 I 6 B 4 E 30 F 11 A 2

82. Identify the symbols to be inserted to make the expression correct.

16 Δ 2 Δ 9 Δ 3 Δ 4 Δ 5

(*a*) ÷ = + × + (*b*) + = ÷ + ×
(*c*) ÷ + = × + (*d*) + + ÷ = ×

Directions : In question nos. **83** to **88**, find related word/number from the given alternatives.

83. Planet : Mercury :: Bone : ?

(*a*) Femur (*b*) Stapes
(*c*) Scapula (*d*) Tibia

84. Plants : Cell Wall :: Animals : ?

(*a*) Ribosome

(*b*) Nucleus

(*c*) Lysosome

(*d*) Cell Membrane

85. Bear : Growl :: Doors : ?

(*a*) Chime (*b*) Creak
(*c*) Splatter (*d*) Jangle

86. Battery : Terminals :: Magnet : ?

(*a*) Charger (*b*) Poles
(*c*) Power (*d*) Ends

87. Lion : Pride :: Geese : ?

(*a*) Herd (*b*) Pack
(*c*) Gaggle (*d*) Flock

88. 15 : 65 : : 24 : ?

(*a*) 76 (*b*) 96
(*c*) 126 (*d*) 144

Directions : In question nos. **89** to **90**, find the odd word/number/letters/number pair from the given alternatives.

89. (*a*) $\frac{4}{5}$ (*b*) $\frac{6}{11}$
(*c*) $\frac{3}{2}$ (*d*) $\frac{8}{13}$

90. (*a*) Father (*b*) Wife
(*c*) Mother (*d*) Son

GENERAL AWARENESS

91. Who founded the 'Indian National Union' during 1884?

(*a*) Firoz shah Mehta (*b*) A O Hume
(*c*) Mary Carpenter (*d*) Anand Mohan Bose

92. The study of skin is called

(*a*) Dermatology (*b*) Paleontology
(*c*) Entomology (*d*) Etiology

93. The Caribbean Sea is situated around

(*a*) North, Central and South America

(*b*) North, Central and South Africa

(*c*) North, Central and South Asia

(*d*) North, Central and South Australia

94. Which of the following treaty was signed at the end of the First World War?

(*a*) Treaty of Versailles

(*b*) Maastricht Treaty

(*c*) Treaty of Lisbon

(*d*) Treaty of Rome

95. Who gave the theoretical concept 'Survival of the Fittest'?

(*a*) Louis Pasteur (*b*) Charles Darwin

(*c*) Ernst Haeckel (*d*) Edward Jenner

96. Which of the following does not belong to the reptilian class?

(*a*) Lizard (*b*) Snake

(*c*) Crocodile (*d*) Frog

97. Which of the following hormone is produced during pregnancy?

(*a*) Human Chorionic Gonadotropin

(*b*) Testosterone

(*c*) Aldosterone

(*d*) Somatotropin

98. Who discovered the relationship between a changing magnetic field and the electric field?

(*a*) Michael Faraday (*b*) Graham Bell

(*c*) Albert Einstein (*d*) Isaac Newton

99. Ghughua Fossil National Park is located in which of the following states?

(*a*) Madhya Pradesh

(*b*) Andhra Pradesh

(*c*) Himcahal Pradesh

(*d*) Nagaland

100. Lokpriya Gopinath Bordoli International Airport is located in which of the following cities?

(*a*) Tejpur (*b*) Silchar

(*c*) Guwahati (*d*) Gangtok

101. Which of the following is not a part of an ecosystem?

(*a*) Metro train (*b*) Trees

(*c*) Soil (*d*) Birds

102. Which of the following element is used in the filament of bulb?

(*a*) Mercury (*b*) Tungsten

(*c*) Cadmium (*d*) Silicon

103. Silver iodide is used in

(*a*) Photography (*b*) Artificial rain

(*c*) Antiseptic (*d*) All of the above

104. A row of trees planted to protect an area from the wind is known as

(*a*) Windbreak (*b*) Shelterbelts

(*c*) Both (*a*) and (*b*) (*d*) Contour farming

105. Which of the following airport is not situated in Asia?

(*a*) Aden International Airport

(*b*) Amami Airport

(*c*) Ahwaz airport

(*d*) El Dorado International Airport

QUANTITATIVE APTITUDE

106. Find the least number which must be added to 21590 to make it a perfect square.

(*a*) 20 (*b*) 16

(*c*) 24 (*d*) 19

107. A and B together can complete a piece of work in $\frac{27}{2}$ days, B and C together can do it in $\frac{21}{4}$ days while A and C together can complete it in $\frac{9}{2}$ days. In how many days can the three working together complete the same work?

(*a*) $\frac{169}{46}$ days (*b*) $\frac{189}{23}$ days

(*c*) $\frac{189}{46}$ days (*d*) $\frac{159}{46}$ days

108. Two adjacent sides of a parallelogram are 8 cm and 15 cm while the angle between them is 30°. Find the area of the parallelogram.

(*a*) 30 cm^2 (*b*) 60 cm^2

(*c*) 90 cm^2 (*d*) 120 cm^2

109. A dealer marks up an article in such a way that he earns a profit of 16.67% by selling the article at a discount of 16.67%. By what percent above the cost price should he mark up the article?

(*a*) 40%

(*b*) 37.5%

(*c*) 28%

(*d*) 25%

110. The ratio of the sum and the difference of two numbers is 28 : 17, and their HCF is 2. Find the product of these numbers.

(*a*) 475 (*b*) 465
(*c*) 485 (*d*) 495

111. The average age of 78 students in a group is 25.5 years. When the ages of some teachers were included in the group the new average became 29 years. Find the number of teachers included in the group if the average age of the teachers is 68 years.

(*a*) 8 (*b*) 7
(*c*) 9 (*d*) 6

112. A dishonest dealer uses a weight of 1200 grams instead of 1 kg at the time of buying while a weight of 900 grams instead of 1 kg at the time of selling. Find the percentage of profit made by him if he professes to sell his goods at the cost price.

(*a*) 20% (*b*) 37.5%
(*c*) 25% (*d*) 33.33%

113. If A's salary is 37.5% more than B's salary, C's salary is 20% less than A's salary and D's salary is 50% more than C's salary, then by what percent is A's salary more or less than that of D's salary?

(*a*) 16.67% Less (*b*) 15.38% Less
(*c*) 18.18% More (*d*) 15.38% More

114. A boat covers a distance of 375 m against the current in 25 seconds while it covers a distance of 675 m along the current in 27 seconds. Find the distance travelled by the boat in still water in 4.5 hrs.

(*a*) 324 km (*b*) 342 km
(*c*) 243 km (*d*) 423 km

115. If the simple interest on Rs. 200 for 2 months is Rs. 2, then find the compound interest on Rs. 27,500 for one year at the same rate of interest compounded semi-annually.

(*a*) Rs. 1,684.75 (*b*) Rs. 1,674.75
(*c*) Rs. 1,664.75 (*d*) Rs. 1,674.25

116. Find the maximum value of $3.75 + 2x - 4x^2$.

(*a*) 3.75 (*b*) 4
(*c*) 4.25 (*d*) 3.5

117. If $a^3 + b^3 + c^3 = -3$, $a^2 + b^2 + c^2 = ab + bc + ca$ and $(a + b)(b + c)(c + a) = -8$, then find the value of $(a - b - c)^3 + 3abc$.

(*a*) 0 (*b*) 2
(*c*) 1 (*d*) –2

118. If $a + b = 8$ and $a^3 + b^3 = 152$, then the value of $a - b$ will be

(*a*) 4 (*b*) 2
(*c*) 6 (*d*) 8

119. If $\frac{2x}{a+b} + 1 = \frac{2x}{a-b} + \frac{a-b}{a+b}$, then find the value of x.

(*a*) $2a - b$ (*b*) $\frac{a+b}{2}$
(*c*) $a - 2b$ (*d*) $\frac{a-b}{2}$

120. What will be the ratio of the circumradius and the median of an equilateral triangle the length of whose side is 14 cm?

(*a*) 1 : 2 (*b*) 2 : 1
(*c*) 2 : 3 (*d*) 3 : 2

GENERAL ENGLISH

Directions: In question nos. 121 to 123, out of the four alternatives, choose the one which can be substituted for the given words/sentences and indicate it.

121. Study of Fossils

(*a*) Zoology (*b*) Mythology
(*c*) Anthropology (*d*) Palaeontology

122. Excessive desire to work

(*a*) Ergomania (*b*) Hippomania
(*c*) Phaneromania (*d*) Sophomania

123. Governance in which a country is ruled by significantly old people

(*a*) Plutocracy (*b*) Oligarchy
(*c*) Gerontocracy (*d*) Republic

Directions: In question nos. 124 to 128, a sentence/a part of the sentence is underlined. Below are given alternatives to the underlined part which may improve the sentence. Choose the correct alternative. In case no improvement is needed, choose "No Improvement". Mark your answer.

124. Winter in New Zealand sets on June.

(*a*) No Improvement (*b*) sets at
(*c*) sets upon (*d*) sets in

125. That fellow trumped up a story.

(*a*) trumped at (*b*) No Improvement
(*c*) trumped on (*d*) trumped in

126. At present, I am reading Shakespeare, but it is sometimes difficult to take on his meaning.

(*a*) take off (*b*) take in

(*c*) take about (*d*) No Improvement

127. Very heavy rain fell so that the rivers were soon on flood.

(*a*) No Improvement (*b*) at

(*c*) above (*d*) in

128. Protestors burned railway stations and car showrooms, blocked road traffic and force the cancellation of hundreds of trains.

(*a*) forcibly the cancellation

(*b*) force the cancelling

(*c*) No Improvement

(*d*) forced the cancellation

Directions: In question nos. 129 to 133, you have a passage with 5 questions following. Read the passage carefully and choose the best answer to each question out of the four alternatives and mark it.

Passage

After decades of strict enforcement, China has finally scrapped its controversial one-child policy. In a move that can have huge implications globally, China recently announced that it will allow all married couples to have two children. In the wake of realisation that the policy is leaving the country with an ageing population and shrinking workforce, the move comes as an encouraging step towards greater personal freedom in China. China's working-age population fell in 2012 and there was the realisation that it could be the first country in the world to get old before it gets rich.

The controversial one-child policy, which came into effect in 1980 restricted most couples to only a single offspring and after its implementation authorities argued that it was a key contributor to China's economic success. The policy was initially successful in its implementation with prevention of 400 million births. However, it met with a backlash and was criticized as being overly harsh and even inhumane. Couples who violated the one-child policy faced many punishments, from forced abortions to harassment to fines and loss of employment.

129. What according to authorities were the initial consequences of China's one child policy?

(*a*) It led to numerous deaths throughout the country

(*b*) It gave birth to a number of dissents

(*c*) It was severely objected by the masses

(*d*) It brought economic success to the country

130. What has made China allow couples to bear two kids now?

(*a*) Shrinking workforce and ageing population

(*b*) Harsh criticism by the citizens

(*c*) Receding economic success

(*d*) Need to give personal freedom to individuals

131. What did the general public think of one-child policy?

(*a*) They lauded the policy for its success

(*b*) They tried to overthrow the government

(*c*) They criticized the policy for being extremely harsh

(*d*) They decided to adhere to it unquestioningly

132. How did China deal with couples who bore more than one child post 1980?

(*a*) They were warned against any future transgressions

(*b*) They were severely punished for violating the law

(*c*) They were severely reprimanded in public

(*d*) They were meted with capital punishments

133. How many births have been prevented by one-child policy?

(*a*) Less than a million

(*b*) More than a billion

(*c*) More than 100 million

(*d*) More than 100 billion

Directions: In question no. 134, out of the four alternatives, find the one which best expresses the meaning of the given word.

134. REPRIMAND

(*a*) Appraisal (*b*) Accord

(*c*) Reproof (*d*) Equate

Directions: In question no. 135, out of the four alternatives, choose the opposite in meaning of the given word.

135. ELATED

(*a*) Convivial (*b*) Peppy

(*c*) Exultant (*d*) Morose

136. In the following question, four words are given in each question, out of which only one word is correctly spelt. Find the correctly spelt word and mark your answer.

(*a*) examplary (*b*) consumate

(*c*) quintesential (*d*) impeccable

Directions: In question nos. 137 to 139, some parts of the sentence have errors and some are correct. Find out which part of a sentence has an error and blacken the circle corresponding to the appropriate correct option. If a sentence is free from error, mark the "No Error" option.

137. With a laudatory performance on the stage, the Delhi Branch made a befitting entry into the dance competition.

(*a*) With a laudatory performance

(*b*) on the stage,

(*c*) the Delhi Branch made a befitting entry into the dance competition

(*d*) No Error

138. The police reported that the most renowned actor of the century was not only a con but a lecher.

(*a*) The police reported that

(*b*) the most renowned actor of the century

(*c*) was not only a con but a lecher.

(*d*) No Error

139. She felt great after she had had a good night's sleep.

(*a*) She felt great

(*b*) after she had had a

(*c*) good night's sleep.

(*d*) No Error

Directions: In question nos. 140 to 142, sentences are given with blanks to be filled with an appropriate word(s). Four alternatives are suggested for each question. Choose the correct alternative out of the four and indicate it.

140. You have _____ been a spoilt child, and have been used to have your own way a good deal, both in the house and among your playfellows, with whom you were too fond of being a leader.

(*a*) become (*b*) their

(*c*) would (*d*) hitherto

141. His death brought more tributes than have been paid at the _____ of any other human being in history.

(*a*) pass out (*b*) passed

(*c*) passing (*d*) pass

142. He has no liking ___ cats.

(*a*) on (*b*) at

(*c*) to (*d*) for

Directions: In question no. 143 to 145, four alternatives are given for the idiom/phrase underlined in the sentence. Choose the alternative which best expresses the meaning of the idiom/phrase and mark it.

143. The teacher took the erring student to task.

(*a*) praised him

(*b*) patted him

(*c*) took him on a tour

(*d*) rebuked him

144. He kicked up a row at the meeting.

(*a*) made a fuss

(*b*) assaulted somebody

(*c*) shot somebody

(*d*) entertained everyone

145. He seems ill at ease.

(*a*) uncomfortable (*b*) healthy

(*c*) prejudiced (*d*) flexible

Directions: In question nos. **146** to **148**, four alternatives are given for the Idiom/Phrase underlined in the sentence. Choose the alternative which best expresses the meaning of the Idiom/Phrase and mark it.

146. He sported a sunny smile when we met.

(*a*) Appeared sick (*b*) Appeared morose

(*c*) A sad smile (*d*) Very happy

147. We hit the road in a land rover.

(*a*) Meet with an accident

(*b*) Get injured in an accident

(*c*) End of a journey

(*d*) To leave

148. My cousins can be often seen chewing the fat at the local club.

(*a*) To chat idly

(*b*) having red meat

(*c*) indulging in some intellectual discussions

(*d*) having a lavish meal

Directions: In question nos. **149** to **151**, out of the four alternatives, choose the one which can be substituted for the given words/sentences and indicate it.

149. Merciful treatment of someone who could be given harsh punishment

(*a*) absolution (*b*) clemency

(*c*) detention (*d*) circumvention

150. A short interesting story based on a real incident.

(*a*) novella (*b*) fable

(*c*) parable (*d*) anecdote

151. Too sweet or emotional

(*a*) Flamboyant (*b*) Saccharine

(*c*) Piquant (*d*) Acrid

Directions: In question nos. **152** to **156**, a sentence/a part of the sentence is underlined. Below are given alternatives to the underlined part which may improve the sentence. Choose the correct alternative. In case no improvement is needed, choose **"No improvement"**. Mark your answer.

152. One way could be to view the university as a hierarchical place, where certain structures are followed in order to prepare the student for the widest world.

(*a*) No improvement (*b*) biggest

(*c*) wider (*d*) largest

153. It is very important that in such a troubled political climate, we take a non-idealistic view of the university.

(*a*) No improvement (*b*) hydrosphere

(*c*) troposphere (*d*) weather

154. The air pollution in Calcutta is getting worst with every passing day.

(*a*) growing worse (*b*) getting nuisance

(*c*) indulging worse (*d*) No improvement

155. The eastern churches were exempted to the decree.

(*a*) No improvement (*b*) by

(*c*) at (*d*) from

156. The government must do everything in its power to curb tobacco consumption and must not encourage to pressure from tobacco companies.

(*a*) validate to (*b*) No improvement

(*c*) succumb to (*d*) express to

Directions: In question nos. **157** to **161**, you have a passage with 5 questions following. Read the passage carefully and choose the best answer to each question out of the four alternatives and mark it.

Passage

As Twitter marks its 10th birthday on Monday, it is somewhat of an awkward child -- having become a powerful communication tool but still struggling to win users and reach profitability.Since making a star-quality entrance a decade ago, Twitter has become a must-have tool for journalists, activists and celebrities but has struggled to show it can expand beyond its devoted "twitterati" to become a mainstream hit.While Internet lovers might have trouble envisioning life without Twitter, the San Francisco-based company has seen its stock tank, a chief executive leave, and its staff cut. Twitter's woes include a slump in its stock price to all-time lows this year -- down nearly half from its 2013 stock market debut -- and ongoing losses, even as its revenue grows.Twitter's base of monthly active users remained stuck at 320 million at the end of 2015. While that is a big accomplishment, Twitter has failed to keep pace with fast-growing rivals and to expand beyond its base.The troubles have forced Twitter to bring back co-founder Jack Dorsey as chief executive, but that has not stemmed rumors about a possible buyout or merger."It's not dead yet," independent analyst Rob Enderle of the Enderle Group said of Twitter.Some analysts believe Twitter's true value is being demonstrated in the US presidential race, especially by the Republican frontrunner, Donald Trump.Twitter and other social media have been important tools in movements such as the Arab Spring, and in protests in Turkey, where the government has sought to ban it.An advantage of Twitter is that it lets users "amplify" messages with tweets that echo on the Internet, and assess public sentiment in real-time by getting quick feedback.

157. Jack Dorsey, co-founder of Twitter has been brought back as _________.

(*a*) COO (*b*) Chief Executive

(*c*) President (*d*) Chief Manager

158. Where is the head office of Twitter located?

(*a*) Los Angeles (*b*) Paris

(*c*) New York (*d*) San Francisco

159. Twitter has been instrumental in

(*a*) character assassination

(*b*) spreading movements such as Arab Spring

(*c*) endangering democracy

(*d*) spreading wars in South Asia

160. For how long has Twitter has been in use?

(*a*) 9 years (*b*) Half a decade

(*c*) A decade (*d*) 20 years

161. How many active users of Twitter were recorded at the end of 2015?

(*a*) 320 million (*b*) 220 million

(*c*) 310 million (*d*) 330 million

Directions: In question no. **162** , out of the four alternatives, choose the one which best expresses the meaning of the given word and mark it.

162. Pallor

(*a*) Lucid (*b*) Description

(*c*) Pallid (*d*) Vehement

Directions: In question nos. **163**, choose the word opposite in meaning to the given word and mark it.

163. Ennui

(*a*) Brilliance (*b*) Glitzy

(*c*) Lullaby (*d*) Vivacity

164. In the following question, four words are given in each question, out of which only one word is correctly spelt. Find the correctly spelt word and mark your answer.

Find the correctly spelt word:

(*a*) connoissur (*b*) conoisseur

(*c*) conoiseur (*d*) connoisseur

Directions: In question nos. **165** to **167**, some parts of the sentences have errors and some are correct. Find out which part of a sentence has an error and blacken the circle corresponding to the appropriate correct option. If a sentence is free from error,mark the "No error" option.

165. Making a movie and to write about a movie are two separate domains needing two different sets of abilities.

(*a*) Making a movie and to write about a movie

(*b*) are two separate domains

(*c*) needing two different sets of abilities

(*d*) No error

166. In a city full of Indian Chinese restaurants, it is hard to choose the most perfect destination for authentic Chinese cuisine.

(*a*) No error

(*b*) In a city full of Indian Chinese restaurants,

(*c*) it is hard to choose the most perfect destination

(*d*) for authentic Chinese cuisine.

167. Can a journalist in a conflict area, who's security and career depends on his relationship with the local police, afford to turn down such a request?

(*a*) Can a journalist in a conflict area,

(*b*) who's security and career depends on his relationship

(*c*) with the local police, afford to turn down such a request?

(*d*) No error

Directions: In question nos. **168** to **170**, sentences are given with blanks to be filled in with an appropriate word(s). Four alternatives are suggested for each question. Choose the correct alternative out of the four and indicate it.

168. So long as the rain _____, I _____ at home.

(*a*) continued; stayed

(*b*) had continued; stay

(*c*) has continued; can stayed

(*d*) will continue; stayed

169. Galileo _______ that the earth _____ round the sun.

(*a*) maintains; moved

(*b*) was maintaining; was moving

(*c*) maintained; can be moving

(*d*) maintained; moves

170. Apart from the injustice and impracticability of a course that is bound to aggravate resentment, discrimination will also play ________ the hands of the jihadists.

(*a*) below (*b*) after

(*c*) into (*d*) above

Directions: In question nos. **171** to **173**, some parts of the sentences have errors and some are correct. Find out which part of a sentence has an error and mark the appropriate option. If a sentence is free from error, mark the "No error" option.

171. My mother vetoed the idea of me going to Poona to join the Film and TV institute.

My mother vetoed (*a*)/ the idea of me going (*b*)/ to Poona to join the Film and TV institute. (*c*)/ No error (*d*)

172. Even before I asked him questions, Gangaram had answers ready, almost as if he rehearsed them.

Even before I asked (*a*) / him questions, Gangaram had (*b*) / answers ready, almost as if he rehearsed them. (*c*) / No error (*d*)

173. In fact, when I'm sometimes on the road, random people have asked me that when I would be dubbing my next film.

In fact, when I'm sometimes (*a*) / on the road, random people have asked me that when (*b*) / I would be dubbing my next film. (*c*) / No error (*d*)

Directions: In question nos. **174** to **176**, sentences given with blanks are to be filled in with appropriate word(s). Four alternatives are suggested for each question. Choose the correct alternative out of the four.

174. He has lost all his teeth, _______ he cannot eat hard food.

(*a*) consequently (*b*) but

(*c*) however (*d*) although

175. Scientists believe the methane could stem from micro-organisms, called methanogens, that _____ became extinct millions of years ago or left gas frozen below the planet's surface, or that some methane producing organisms still survive.

(*a*) has (*b*) neither

(*c*) either (*d*) neither more

176. The ancient Greeks, though born in a warm climate, seem to have been much addicted _____ the bottle.

(*a*) in

(*b*) for

(*c*) on

(*d*) to

Directions: In question nos. **177** to **179**, four alternatives are given for the Idiom/Phrase underlined in the sentence. Choose the alternative which best expresses the meaning of the Idiom/Phrase.

177. The belief in witchcraft is slowly <u>losing ground</u> in India.

(*a*) gaining popular support

(*b*) gaining power

(*c*) moving slowly

(*d*) becoming less powerful

178. Recently, he has been <u>giving himself airs</u>.

(*a*) gone on a vacation

(*b*) been ill

(*c*) behaving arrogantly

(*d*) behaving politely

179. I am prepared to <u>meet you half-way</u>.

(*a*) come to a compromise with you

(*b*) fight with you

(*c*) meet you on the road

(*d*) beat you

Directions: In question nos. **180**, out of the four alternatives, choose the one which can be substituted for the given words/sentences.

180. Set in wrong period of time.

(*a*) Anachronistic

(*b*) Aesthetic

(*c*) Pluralistic

(*d*) Diagnostic

OMR SHEET

1.	A	B	C	D	26.	A	B	C	D	51.	A	B	C	D	76.	A	B	C	D
2.	A	B	C	D	27.	A	B	C	D	52.	A	B	C	D	77.	A	B	C	D
3.	A	B	C	D	28.	A	B	C	D	53.	A	B	C	D	78.	A	B	C	D
4.	A	B	C	D	29.	A	B	C	D	54.	A	B	C	D	79.	A	B	C	D
5.	A	B	C	D	30.	A	B	C	D	55.	A	B	C	D	80.	A	B	C	D
6.	A	B	C	D	31.	A	B	C	D	56.	A	B	C	D	81.	A	B	C	D
7.	A	B	C	D	32.	A	B	C	D	57.	A	B	C	D	82.	A	B	C	D
8.	A	B	C	D	33.	A	B	C	D	58.	A	B	C	D	83.	A	B	C	D
9.	A	B	C	D	34.	A	B	C	D	59.	A	B	C	D	84.	A	B	C	D
10.	A	B	C	D	35.	A	B	C	D	60.	A	B	C	D	85.	A	B	C	D
11.	A	B	C	D	36.	A	B	C	D	61.	A	B	C	D	86.	A	B	C	D
12.	A	B	C	D	37.	A	B	C	D	62.	A	B	C	D	87.	A	B	C	D
13.	A	B	C	D	38.	A	B	C	D	63.	A	B	C	D	88.	A	B	C	D
14.	A	B	C	D	39.	A	B	C	D	64.	A	B	C	D	89.	A	B	C	D
15.	A	B	C	D	40.	A	B	C	D	65.	A	B	C	D	90.	A	B	C	D
16.	A	B	C	D	41.	A	B	C	D	66.	A	B	C	D	91.	A	B	C	D
17.	A	B	C	D	42.	A	B	C	D	67.	A	B	C	D	92.	A	B	C	D
18.	A	B	C	D	43.	A	B	C	D	68.	A	B	C	D	93.	A	B	C	D
19.	A	B	C	D	44.	A	B	C	D	69.	A	B	C	D	94.	A	B	C	D
20.	A	B	C	D	45.	A	B	C	D	70.	A	B	C	D	95.	A	B	C	D
21.	A	B	C	D	46.	A	B	C	D	71.	A	B	C	D	96.	A	B	C	D
22.	A	B	C	D	47.	A	B	C	D	72.	A	B	C	D	97.	A	B	C	D
23.	A	B	C	D	48.	A	B	C	D	73.	A	B	C	D	98.	A	B	C	D
24.	A	B	C	D	49.	A	B	C	D	74.	A	B	C	D	99.	A	B	C	D
25.	A	B	C	D	50.	A	B	C	D	75.	A	B	C	D	100.	A	B	C	D

OMR SHEET

101. (A) (B) (C) (D)	121. (A) (B) (C) (D)	141. (A) (B) (C) (D)	161. (A) (B) (C) (D)
102. (A) (B) (C) (D)	122. (A) (B) (C) (D)	142. (A) (B) (C) (D)	162. (A) (B) (C) (D)
103. (A) (B) (C) (D)	123. (A) (B) (C) (D)	143. (A) (B) (C) (D)	163. (A) (B) (C) (D)
104. (A) (B) (C) (D)	124. (A) (B) (C) (D)	144. (A) (B) (C) (D)	164. (A) (B) (C) (D)
105. (A) (B) (C) (D)	125. (A) (B) (C) (D)	145. (A) (B) (C) (D)	165. (A) (B) (C) (D)
106. (A) (B) (C) (D)	126. (A) (B) (C) (D)	146. (A) (B) (C) (D)	166. (A) (B) (C) (D)
107. (A) (B) (C) (D)	127. (A) (B) (C) (D)	147. (A) (B) (C) (D)	167. (A) (B) (C) (D)
108. (A) (B) (C) (D)	128. (A) (B) (C) (D)	148. (A) (B) (C) (D)	168. (A) (B) (C) (D)
109. (A) (B) (C) (D)	129. (A) (B) (C) (D)	149. (A) (B) (C) (D)	169. (A) (B) (C) (D)
110. (A) (B) (C) (D)	130. (A) (B) (C) (D)	150. (A) (B) (C) (D)	170. (A) (B) (C) (D)
111. (A) (B) (C) (D)	131. (A) (B) (C) (D)	151. (A) (B) (C) (D)	171. (A) (B) (C) (D)
112. (A) (B) (C) (D)	132. (A) (B) (C) (D)	152. (A) (B) (C) (D)	172. (A) (B) (C) (D)
113. (A) (B) (C) (D)	133. (A) (B) (C) (D)	153. (A) (B) (C) (D)	173. (A) (B) (C) (D)
114. (A) (B) (C) (D)	134. (A) (B) (C) (D)	154. (A) (B) (C) (D)	174. (A) (B) (C) (D)
115. (A) (B) (C) (D)	135. (A) (B) (C) (D)	155. (A) (B) (C) (D)	175. (A) (B) (C) (D)
116. (A) (B) (C) (D)	136. (A) (B) (C) (D)	156. (A) (B) (C) (D)	176. (A) (B) (C) (D)
117. (A) (B) (C) (D)	137. (A) (B) (C) (D)	157. (A) (B) (C) (D)	177. (A) (B) (C) (D)
118. (A) (B) (C) (D)	138. (A) (B) (C) (D)	158. (A) (B) (C) (D)	178. (A) (B) (C) (D)
119. (A) (B) (C) (D)	139. (A) (B) (C) (D)	159. (A) (B) (C) (D)	179. (A) (B) (C) (D)
120. (A) (B) (C) (D)	140. (A) (B) (C) (D)	160. (A) (B) (C) (D)	180. (A) (B) (C) (D)

ANSWERS

1. (a)	2. (b)	3. (a)	4. (c)	5. (d)	6. (c)	7. (a)	8. (b)	9. (d)	10. (d)
11. (c)	12. (c)	13. (a)	14. (d)	15. (d)	16. (b)	17. (c)	18. (a)	19. (d)	20. (c)
21. (d)	22. (a)	23. (a)	24. (b)	25. (a)	26. (b)	27. (b)	28. (b)	29. (a)	30. (a)
31. (c)	32. (c)	33. (b)	34. (c)	35. (b)	36. (d)	37. (a)	38. (a)	39. (c)	40. (a)
41. (d)	42. (d)	43. (c)	44. (a)	45. (a)	46. (a)	47. (a)	48. (a)	49. (c)	50. (b)
51. (a)	52. (a)	53. (c)	54. (a)	55. (a)	56. (b)	57. (a)	58. (d)	59. (b)	60. (b)
61. (a)	62. (c)	63. (a)	64. (c)	65. (d)	66. (a)	67. (c)	68. (c)	69. (b)	70. (b)
71. (b)	72. (c)	73. (c)	74. (d)	75. (a)	76. (c)	77. (d)	78. (d)	79. (b)	80. (d)
81. (a)	82. (c)	83. (b)	84. (d)	85. (b)	86. (b)	87. (c)	88. (c)	89. (c)	90. (b)
91. (b)	92. (a)	93. (a)	94. (a)	95. (b)	96. (d)	97. (a)	98. (a)	99. (a)	100. (c)
101. (a)	102. (b)	103. (d)	104. (c)	105. (d)	106. (d)	107. (c)	108. (b)	109. (a)	110. (d)
111. (b)	112. (d)	113. (a)	114. (a)	115. (b)	116. (b)	117. (d)	118. (b)	119. (d)	120. (c)
121. (d)	122. (a)	123. (c)	124. (d)	125. (b)	126. (b)	127. (d)	128. (d)	129. (d)	130. (a)
131. (c)	132. (b)	133. (c)	134. (c)	135. (d)	136. (d)	137. (a)	138. (c)	139. (d)	140. (d)
141. (c)	142. (d)	143. (d)	144. (a)	145. (a)	146. (d)	147. (d)	148. (a)	149. (b)	150. (d)
151. (b)	152. (c)	153. (a)	154. (a)	155. (d)	156. (c)	157. (b)	158. (d)	159. (b)	160. (c)
161. (a)	162. (c)	163. (d)	164. (d)	165. (a)	166. (c)	167. (b)	168. (a)	169. (d)	170. (c)
171. (b)	172. (c)	173. (b)	174. (a)	175. (c)	176. (d)	177. (d)	178. (c)	179. (a)	180. (a)

EXPLANATIONS

1. In pure Germanium,

   $n_i = n = p = 2.5 \times 10^{13}$

   When it is doped with arsenic, it donates an extra 10^{17} free electrons. So, number of electrons now are

   $n = (2.5 \times 10^{13} + 10^{17})$

   n_i (intrinsic concentration) will remain same. Using mass action law,

   $n \cdot p = n_i^2$

   $(2.5 \times 10^{13} + 10^{17}) \times p = (2.5 \times 10^{13})^2$

   $P = 6.25 \times 10^9$

2. Hall experiment is used to determine whether a semiconductor is n-type or p-type and the carrier concentration. Also by simultaneously measuring the conductivity σ, the mobility μ can be calculated.

3. Since silicon is an indirect band gap semiconductor so electron cannot fall directly to the valence band maximum but must undergo a momentum change as well as change in energy. So, senergy is released as heat along with light, so not used.

4. Band gap of elements at 300 k

   Si → 1.1 eV

   Ge → 0.67 eV

   Diamond (C) → 5.5 eV

7. $A_{\text{non-inverting}} = \left(1+\frac{R_F}{R_i}\right) = \left(1+\frac{10^5}{10^3}\right) = 101$

   $A_{\text{inverting}} = -\frac{R_F}{R_i} = -\frac{10^5}{10^3} = -100$

   ∴ The magnitude of non-inverting and inverting gains are nearly same.

8. Advantage of crystal oscillator
   (i) It is one of the most stable oscillator.
   (ii) High Q-factor of the range of 10^4 to 10^6 compared to 10^2 for LC oscillator.
   (iii) No ageing affect. Phase shift oscillator is used in audio range frequency.

9. **Difference between Butterworth and Chebyshev filter**

   The magnitude response (H | ω |) of Butterworth filter decreases with increase in frequency from 0 to ∞, while the magnitude response of the chebyshev filter fluctuates or show ripples in the pass band and stop band depending on the type of the filter.

10. In regulator circuit, output does not change due to the change in the load voltage or the input. But the output can change due to change of reference voltage.

11. Applying KVL in outor loop

$10 - 2.2K\, I_T - 3.5 = 0$

$$I_T = \frac{6.5}{2.2K} = \frac{65}{22} mA$$

Current through Zener Diode

$10 - 2.2K\, I_T + I_Z R_Z - 3.3 = 0$

$6.7 - 6.5 - 0.1 I_z = 0$

$0.2 - 0.1\, I_z = 0$

$$I_Z = \frac{0.2}{0.7},\ I_Z = 2\ mA$$

12. From the ckt, if 'D_1' is 'ON' Kirchoff's law is not satisfied

$\therefore$ D_1 is OFF and D_2 is ON

$\therefore$ I = 0 mA

13. Current will pass through its simplest path (or) low Resistance path. D_1 is become forward biased and D_2 is reverse biased

I = 0 mA

15. Max. Zener Current $= \frac{600}{24} = 25mA$

$$R = \frac{32 - 24}{25 \times 10^{-3}} = 320 \text{ ohm}$$

Maximum load current = 25 – 10 = 15 mA

16. $0 - 0.7 - I_b(\beta + 1) \times 6.3 = -10$ V

$$I_b = \frac{10 - 0.7}{81 \times 6.3} mA = 18.452\ \mu A$$

17. $$g_m = -\frac{2\, I_{DSS}}{V_P}\left(1 - \frac{V_{GS}}{V_P}\right)$$

$$1\ mAN = \frac{-2\, I_{DSS}}{-5}\left(1 - \frac{-3}{-5}\right)$$

$$= \frac{2}{5} \times \left(1 - \frac{3}{5}\right) I_{DSS}$$

$$\Rightarrow \quad I_{DSS} = \left(\frac{5}{2}\right)^2 mA$$

$$(g_m)_{max} = -\frac{2\, I_{DSS}}{V_P} = -\frac{2 \times 25}{4 \times (-5)} mAN$$

$= 2.5$ mAN

18. S.R. $= \omega V_m$

$62.8\ V/\ s = 2\pi f \times 10$

$$f = \frac{62.8 \times 10^6}{2\pi \times 10} = 1\ MHz$$

19. For emitter follower

$A_V \cong 1;\ A_I$ is high

$\Rightarrow A_p = A_V A_I$ is high $\Rightarrow P_{out} > P_{in}$

For transformer $P_{in} = P_{out}$

20.

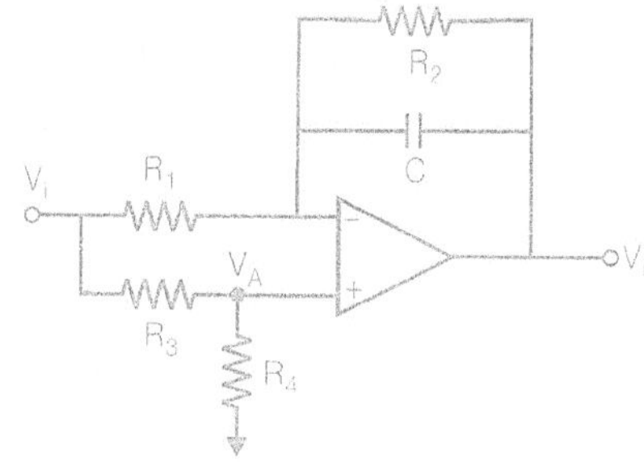

$$V_A = \frac{R_4}{R_4 + R_3} \times V_p$$

$$V_A = \frac{R_B}{R_B + R_B} \times V_i = \frac{V_i}{2}$$

Let V_i only on inverting terminal

$$V_{01} = \frac{-R_f}{R_1} V_i$$

Here, $$R_f = R_2 || \frac{1}{sC} = \frac{R_2}{1 + sCR_2}$$

Let V_i only on non-inverting terminal

$$V_{02} = \left(1 + \frac{R_f}{R_1}\right)\frac{V_i}{2}$$

$$V_0 = V_{01} + V_{02}$$

$$= \left(1 - \frac{R_f}{R_1}\right)\frac{V_i}{2}$$

Putting the value of R_f, we get

$$V_0 = \left(1 - \frac{R_2}{R_1(1 + sCR_2)}\right)\frac{V_i}{2}$$

Here $(R_1 = R_2 = R_A)$

$$\therefore \quad V_0 = \frac{1}{1 + \frac{1}{sCR_A}} \times \left(\frac{V_i}{2}\right)$$

So, it is a high pass filter.

21.

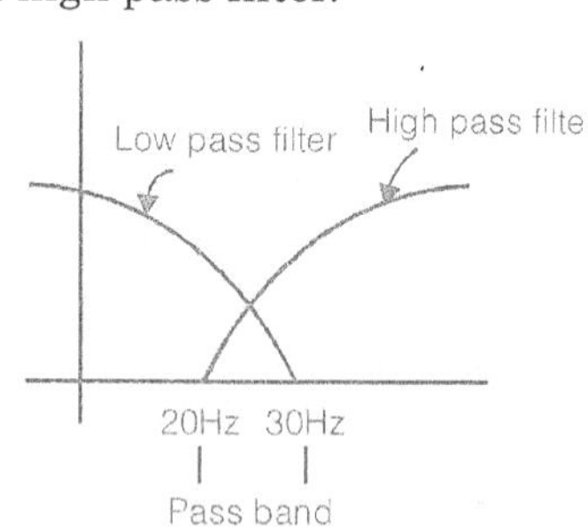

It is a band pass filter

23.

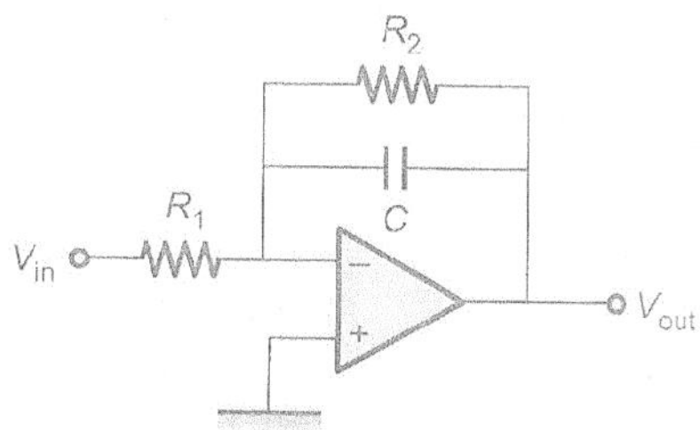

$$\frac{V_{out}}{V_{in}} = -\left[\frac{R_2 \cdot \frac{1}{j\omega C} \Big/ R_2 + \frac{1}{j\omega C}}{R_1}\right]$$

$$\frac{V_{out}}{V_{in}} = -\left[\frac{R_2}{R_1(R_2 j\omega C + 1)}\right]$$

So the system is a low pass filter.

25. Type = no. of poles at origin = 1

order = total no. of poles = 3

26. To achieve maximum power transfer.

27. $Z_i = \infty \rightarrow \boxed{\text{Voltage amplifier}} \rightarrow Z_0 = 0$

28. The collector resistance R_C is replaced by inductor where the dc resistance is low.

29. The coupling capacitor limits the low frequency response.

30. Cascode amplifier (CE-CB) is a special type which improves Z_i and A_V.

31. Transformer coupling will have higher voltage gain and impedance matching .

32. Cascading of several amplifiers result in increase in voltage gain and decrease in bandwidth.

37. Given data is,

$\beta = 300$

$I_e = 7.5\ \mu A$

$R_C = 1\ m\Omega$

$V_T = 25\ mV$

$$r_e = \frac{V_T}{I_E} = \frac{25\ mV}{7.5\ \mu A}$$

$$\boxed{r_e = 3.33\ k\Omega}$$

The voltage gain for the single ended output is

$$A = \frac{R_C}{2\,r_e}$$

$$A = \frac{1\ m\Omega}{2(3.33\ k\Omega)}$$

$$\boxed{A = 150}$$

The ac output voltage is

$V_{out} = AV_1$

$V_{out} = 150\ (7\ mV) = 1.05\ V$

The input impedance is

$Z_{in} = 2\ \beta r_e$

$Z_{in} = 2 \times 300 \times 3.33$

$Z_{in} = 2\ m\Omega$

38. CMRR(dB) = 20 log CMRR

$$\text{CMRR} = \text{anti}\log\frac{\text{CMRR dB}}{20}$$

$$\text{CMRR} = \text{anti}\log\frac{90\text{dB}}{20} = 31{,}600$$

$$A_{cm} = \frac{A}{\text{CMRR}} = \frac{200{,}000}{31{,}600}$$

$$\boxed{A_{cm} = 6.32}$$

The desired output component is

$V_{out\,1} = AVi$

$= 2000{,}000\ (1\ \mu A)$

$= 0.2\ V$

The common-mode output is

$V_{out\,2} = A_{cm}\ V_{icm}$

$= 6.32\ (1\ \mu V)$

$= 6.32\ mV$

V_{icm} = common-mode signal

V_i = Desired signal

The desired output is much larger than the common-mode otuput.

39. Given $A = 100$

$Z_i = 1\ k\Omega$

$Z_0 = 100\ \Omega$

$\beta = 0.99$

In voltage series, $Z_{if} = Z_i(1 + A\beta) = 1\ k\Omega\ (1 + 100 \times 0.99) = 1\ k\Omega\ (1 + 99) = 100\ k\Omega$

$$Z_{of} = \frac{Z_0}{1 + A\beta} = \frac{100\Omega}{(1 + 100 \times 0.99)} = \frac{100}{100} = 1\Omega$$

40. We have $R_i = 1\ k\Omega$, $\beta = 0.2$, $A = 50$

$$\text{Thus } R_{if} = \frac{R_i}{1 + A\beta} = \frac{1\ k\Omega}{1 + 50 \times 0.2} = \frac{1}{11}\ k\Omega$$

41. In current shunt $R_{if} = \dfrac{R_i}{1 + A\beta}$

$R_{of} = R_o(1 + A\beta)$

42. $\dfrac{\partial A_f}{A_f} = \dfrac{1}{1+A\beta}\dfrac{\partial A}{A}$;

$\dfrac{0.1}{100} = \dfrac{10}{1000}\left[\dfrac{1}{1+1000\beta}\right]$

$1 + 1000\beta = 10, \beta = \dfrac{9}{1000}$

43. Open loop gain $A = -100$

Negative feedback $\beta = \dfrac{-4}{100}$

Closed loop gain $A_f = \dfrac{A}{1+A\beta}$

$= \dfrac{-100}{1+(-100)\left(\dfrac{-4}{100}\right)} = -20$

44. Barkhausen cirterion for sustained oscillations $|A\beta| \geq 1$ and phase shift must be 2π.

$A = 1 + \dfrac{R_2}{R_1}$; $\beta(f) = \dfrac{1}{6}\angle 0°$

For sustained oscillation

$6 = 1 + \dfrac{R_2}{R_1}$

(or) $R_2 = 5R_1$

45. Given that

$C_1 = 0.005\ \mu F$,

$C_2 = 0.01\ \mu F$,

$L = 100\ \mu H$

The frequency of oscillations of a colpitts oscillator is given by

$$f = \frac{1}{2\pi\sqrt{LC}}$$

$$C = \frac{C_1C_2}{C_1+C_2}$$

Now, $C = \dfrac{0.005 \times 0.01}{(0.005+0.01)}$

$\boxed{C = 0.00333\ \mu F}$

$$f = \frac{1}{2\pi\sqrt{LC}}$$

$$f = \frac{1}{2\pi[100\times10^{-6}\times0.0033\times10^{-6}]^{1/2}}$$

$\boxed{f = 277.05\ kHz}$

46. Given that:

$C_1 = C_2 = 500$ PF and $L = 1$ mH

The equivalent capacitance for a colpitt's oscillator is given by

$C = \dfrac{C_1C_2}{C_1+C_2}$, $C = \dfrac{500\times500}{500+500}$

$\boxed{C = 250\ PF}$

The frequency of oscillations is given by

$$f = \frac{1}{2\pi\sqrt{LC}}$$

$$f = \frac{1}{2\pi\sqrt{1\times10^{-3}\times250\times10^{-12}}}$$

$\boxed{f = 318.3\ kHz}$

47. Here, effective inductance

$L = L_1 + L_2 + 2M$

$L = (1\times10^{-3}) + (100\times10^{-6}) + (2\times50\times10^{-6})$

$C = 100\ pF = 100^{-10}\ F$

Frequency of oscillation,

$$f_0 = \frac{1}{2\pi\sqrt{LC}}$$

or $f_0 = \dfrac{1}{2\pi\sqrt{1200\times10^{-6}\times10^{-6}}}$

$\boxed{f_0 = 459\ kHz}$

48. Given that series resistance,

$R_S = 5k\Omega = 5{,}000\ \Omega$

Series inductance, $L_S = 0.8$ H

series capacitance, $C_S = 0.08$ pF

$= 8 \times 10^{-14}$ F

Parallel capacitance, C_p or C_m

= 1.0 pF = 1 × 10–12 F

Series Resonant frequency,

$$f_s = \frac{1}{2\pi\sqrt{L_SC_S}}$$

$$f_s = \frac{1}{2\pi\sqrt{0.8\times8\times10^{-4}}}$$

$\boxed{f_s = 629\ kHz}$

49. We have $f_s = \dfrac{1}{2\pi\sqrt{L_SC_S}}$

Given that

$L_S = 800\ mH = 0.8\ H$

$C_S = 0.01\ pF = 0.01 \times 10^{-12}F$;

$C_P = 290\ pF$

therefore,

$$f_s = \frac{1}{2\pi\sqrt{0.8 \times 0.01 \times 10^{-12}}}$$

$$\boxed{f_s = 1779.40 \text{ kHz}}$$

50. Given that

$C_1 = 120$ pF

$C_2 = 1.5$ nF

$C_3 = 15$ pF

$L_3 = 10$ mH.

Equivalent capacitance,

$$\frac{1}{C} = \frac{1}{C_1} + \frac{1}{C_2} + \frac{1}{C_3}$$

$$\frac{1}{C} = \frac{1}{120} + \frac{1}{1500} + \frac{1}{15}$$

The frequency of oscillations is given by,

$$f = \frac{1}{2\pi\sqrt{CL_3}}$$

$$f = \frac{1}{2\pi\sqrt{13.2 \times 10^{-12} \times 10 \times 10^{-6}}}$$

$$\boxed{f = 13.84 \text{ MHz}}$$

If we calculate the approximate value of oscillator frequency then

$C = C_3 = 15$ pF

Hence,

$$f = \frac{1}{2\pi\sqrt{C_3L_3}}$$

$$f = \frac{1}{2\pi\sqrt{15 \times 10^{-12} \times 10 \times 10^{-6}}}$$

$$\boxed{f = 12.99 \text{ MHz}}$$

51. The frequency of oscillations is given by

$$f = \frac{1}{2\pi\sqrt{(L_1 + L_2)C_3}}$$

$$f = \frac{1}{2\pi\sqrt{100 \times 10^{-6} \times 100 \times 10^{-12}}}$$

$$\boxed{f = 1.59 \text{ MHz}}$$

52. Given that:

$R_1 = R_2 = R_3 = 800 \text{ k}\Omega = 8 \times 10^5\ \Omega$

$C_1 = C_2 = C_3 = C = 100 \text{ pF} = 10^{-10}$ F

Frequency of oscillation,

$$f_0 = \frac{1}{2\pi RC\sqrt{6}} = \frac{1}{2\pi \times 8 \times 10^5 \times 10^{-10} \times \sqrt{6}} = 812 \text{ Hz}$$

54. This is due to heavy doping.

55. Tunnel diode has heavily doped p and n regions.

56. Impedance matching is required to achieve maximum power transfer, because if there is no impedance matching then power transfer from one stage to other will be poor.

57. In PIN diode intrinsic layer decides the reverse blocking capacity.

58. Negative feedback increases the BW by a factor $(1+A\beta)$. Where as it reduces the gain.

73. $T = (x+y)(x+\bar{y})(\bar{x}+y)$

$= (xx + x\bar{y} + xy + y\bar{y})(\bar{x}+\bar{y})$

$= (x + x\bar{y} + xy + 0)(\bar{x}+y) \quad (\because xx = x \text{ and } y\bar{y} = 0)$

$= (x + x\bar{y} + xy)(\bar{x}+y)$

$= x(1 + \bar{y} + y)(\bar{x}+y)$

$= x(1+1)(\bar{x}+y) \qquad (\because 1+1=0)$

$= x(\bar{x}+y) = x\bar{y} + xy$

$= 0 + xy \qquad (\because x\bar{y} = 0)$

$= xy$

74. For low frequency $Z_C = \frac{1}{j\omega C} \Rightarrow \infty$

So, output $V_0 = -\frac{R_f}{R_1}V_i$

We will get the output for high frequency

$$Z_C = \frac{1}{j\omega C} \Rightarrow 0$$

$\therefore$ output $V_0 \cong 0$

so cut the high frequency

so it is a low pass filter

75. According to the question $V_{CB} = 0.2$V i.e. the collector to base junction is reverse bias and $V_{BE} = 0.7$ i.e. base to emitter junction is forward bias. this condition meet in active region.

76. In the coding, letters of the word are reversed and then first two letters and last two letters exchange their positions. Hence, the code for UNDER is ERDUN.

77. Coding follows as:

N → 5

O → 2

V → 7

E → 0

L → 8

T $\rightarrow$ 1

Y $\rightarrow$ 3

and

E $\rightarrow$ 0

S $\rightarrow$ 4

S $\rightarrow$ 4

A $\rightarrow$ 6

Y $\rightarrow$ 3

Hence, the code for SEVEN is 40705.

78. Last two letters in reverse order become first two letters and remaining 3 letters come in the same order.

Hence, the code for FLIRT is TRFLI.

79. Let 'x' be the required percentage. Then,

$$\frac{35}{100} \times 1180 + \frac{65}{100} \times 860 = 413 + 559$$

$$= 972 = \frac{x}{100} \times 8100$$

$$\Rightarrow x = \frac{97200}{8100} = 12.$$

80. $30 \div 5 \times 4 + 3 - 2 = 6 \times 4 + 3 - 2 = 25.$

81. Option (A): $18 \div 6 + 4 < 30 - 11 \times 2$ (Correct)

Option (B): $18 - 6 \div 4 > 30 + 11 \times 2$ (Not correct)

Option (C): $18 - 6 \div 4 + 30 < 11 \times 2$ (Not correct)

Option (D): $18 \div 6 > 4 + 30 - 11 \times 2$ (Not correct)

82. $16 \div 2 + 9 = 3 \times 4 + 5$

83. The smallest planet in our solar system is 'mercury'. Similarly, the smallest bone in the human body is 'stapes'. Hence, option (B) is the correct answer.

84. The outermost layer of a plant cell is called a cell wall. The outermost layer of an animal cell is called a cell membrane. Hence, option (D) is the correct answer.

85. The sound made by a bear is called 'growl'. 'Creak' refers to the sound made by an old door when it is opened. Hence, option (B) is the correct answer. 'Chime' is the sound made by bells. 'Splatter' is the sound of wet object hitting something hard. 'Jangle' refers to a harsh ringing sound.

86. The two ends of a battery are called terminals. Similarly, the two ends of a magnet are called poles. Hence, option (*b*) is the correct answer.

87. A group of lions is called 'pride'. Similarly, a group of geese on land is called 'gaggle'. Hence, option (*c*) is the correct answer. When in flight, it is called 'skein'.

88. $(4 \times 4 - 1) : (4 \times 4 \times 4 + 1) :: (5 \times 5 - 1) : (5 \times 5 \times 5 + 1)$

Hence, required number is 126.

89. In all the fractions except (C), the numerator is less than the denominator.

90. Other than wife all are blood relatives. Hence, option (B) is the correct answer.

91. A O Hume formed the Indian National Union in 1884.

92. Dermatology focuses on the diagnosis and treatment of conditions related to the skin, hair and nails.

93. The Caribbean Sea is an area around North, Central and South America and the island arcs Antilles.

94. The Treaty of Versailles was one of the peace treaties at the end of World War I. It ended the state of war between Germany and the Allied Powers.

95. Charles Darwin suggested that "survival of the fittest" was the basis for organic evolution (the modification of living things with time).

96. Frogs are members of the zoological class called Amphibia.

97. The hormone called as Human Chorionic Gonadotropin (better known as HCG) is produced during pregnancy. It is also known as pregnancy test hormone.

98. Michael Faraday's law of induction establishes a relationship between a changing magnetic field and the electric field.

99. The Ghughua Fossil National Park of Madhya Pradesh is a unique destination in India, which has been blessed with a priceless treasure trove of plant fossils.

102. Light bulbs are manufactured with tungsten filaments because tungsten has an abnormally high melting temperature.

103. AgI that darkens on exposure to light and is used in photographic emulsions, rainmaking and medicine, especially as an antiseptic.

104. A windbreak or shelterbelt is a plantation usually made up of one or more rows of trees or shrubs planted in such a manner as to provide shelter from the wind.

105. Aden International Airport Country: Yemen

Ahwaz Airport Country: Iran

Amami Airport Country: Japan

106. $147^2 = 21609$

Hence, the required number is 19.

Note: In order to find the square root of a five digit number, try to find the nearest integral square root of the first three digits which is 14. Thus, the required square root will lie between 140 and 150. Now, the square of 145 is nothing but |14 × 15|25 = 21025 and that is how we reach the required number easily which is 147.

107. Let the total work be $\frac{27 \times 7}{2}$ units i.e. LCM of $\frac{27}{2}$, $\frac{21}{4}$ and $\frac{9}{2}$.

⇒ Work done by A and B in one day = 7 units, B and C in one day = 18 units and A and C in one day = 21 units.

So, work done by A, B and C in one day = 23 units.

Hence, required time = $\frac{189}{46}$ days.

108.

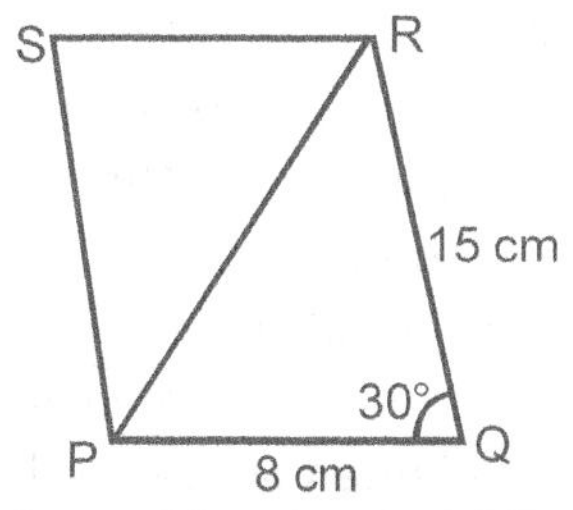

Area of the triangle PQR

$= \frac{1}{2} \times PQ \times QR \times \sin 30°$

$= \frac{1}{2} \times 8 \times 15 \times \frac{1}{2} = 30 \text{ cm}^2.$

Hence, area of the parallelogram = $30 \times 2 = 60 \text{ cm}^2$.

109. $\frac{5}{6} MP = \frac{7}{6} CP \Rightarrow MP = \frac{7}{5} CP$

Hence, required percent = 40%.

110. $a + b = 28 \times 2$ and

$a - b = 17 \times 2$

$\Rightarrow a = 45$ and $b = 11$

Hence, $a \times b = 45 \times 11 = 495$.

111. Let the number of teachers who joined the group be x.

Sum of the ages of 78 students = 78 × 25.5 = 1989

$\Rightarrow 1989 + 68x = (78 + x) \times 29$

Hence, x = 7.

112. Required percent = $\frac{1200 - 900}{900} \times 100$

= 33.33%.

113. Let the salary of B is Rs. 80

⇒ A's salary = Rs. 110, C's salary = Rs. 88 and D's salary = Rs. 132

Hence, A's salary is 16.67% less than that of D's salary.

114. Speed of the boat upstream = $\frac{375}{25} \times \frac{18}{5} = 54$ km/hr

and speed of the boat downstream

$= \frac{675}{27} \times \frac{18}{5} = 90$ km/hr.

⇒ Speed of the boat in still water

$= \frac{90 + 54}{2} = 72$ km/hr.

Hence, required distance = 72 × 4.5 = 324 km.

115. Simple interest on Rs. 200 for 2 months is Rs. 2 or Rs. 12 for 12 months which gives the rate of interest as 6%.

Hence, required interest = $\frac{27500 \times 6.09}{100}$

= Rs. 1,674.75.

116. $\frac{15}{4} + 2x - 4x^2 = 4 - \left(2x - \frac{1}{2}\right)^2$

Hence, required value is 4.

117. Satisfying each of the equations given in the question with a = b = c = – 1, we get the required answer as – 2.

118. $(a + b)^3 = 8^3$

$\Rightarrow a^3 + b^3 + 3ab(a + b) = 512$

$\Rightarrow 152 + 3 \times 8 \times ab = 512$

$\Rightarrow ab = 15$

Again, $(a + b)^2 = 64$

$\Rightarrow a^2 + b^2 = 34$

$\Rightarrow a - b = 2.$

119. $2x\left(\frac{2b}{a^2 - b^2}\right) = \frac{2b}{a + b} \Rightarrow x = \frac{a - b}{2}$

Also, going through the options we get the required value of x as $\frac{a-b}{2}$.

120.

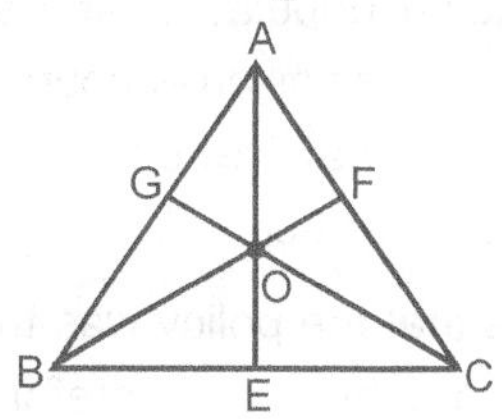

O is the centroid of the triangle and AO : OE = 2 : 1

Hence, the required ratio = 2 : 3.

121. Palaeontology is the branch of science concerned with fossil animals and plants. Zoology is the scientific study of the behaviour, structure, physiology, classification and distribution of animals. Mythology is a collection of myths, especially one belonging to a particular religious or cultural tradition. 1nthropology is the study of humankind.

122. 'Ergo' means work and 'mania' means crazy. Therefore, ergomania means excessive desire to work; workaholism. Hippomania means obsession with horses, phaneromania is the habit of biting one's nails and sophomania is a delusion that one is incredibly intelligent.

123. Plutocracy is the governance of rich people. Oligarchy is a kind of governance where power resides in the hand of a few people or the members of a family. A republic is a state in which supreme power is held with the people and their elected representatives. Gerontocracy is the kind of rule where power is with people who are relatively older than most adult population of the country.

124. 'Sets in' means 'begins'.

125. The sentence needs no improvement. 'Trumped up' refers to a concocted story.

126. 'Take in' means to comprehend.

127. 'In' is the correct preposition.

128. The verb should be 'forced' because the other verbs are in the past tense.

129. Refer to the following line from the passage – "The controversial one-child policy, which came into effect in 1980 restricted most couples to only a single offspring and after its implementation authorities argued that it was a key contributor to China's economic success." This makes option (d) correct. Option (c) is incorrect because it is not the opinion of the masses.

130. Refer to the following line from the passage – "In the wake of realisation that the policy is leaving the country with an ageing population and shrinking workforce, the move comes as an encouraging step towards greater personal freedom in China." Therefore, option (A) is correct.

131. The passage states that the policy was thought to be extremely harsh or inhuman. Therefore, option (c) is correct.

132. Option (b) is the answer because the passage states that ". 3ouples who violated the one-child policy faced many punishments, from forced abortions to harassment to fines and loss of employment."

133. According to the passage, it saved 400 million births. Therefore, option (c) is the answer.

137. The word 'laudatory' is used to describe a speech or a piece of writing that is expressing praise and commendation. The apt adjective in the above context is 'laudable', which is used to describe something that deserves praise or admiration instead. Therefore, the correct sentence would be, "With a laudable performance on the stage, the Delhi Branch made a befitting entry into the dance competition."

138. 'Not only' should be followed by 'but also' and not 'but' to combine two negative or two positive qualities of a person. In this case, the two negative qualities referred to are – a con and a lecher. Therefore, the correct sentence would be, "The police reported that the most renowned actor of the century was not only a con but also a lecher."

139. The given sentence has no error.

149. Detention means holding a person in temporary custody while awaiting trial. Circumvention means to avoid defeat, capture or any other unpleasantness by outwitting the problem with artfulness and deceit. Absolution means complete forgiveness while clemency means lessening someone's punishment by dealing with that person leniently. Therefore, option (b) is the answer.

150. A novella is a fictional narrative prose longer than a short story and shorter than a novel. A fable is a short fictional story, typically with animals as characters, conveying a moral. A parable is a simple short story used to illustrate a moral or a spiritual lesson, as told by Jesus in the Gospel. An anecdote is a short amusing story about a real incident or person.

151. Saccharine means excessively sweet or sentimental. Flamboyant describes something that is very confident, stylish and attractive. Piquant describes something having pleasantly sharp flavor. Acrid means unpleasantly bitter or pungent.

152. The comparative degree should be used.

153. The sentence needs no improvement.

154. The comparative degree of 'bad' should be used.

155. 'Exempted' is always followed by the preposition 'from'.

156. 'Succumb to' refers to 'yielding to.'

157. Refer to the fourth sentence of the second paragraph.

158. Refer to the last sentence of the first paragraph.

159. Refer to the second last sentence of the passage.

160. Refer to the first sentence of the passage.

161. Refer to the second sentence of the second paragraph of the passage.

165. According to parallelism, here the structure of the two actions given before and after 'and' should be the same. Therefore, the correct sentence should be, "Making a movie and writing about a movie are two separate domains needing two different sets of abilities."

168. The word 'perfect' itself is a superlative adjective. To qualify it with 'most' is redundant and therefore incorrect.

167. The word 'who's' is a short form of 'who is' and therefore inappropriate in this context. The correct word that should replace it is 'whose', a possessive pronoun, which in this context denotes 'journalist's'.

168. 'Continued' and 'stayed' correctly fill the blanks.

169. A past tense in the principal clause may be followed by a present tense in the subordinate clause when the subordinate clause expresses a universal truth.

171. Since the given sentence means that the mother hated the person's action of going to Poona and so, therefore the word 'going' should be preceded by the possessive pronoun 'my' and not 'me'. The correct sentence would be, "My mother vetoed the idea of my going..."

172. Since Gangaram's action of rehearsing the answers would have occurred before he was asked the questions, suggests that the former action - of rehearsing - should be written in past perfect tense to show that order. Hence, the correct sentence would be - "Even before I asked him questions, Gangaram had answers ready, almost as if he had rehearsed them."

173. It is inappropriate to use a question word in a sentence that is not interrogative. Hence, the correct phrase would be, "...random people have asked me when I would be dubbing my next film."

PRACTICE PAPER-4

ELECTRONICS ENGINEERING

Time Allowed : 2:15 Hours **Maximum Marks : 180**

Read the following instructions carefully before you begin to answer the questions. This Booklet contains questions in English.

INSTRUCTIONS

1. This Booklet contains 180 questions in total comprising of the following five parts:

Part – I:	Domain Knowledge	[1 – 75 Questions]
Part – II:	General Intelligence and Reasoning	[76 – 90 Questions]
Part – III:	General Awareness	[91 – 105 Questions]
Part – IV:	Quantitative Aptitude	[106 – 120 Questions]
Part – V:	General English	[121 – 180 Questions]

2. All questions are compulsory and carry 1 mark each.

3. Before you start to answer the questions you must check this Booklet to ensure that it contains all the pages and see that no page is missing or repeated.

4. For every wrong answer 1/3 marks will be deducted.

5. You are supplied with the OMR-Sheet separately.

6. Answers must be indicated by completely blackening one of the corresponding ovals on OMR-Sheet against the relevant question number, by black/blue ball pen only. Answers which are not indicated by black/blue ball pen will not be awarded any mark.

7. The OMR-Sheet must be handed over to the Invigilator before you leave the Examination Hall.

8. Failure to comply with any of the above instructions will render a candidate liable to such action/penalty as may be deemed fit.

PRACTICE PAPER – 4

DOMAIN KNOWLEDGE

1. In the given circuit shown the three voltmeter readings are

   $V_1 = 220V, V_2 = 122V, V_3 = 136V$

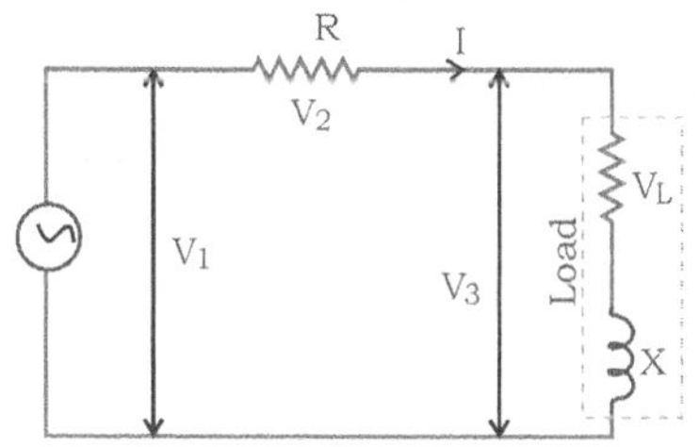

   The power factor of the load is

   (a) 0.45 (b) 0.50
   (c) 0.55 (d) 0.60

2. Calculate the power consumed by the load the given data is

   Power factor = 0.45

   Voltage across the load = 136 V

   Load Resistance $R_L = 5\ \Omega$

   (a) 360 W (b) 850 W
   (c) 750 W (d) 700 W

3. A moving coil of a meter has 100 turns and a length and depth of 10 mm and 20 mm respectively. It is positioned in a uniform radial flux density of 200 mT. The coil carries a current of 50 mA. the torque on the coil is

   (a) 200 μN-m (b) 100 μN-m
   (c) 2 μN-m (d) 1 μN-m

4. A moving iron ammeter produces a full scale torque of 240 μN-m with a deflection of 120º at a current of 10A. the rate of change of self inductance (μH/radian) of the instrument of full scale is

   (a) 2.0 μH/radian (b) 4.8 μH/radian
   (c) 12.0 μH/radian (d) 114.6 μH/radian

5. A metal strain gauge has factor of two Its nominal resistance is 120 ohms. It under goes a strain of 10^{-5} the value of change in resistance in response to the strain is

   (a) 240 ohms (b) 2×10^{-5} ohms
   (c) 2.4×10^{-3} ohms (d) 1.2×10^{-3} ohms

6. A (0-10) mA PMMC ammeter reads 4 mA in a circuit. Its bottom control spring snaps suddenly, the meter will now read nearly

   (a) 10 mA (b) 8 mA
   (c) 2 mA (d) 0

7. Two 100 mA full scale PMMC meters are employed to construct a 10V and a 100V full scale voltmeter. These meter will have figures of merits (sensitivities) as

   (a) 10 kΩ/V and 10 kΩ/V
   (b) 100 kΩ/V and 10 kΩ/V
   (c) 10 kΩ/V and 100 kΩ/V
   (d) 10 kΩ/V and 1 kΩ/V

8. The bridge method commonly used for finding mutual inductance is

   (a) Heaviside Campbell bridge
   (b) Schering bridge
   (c) De sauty bridge
   (d) Wien bridge

9. The voltage phasor of a circuit is $10\angle 15°$ V and the current phasor is $2\angle -45$ A. The active and the reactive powers in the circuit are

   (a) 10 W and 17.32 VAR
   (b) 5 W and 8.66 VAR
   (c) 20 W and 60 VAR
   (d) $20\sqrt{2}$ W and $10\sqrt{2}$ VAR

10. The minimum number of wattmeter(s) required to measure 3-phase, 3-wire balance or unbalanced power is

    (a) 1 (b) 2
    (c) 3 (d) 4

11. An energy meter having meter constant of 1200 revolutions kWh, makes 20 revolutions in 30 seconds for a constant load. the load, in kW is

    (a) 2 kW (b) 5 kW
    (c) 1 kW (d) 10 kW

12. Which one of the following is the most suitable definition of ARRAY?

    (a) It is a collection of items which share a common name
    (b) It is a collection of items which share a common name
    (c) It is a collection of items of the same type and storage class which share common name and occupy consecutive memory locations
    (d) It is just a collection of unordered items

13. Handshaking mode of data transfer is
(*a*) synchronous data transfer
(*b*) asynchronous data transfer
(*c*) interrupt driven data transfer
(*d*) level mode of DMA data transfer

14. Output of the assemble in machine code is referred to as
(*a*) object program (*b*) source program
(*c*) Macroinstruction (*d*) symbolic addressing

15. In a microprocessor the address of the next instruction to be excited is stored in
(*a*) stack pointer
(*b*) address latch
(*c*) program counter
(*d*) general purpose register

16. In Intel 8085 microprocessor ALE signals mode high to
(*a*) enable data bus to be used as low order address bus
(*b*) to latch data D_0-D_7 from data bus
(*c*) to disable data bus
(*d*) to achieve all the functions listed above

17. In PAM the carrier wave consists of a periodic train of rectangular pulses and the carrier frequency is equal to
(*a*) the bandwidth of the modulating signal
(*b*) the sampling rate of the modulating signal
(*c*) at least ten times greater than the bandwidth of the modulating signal
(*d*) twice the bandwidth of the modulating signal

18. A 25 MHz carrier is modulated by a 400 Hz audio sine wave. if the carrier voltage is 4V and the maximum frequency deviation is 10 kHz. calculated the modulation index.
(*a*) 25 (*b*) 30
(*c*) 40 (*d*) 55

19. The antenna current of an AM transmitter is 7.5 A, when only the carrier is sent, but it is increases to 8.65A , when the carrier is sinusoidally modulated. Find the percentage modulation.
(*a*) 95.8% (*b*) 55%
(*c*) 81.2% (*d*) 74.5%

20. If the frequency deviation is 30 KHz and modulating signal is 5 KHz. calculated percent modulation
(*a*) 200% (*b*) 600%
(*c*) 100% (*d*) 350%

21. What is the time constant of the circuit below?

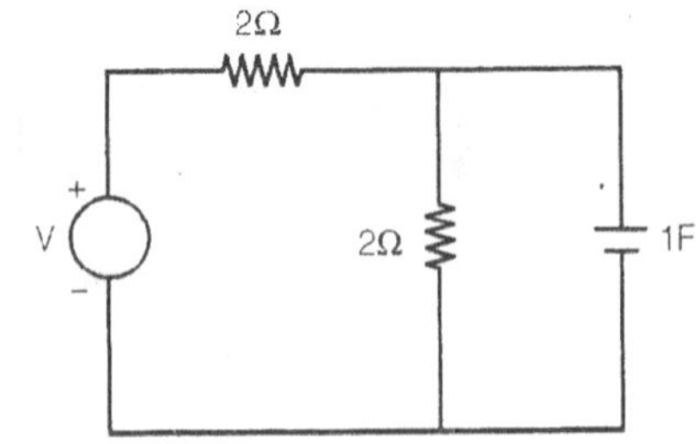

(*a*) 0.5s (*b*) 1s
(*c*) 2s (*d*) 4s

22. The open circuit impedance of a certain length of a loss less line is 50Ω. the short circuit impedance of the same line is also 50Ω. The characteristic impedance of the line is
(*a*) $100\sqrt{2}\Omega$ (*b*) 5Ω
(*c*) $100/\sqrt{2}\Omega$ (*d*) 100Ω

23. A medium behaves like dielectric when the
(*a*) displacement current is just equal to the conduction current
(*b*) displacement current is less than the conduction current
(*c*) displacement current is much greater than the conduction current
(*d*) displacement current is almost negligible

24. A loss-less transmission line with characteristic impedance of 600 ohms is terminated in a purely resistive load of 900 ohms. The reflection coefficient is
(*a*) 0.2 (*b*) 0.5
(*c*) 0.667 (*d*) 1.5

25. For a conducting medium with conductivity σ, permeability μ and permittivity ε, the skin depth for an electromagnetic signal at an angular frequencyω is proportional to
(*a*) σ (*b*) $1/\omega$
(*c*) $1/\sqrt{\sigma}$ (*d*) $1/\mu$

26. An electromagnetic field is said to be conservative when
(*a*) $\nabla^2 E = \mu \in (\partial^2 E/\partial t^2)$
(*b*) $\nabla^2 H = \mu \in (\partial^2 H/\partial t^2)$
(*c*) Curl of the field is zero
(*d*) Divergence of the field is zero

27. If $\overline{A}$ and $\overline{J}$ are the vector potential and current density vectors associated with a coil, then $\int(\overline{A}.\overline{J})dv$ has the units of
(*a*) flux-linkage (*b*) power
(*c*) energy (*d*) inductance

28. What is the value of total electric flux coming out of a closed surface?

(*a*) Zero

(*b*) Equal to volume charge density

(*c*) Equal to the total charge enclosed by the surface

(*d*) Equal to the surface charge density

29. Which one of the following statements is correct? For a lossless dielectric medium, the phase constant for a travelling wave, β is proportional to

(*a*) ε_r (*b*) $\sqrt{\varepsilon_r}$

(*c*) $1/\varepsilon_r$ (*d*) $1/\sqrt{\varepsilon_r}$

30. In a lossless medium, the intrinsic impedance $\eta = 60\pi$ and $\mu_r = 1$. What is the value of the dielectric constant ε_r?

(*a*) 2 (*b*) 1

(*c*) 4 (*d*) 8

31. What is the attenuation constant α for distortionless transmission line?

(*a*) $\alpha = 0$ (*b*) $\alpha = R\sqrt{\frac{C}{L}}$

(*c*) $\alpha = R\sqrt{\frac{L}{C}}$ (*d*) $\alpha = \sqrt{\frac{RL}{C}}$

32. Which one of the following is the correct expression for the propagation constant in a transmission line?

(*a*) $(R - j\omega L)(G - j\omega C)$ (*b*) $\sqrt{\frac{(R - j\omega L)}{(G - j\omega C)}}$

(*c*) $\sqrt{\frac{(R + j\omega L)}{(G + j\omega C)}}$ (*d*) $\sqrt{(R + j\omega L)(G + j\omega C)}$

33. Which one of the following statements is correct? If the Fermi level lies midway between the conduction and valence bands, then the semiconductor is

(*a*) intrinsic (*b*) extrinsic

(*c*) p-type (*d*) n-type

34. Six capacitors of different capacitances C_1, C_2, C_3, C_4, C_5 and C_6 are connected in series. $C_1 > C_2 > C_3 > C_4 > C_5 > C_6$. What is the total capacitance almost equal to?

(*a*) C_1 (*b*) C_3

(*c*) C_4 (*d*) C_6

35. Where is the Laplace's equation valid?

(*a*) Only in free space

(*b*) Only in conductors

(*c*) Only in charge free dielectric regions

(*d*) Only in cavities bounded on all sides by conducting walls

36. The latching current in the below circuit is 4 mA The minimum width of the gate pulse required to turn on the thyristor is

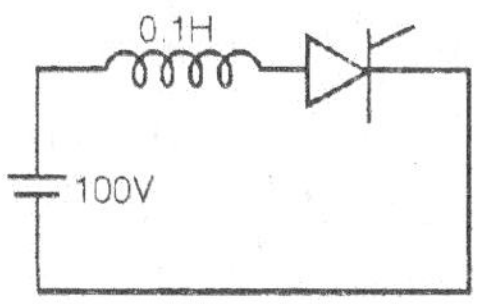

(*a*) 6 μs (*b*) 4 μs

(*c*) 2 μs (*d*) 1 μs

37. The snubber circuit is used in thyristor circuits for

(*a*) triggering (*b*) dv/dt protection

(*c*) di/dt protection (*d*) phase shifting

38. It is preferable to use a train of pulse of high frequency for gate triggering of SCR in order to reduce

(*a*) dv/dt problem

(*b*) di/dt problem

(*c*) the size of the pulse transformer

(*d*) the complexity of the firing circuit

39. Output of the assembler in machine codes is referred to as

(*a*) Object program (*b*) Source program

(*c*) flag instruction (*d*) Symbolic addressing

40. The cycle required to fetch and execute an instruction in a 8085 micro-processor is which one of the following?

(*a*) Clock cycle (*b*) Memory cycle

(*c*) Machine cycle (*d*) Instruction cycle

41. Assume that D_1 and D_2 in figure are ideal diodes. The value of the current I is

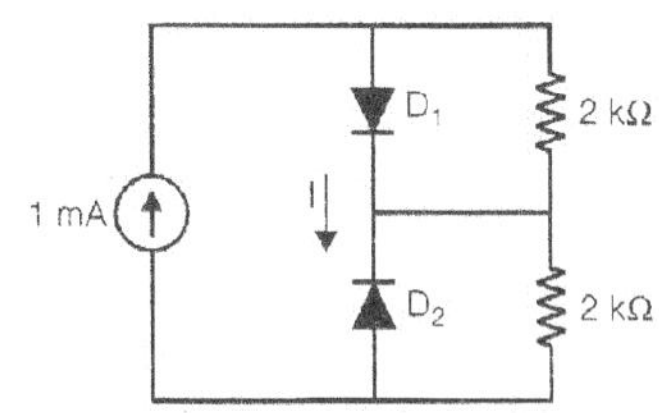

(*a*) 0 mA (*b*) 0.5 mA

(*c*) 1 mA (*d*) 2 mA

42. The current through the Zener diode in figure is

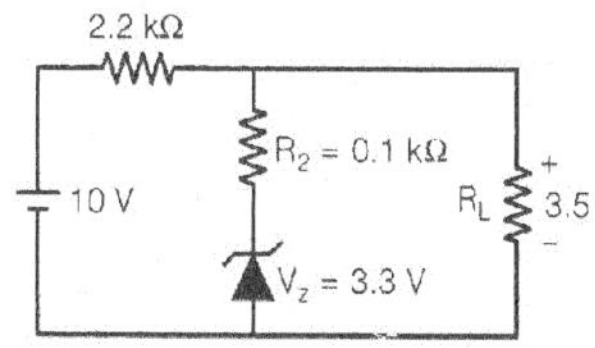

(*a*) 33 mA (*b*) 3.3 mA

(*c*) 2 mA (*d*) 0 mA

43. In a half-wave rectifier. If an ac supply is 60 Hz, then what is the ac ripple at output?

(a) 30 Hz (b) 60 Hz

(c) 120 Hz (d) 15 Hz

44. A half-wave rectifier having a resistance load of 1 kΩ rectifiers an ac voltage of 325 V peak value and the diode has a forward resistance of 100Ω. What is the RMS value of the current?

(a) 295.4 mA (b) 94 mA

(b) 147.7 mA (d) 208 mA

45. In a centre tap full-wave rectifier, 100 V is the peak voltage between the centre tap and one end of the secondary. What is the maximum voltage across the reverse biased diode?

(a) 200 V (b) 141V

(c) 100V (d) 86V

46. The circuit shown below, the thevenin's equivalent circuit of a centre tapped full-wave rectifier with diode forward resistance R_f = 100Ω, transformer secondary coil resistance R_s = 30Ω, peak input voltage V_m = 10 V. What are the values of V and R_0 respectively.

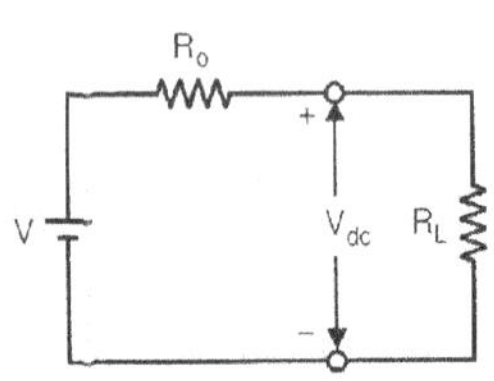

(a) 10 V, 100Ω (b) 6.36 V, 130Ω

(c) 6.36 V, 115Ω (d) 4.54 V, 130Ω

47. Which one of the following statements is correct. An ideal regulated power supply should have

(a) 100% regulation (b) 50% regulation

(c) 0% regulation (d) 75% regulation

48. In the circuit shown in the figure the, current flowing through resistance of 100Ω would be

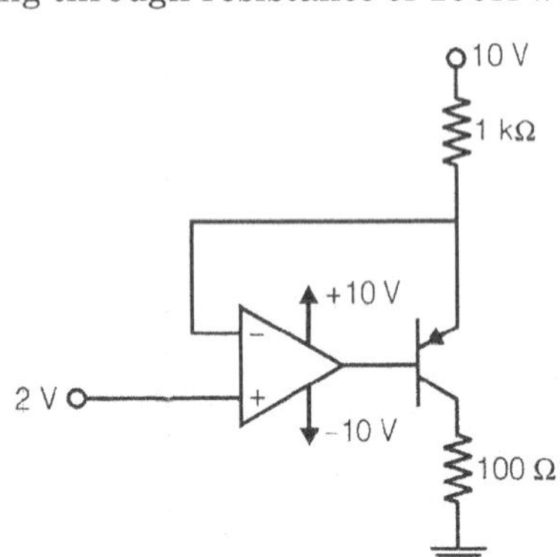

(a) 8 mA (b) 10 mA

(c) 20 mA (d) 100 mA

49. Z_L = 200Ω and it is desired that Z_i = 50Ω The quarter wave transformer should have a characteristic impedance of

(a) 100Ω (b) 40Ω

(c) 10000Ω (d) 4Ω

50. Refractive index of glass is 1.5. Find the wavelength of a beam of light with a frequency of 10^{14} Hz in glass. Assume velocity of light is 3 × 10^8 m/sec in vacuum.

(a) 4 m (b) 3 m

(c) 2 m (d) 1 m

51. The output resistance of the circuit at port AB is

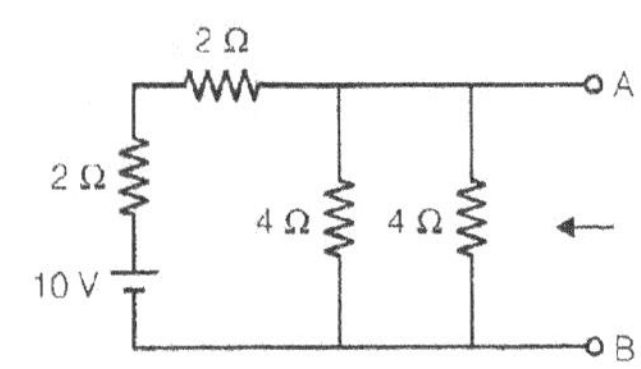

(a) 1Ω (b) 1.2 Ω

(c) 1.33 Ω (d) 1.5 Ω

52. The input impedance of short-circuited line of length l where $\lambda/4 < l < \lambda/2$, is

(a) resistive (b) inductive

(c) capacitive (d) none of the above

53. Find the radiation resistance of an antenna of length λ/10 meter?

(a) 8 Ω (b) 780 Ω

(c) 7.8 Ω (d) 78 Ω

54. The velocity of electromagnetic wave in a good conductor is

(a) 3 × 10^8 m/s

(b) more than 3 × 10^8 m/s

(c) very low

(d) high

55. A (75-j 40) W load is connected to a co-axial line of z_0 = 75ω at 6 MHz. The load matching on the line can be accomplished by connecting

(a) a short circuit stub at the load

(b) an inductance at the load

(c) a short circuit stub at a specific distance from the load

(d) a capacitance at a specific distance from the load

56. The electric field on the surface of a perfect conductor is 2V/m. The conductor is immersed in water with ε =80 ε_0. The surface charge density on the conductor is (($\varepsilon_0 = 10^{-9}/36\pi$)F/m)

(a) 0 c/m^2 (b) 2 c/m^2

(c) 1.8 × 10^{-11} c/m^2 (d) 1.14 × 10^{-9} c/m^2

57. A plane electromagnetic wave travels in dielectric medium of relative permittivity 9. Relative to free space, the velocity of propagation in the dielectric is

(*a*) increased by a factor of 9

(*b*) increased by a factor of 3

(*c*) unchanged

(*d*) reduced by a factor of 1/3

58. For attenuation of high frequencies we should use

(*a*) shunt capacitance (*b*) series capacitance

(*c*) inductance (*d*) resistance

59. VSB modulation is preferred in TV because

(*a*) it reduces the bandwidth requirement to half

(*b*) it avoids phase distortion at low frequencies

(*c*) it results in better reception

(*d*) none of the above

60. In FM signal with a modulation index m_f is passed through a frequency tripler. The wave in the output of the tripler will have a modulation index of

(*a*) m_f (*b*) $3m_f$

(*c*) $m_f/3$ (*d*) $m_f/9$

61. The electrical power to a meggar is provided by

(*a*) battery

(*b*) permanent magnet DC generator

(*c*) AC generator

(*d*) any of the above

62. Basically a potentiometer is a device for

(*a*) comparing two voltages

(*b*) measuring a current

(*c*) comparing two currents

(*d*) none of the above

63. A device which converts BCD to seven segment is called

(*a*) encoder (*b*) decoder

(*c*) multiplexer (*d*) none of these

64. The basic storage element in a digital system is

(*a*) flip flop (*b*) counter

(*c*) multiplexer (*d*) encoder

65. The output of a half adder is

(*a*) SUM (*b*) CARRY

(*c*) SUM and CARRY (*d*) none of the above

66. A transmission line section shows an input impedance of 36 Ω and 64 Ω respectively, when short circuited and open circuited. What is the characteristic impedance of the transmission line?

(*a*) 100 Ω (*b*) 50 Ω

(*c*) 45 Ω (*d*) 48 Ω

67. A signal of 10 V is applied to a 50 ohm coaxial transmission line, terminated in a 100 ohm load. The voltage reflection coefficient is

(*a*) 1/4 (*b*) 1/3

(*c*) 1/2 (*d*) 1

68. The output of an EX-OR gate with A and B as input will be

(*a*) $AB + \overline{AB}$ (*b*) $(A+B)\left(\overline{A+B}\right)$

(*c*) $(A + B)(\bar{A}B)$ (*d*) $\overline{A + B} + AB$

69. If the maximum and minimum voltages on a transmission line are 4 V and 2 V, respectively for a typical load, VSWR is

(*a*) 1.0 (*b*) 0.5

(*c*) 2.0 (*d*) 8.0

70. A loss-less transmission line having characteristic impedance Z_0 is terminated in a load of Z_R. If the value of Z_R is exactly half of Z_0 then reflection coefficient Γ_L is

(*a*) $\frac{1}{3}$ (*b*) $\frac{2}{3}$

(*c*) $-\frac{1}{3}$ (*d*) $-\frac{2}{3}$

71. In a two-wattmeter method of measuring power in a balanced 3-phase circuit, the ratio of the two wattmeter readings is 1:2. The circuit power factor is

(*a*) 0.707 (*b*) 0.5

(*c*) 0.866 (*d*) Indeterminate

72. If the total powers consumed by three identical phase loads connected in delta and star configurations are W_1 and W_2 respectively, then is

(*a*) $3\ W_2$ (*b*) $\frac{W_2}{3}$

(*c*) $\sqrt{3}\ W_2$ (*d*) $\frac{W_2}{\sqrt{3}}$

73. If an energy meter makes 5 revolutions in 100 seconds, when a load of 225 W is connected, the meter constant is

(*a*) 800 rev/kWh (*b*) 222 rev/kWh

(*c*) 147 rev/kWh (*d*) 13 rev/kWh

74. In a balanced 3-phase 200 V circuit, the line current is 115.5 A. When the power is measured by two wattmeter method, one of the wattmeters rreads 20 kW and the other one reads zero. What is the power factor of the load?

(*a*) 0.5 (*b*) 0.6

(*c*) 0.7 (*d*) 0.8

75. The capacitance and loss angle of a given capacitor specimen are best measured by

(*a*) Wheatstone bridge (*b*) Maxwell bridge

(*c*) Anderson bridge (*d*) Schering bridge

GENERAL INTELLIGENCE AND REASONING

Directions: In question no. 76 to 81, select the related word/letters/number from the given alternatives.

76. BKLP : DNPU : : SORT : ?

(*a*) URVY (*b*) VRUX

(*c*) TQVX (*d*) UQRY

77. RDN : PBL : : NOS : ?

(*a*) LNQ (*b*) KMR

(*c*) LMQ (*d*) JNR

78. NIKE : JEGA : : SALT : ?

(*a*) OVHP (*b*) OWHP

(*c*) NWGN (*d*) NVGP

79. 15 : 225 : : 24 : ?

(*a*) 360 (*b*) 496

(*c*) 486 (*d*) 676

80. 475 : 639 : : 382 : ?

(*a*) 576 (*b*) 496

(*c*) 582 (*d*) 676

81. Tasty : Tastiness : : ?

(*a*) Vanturesome : Venturesomely

(*b*) Corrode : Corrodible

(*c*) Bureaucratic : Bureaucracy

(*d*) Hazard : Hazardous

82. Arrange the following words as per the order in the dictionary:

(i) Ascend (ii) Asceses

(iii) Ascots (iv) Ascetic

(v) Asbestos

(*a*) (v), (iv), (iii), (ii), (i)

(*b*) (v), (i), (ii), (iv), (iii)

(*c*) (v), (i), (iv), (ii), (iii)

(*d*) (v), (iv), (ii), (iii), (i)

Directions: In question nos. 83 to 86, find the odd word /number/letters from the given alternatives.

83. (*a*) SVY (*b*) EMU

(*c*) KQW (*d*) IMS

84. (*a*) Telephone (*b*) Newspaper

(*c*) Podcast (*d*) Television

85. (*a*) SUXB (*b*) NPSV

(*c*) LNQU (*d*) DFIM

86. (*a*) 217 (*b*) 126

(*c*) 344 (*d*) 729

87. Which one of the given options would be a meaningful order of the following?

(i) Colour (ii) Hang

(iii) Sketch (iv) Frame

(*a*) (i), (iii), (iv), (ii) (*b*) (iii), (i), (ii), (iv)

(*c*) (iii), (i), (iv), (ii) (*d*) (iv), (iii), (i), (ii)

88. Which one of the following would be a meaningful order of the following states in descending order of area?

(i) Rajasthan (ii) Manipur

(iii) Delhi (iv) Uttar Pradesh

(v) Tamil Nadu

(*a*) (v), (i), (iv), (iii), (ii)

(*b*) (i), (iv), (v), (ii), (iii)

(*c*) (iv), (i), (v), (ii), (iii)

(*d*) (i), (iv), (v), (iii), (ii)

Directions: In question nos. 89 to 90, a series is given, with one term missing. Choose the correct alternative from the given ones.

89.

5	14	21
3	4	2
5	6	7
20	62	?

(*a*) 48 (*b*) 49

(*c*) 52 (*d*) 64

90. MLKS, ONMU, QPOW, ?.

(*a*) SQPY (*b*) SRQX

(*c*) SRQY (*d*) SQRY

GENERAL AWARENESS

91. Which article of Indian Constitution is related to joint session of parliament?

(*a*) Article 108 (*b*) Article 110

(*c*) Article 280 (*d*) Article 370

92. Modern forms of social accounting first produced widespread interest in the

(*a*) 1960 (*b*) 1970

(*c*) 1980 (*d*) 1990

93. Economic situation in which consumers spend less than their disposable income, not because they want to save but because the goods they seek are not available. This is known as

(*a*) Inflation (*b*) Stagflation

(*c*) Forced savings (*d*) None of these

94. Labour demand refers to

(*a*) number of hours of hiring that an employer is willing

(*b*) increase in the demand of labour in the market

(*c*) decrease in the labour during peak period

(*d*) None of these

95. Goods and services tax is

(*a*) Value added tax

(*b*) Corporate tax

(*c*) Modified alternate tax

(*d*) All of these

96. Special purpose entity is

(*a*) A legal entity

(*b*) A loan given to bank employees

(*c*) Low interest rate subsidy

(*d*) None of these

97. Which of the following statements is not true about Bal Gangadhar Tilak?

(*a*) Tilak started the home rule league.

(*b*) Tilak was a strong radical in Indian consciousness.

(*c*) Tilak was one of the first and strongest advocates of "Swaraj" (self-rule).

(*d*) None of the above

98. Name the mobile app for iOS and Windows Phone platforms launched by Ministry of External Affairs.

(*a*) mPort Seva (*b*) myPassport Seva

(*c*) mPassport Seva (*d*) mPass Seva

99. Which of the following article is related to adult franchise in Indian Constitution?

(*a*) Article 320 (*b*) Article 326

(*c*) Article 345 (*d*) Article 278

100. Niccole Machiavelli was a citizen of

(*a*) Italy (*b*) Spain

(*c*) Greece (*d*) Russia

101. Which of the following article of Indian constitution deals with right to education?

(*a*) Article 19 (*b*) Article 20

(*c*) Article 21 (*d*) Article 32

102. Who was the founder of the Kanva dynasty?

(*a*) Vasudeva (*b*) Devabhuti

(*c*) Pushyamitra (*d*) Kanishka

103. Regulating Act 1773 deals with

(*a*) Act limited the dividends of the company

(*b*) Prohibited the servants from engaging in any private trade

(*c*) Both (*a*) and (*b*)

(*d*) None of these

104. Which of the following diseases is related to immune system?

(*a*) AIDS/HIV (*b*) Asthma

(*c*) Both (*a*) and (*b*) (*d*) Cirrhosis

105. Moplah rebellion took place in

(*a*) 1919 (*b*) 1920

(*c*) 1921 (*d*) 1922

QUANTITATIVE APTITUDE

106. A is twice as good as B and thrice as good as C. If A and C can finish a piece of work in 15 days, then find the time taken by B and C to finish twice the work while working together.

(*a*) 16 days (*b*) 24 days

(*c*) 32 days (*d*) 48 days

107. A shopkeeper sold an article at Rs. 816 after allowing three successive discounts of 15%, 25% and 20%. Find the selling price of the article if he had offered a single discount of 35%.

(*a*) Rs. 1,020 (*b*) Rs. 1,040

(*c*) Rs. 1,140 (*d*) Rs. 1,120

108. The numbers x, y, z are respectively proportional to 2, 3, 5 and the product of x, y and z is 15360. If z = by + 4, then find the value of b.

(*a*) 2 (*b*) 2.5

(*c*) 1 (*d*) 1.5

109. If the ratio of the cost price and the marked price of an article is 9 :11 and the discount offered is 10%, then the profit percentage is

(*a*) 10% (*b*) 12%

(*c*) 12.5% (*d*) No profit

110. The price of rice increased by 75% while the expenditure increased by 40%. Find the percentage change in the consumption.

(*a*) 20% increase (*b*) 25% decrease

(*c*) 20% decrease (*d*) 25% increase

111. A car covers a total distance of 450 km. For the first 10 hours, the speed was 25 km/hr and then it was reduced to 40%. Find the average speed (in m/s) of the car during the entire journey.

(*a*) $4\frac{1}{6}$ (*b*) $5\frac{1}{6}$

(*c*) $5\frac{2}{5}$ (*d*) $3\frac{1}{6}$

112. In how many years will the ratio of the money invested at 10% per annum compound interest, compounded semi-annually and compound interest be 8000 : 1261?

(*a*) 1 (*b*) 1.5
(*c*) 2 (*d*) 2.5

113. If $(x-2)(x-p) = x^2 - ax + 6$, then the value of p^a is

(*a*) 125 (*b*) 343
(*c*) 32 (*d*) 243

114. If $x - \frac{1}{x} = 2$, then find the value of $x^6 + \frac{1}{x^6}$.

(*a*) 188 (*b*) 198
(*c*) 186 (*d*) 196

115. In ΔABC, if $\angle C - \angle A = 70°$, $\angle B + \angle C = 140°$ and $\angle A + 2\angle B = 100°$, then find the value of $2\angle B - \angle A$.

(*a*) 10° (*b*) 15°
(*c*) 20° (*d*) 25°

116. If PA and PB are two tangents at a circle with center O such that $\angle AOP : \angle APO = 7 : 11$, then find the measure of $\angle BPA$.

(*a*) 88° (*b*) 99°
(*c*) 110° (*d*) 121°

117. If $\tan(5x - 10°) = \cot(20° - 5y)$, then the value of $6x + 8y$ is

(*a*) 26° (*b*) 24°
(*c*) 28° (*d*) 32°

118. Find the maximum value of the product of two natural numbers if their sum is 24.

(*a*) 135 (*b*) 140
(*c*) 143 (*d*) 144

119. The average marks obtained by a class of 60 students are 66. The average marks obtained by one-third of the students are 59 while the average marks obtained by 40% of the remaining students are 68. Find the average marks obtained by the remaining students of the class.

(*a*) 69 (*b*) 69.5
(*c*) 70 (*d*) 70.5

120. If $x = \sqrt{5} + \frac{1}{\sqrt{5}}$ and $y = \sqrt{5} - \frac{1}{\sqrt{5}}$, then find the value of $x^2 + y^2 - 4xy$.

(*a*) $50 - \frac{6}{5}$ (*b*) $50 - \frac{5}{6}$
(*c*) $10 - \frac{6}{5}$ (*d*) $10 - \frac{5}{6}$

GENERAL ENGLISH

Directions: In question nos. **121** to **123**, four alternatives are given for the Idiom/Phrase underlined in the sentence. Choose the alternative which best expresses the meaning of the Idiom/Phrase and mark it.

121. My Ford car went for a song.

(*a*) Sold at a high price
(*b*) Stolen
(*c*) Sold at an auction
(*d*) Sold at an unexpectedly low price

122. As he drove along the coast, he feasted his eyes on the beautiful scenery.

(*a*) Became ill
(*b*) Delighted and gratified by what one sees
(*c*) Ate at a roadside café
(*d*) Was disgusted

123. I expect you to do the job well but don't try to boil the ocean!

(*a*) Sail across the ocean
(*b*) Waste time on a task or project that is unnecessary, not worth doing or impossible to achieve
(*c*) Build a boat to cross the ocean
(*d*) Try to swim across the ocean

Directions: In question nos. **124** to **126**, out of the four alternatives, choose the one which can be substituted for the given words/sentences and indicate it.

124. An annual calendar with positions of stars

(*a*) Manuscript (*b*) Almanac
(*c*) Schedule (*d*) Time table

125. A person holding an opinion at odds with what is generally accepted.

(*a*) Heretic (*b*) Martial
(*c*) Ascetic (*d*) Sceptic

126. A vehicle which is used to carry a dead body

(*a*) Convoy (*b*) Hearse
(*c*) Phaeton (*d*) Cavalcade

Directions: In question nos. **127** to **129**, a sentence/a part of the sentence is underlined. Below are given alternatives to the underlined part which may improve the sentence. Choose the correct alternative. In case no improvement is needed, choose "**No improvement**". Mark your answer.

127. The cistern is running on.

(*a*) running in (*b*) running about
(*c*) running over (*d*) No improvement

128. I have enough capital to set me up in trade.

(*a*) set on (*b*) set about

(*c*) set by (*d*) No improvement

129. She is struck out with jaundice.

(*a*) struck down (*b*) No improvement

(*c*) struck in (*d*) struck on

130. Envisaging a technological future where computers can learn human language and have conversations with people, Microsoft's India-born CEO Satya Nadella has said companies need to build "respectful" technology which gets the action of humanity and not the worst.

(*a*) advantage (*b*) alliteration

(*c*) best (*d*) No improvement

131. Scientists, using a computer model, have attained the mystery behind the sustained eruptions on Saturn's icy moon Enceladus, which serves as a leading candidate for extraterrestrial life.

(*a*) unraveled (*b*) pilfered

(*c*) obtained (*d*) No improvement

Directions: In question nos. **132** to **136**, you have a passage with 5 questions following. Read the passage carefully and choose the best answer to each question out of the four alternatives and mark it.

Passage

Reliance Defence Ltd, and Rafael Advanced Defence Systems Ltd of Israel have decided to set up a joint venture company in India in the highly specialized areas of air-to-air missiles, air defence systems and large aerostats, that will oversee projects worth USD 10 billion (over Rs 65,000 crore) in ten years. The project will have an initial capital outlay of more than Rs.1,300 crore, excluding the cost of technology and the venture will have 51 per cent holding from Reliance Defence and 49 per cent by Rafael as per the current guidelines of the government. The new venture will be located at Dhirubhai Ambani Land Systems Park at Indore in Madhya Pradesh and will generate more than 3,000 highly-skilled jobs.

Rafael has products like Python and Derby in the air-to-air missiles system, Spyder range in air defence systems and the Barak missiles family of surface-to-air missiles in the short and medium range. Rafael has already provided large aerostat systems to the Indian Air Force for meeting its surveillance, reconnaissance, communication and intelligence needs. The two companies, through the joint venture, will offer the entire range of products in these fields to the Indian Armed Forces. Reliance Defence has industrial licences for the full spectrum of military platforms and is the fastest emerging private sector enterprise in the development, manufacture and supply of defence aerospace, land and naval platforms and equipment.

132. Where will the new joint venture project between Reliance and Rafael be set up?

(*a*) Indore (*b*) Rajkot

(*c*) Pune (*d*) Seemanagar

133. The project will have an initial capital outlay of

(*a*) more than Rs.1,300 crores

(*b*) more than Rs. 1,350 crores

(*c*) less than 1,300 crores

(*d*) less than 1,000 crores

134. How many jobs are expected to be generated in the new project?

(*a*) Less than 3000 jobs

(*b*) More than 3000 jobs

(*c*) More than 3500 jobs

(*d*) More than 3300 jobs

135. What does the Spyder Range stand for?

(*a*) Anti submarine warfare

(*b*) Naval defence system

(*c*) Air defence system

(*d*) Anti tank warfare

136. What kind of a missile system is Python?

(*a*) air to ground (*b*) ground to air

(*c*) sea to air (*d*) air to air

Directions: In question no. **137**, out of the four alternatives, choose the one which best expresses the meaning of the given word and mark it.

137. LUMINARY

(*a*) Panjandrum

(*b*) Nihilism

(*c*) A prominent person

(*d*) Pelican

Directions: In question no. **138**, choose the word opposite in meaning to the given word and mark it.

138. DELIRIOUS

(*a*) Surrogacy (*b*) Sensible

(*c*) Sabbatical (*d*) Detection

139. In the following question, four words are given in each question, out of which only one word is correctly spelt. Find the correctly spelt word and mark your answer.

(*a*) Lascivious (*b*) Lasivious

(*c*) Lacivious (*d*) Lascivius

Directions: In question nos. **140** to **142**, some parts of the sentences have errors and some are correct. Find out which part of a sentence has an error and mark the appropriate correct option. If a sentence is free from error, mark the "No error" option.

140. The latest results of a survey show that newspapers are more preferred to television.

The latest results of (*a*) / a survey show that newspapers are (*b*) / more preferred to television. (*c*) / No error (*d*)

141. As many as 192 countries ushered in the International Yoga day with some deep breathing and coordinated movements.

As many as 192 countries ushered in the International Yoga day (*a*) / with some deep breathing (*b*) / and coordinated movements. (*c*) / No error (*d*)

142. A square blue big box lying broken at the site of murder caught the eye of the officer.

A square blue big box lying broken at the (*a*) / site of murder caught (*b*) / the eye of the officer. (*c*) /No error (*d*)

Directions: In question nos. 143 to 145, sentences given with blanks are to be filled in with appropriate word(s). Four alternatives are suggested for each question. Choose the correct alternative out of the four and indicate it.

143. Kumar passed ______ tremendous harassment in the hands of his boss.

(*a*) for (*b*) over
(*c*) through (*d*) at

144. Darpan set ______ for his gym classes early in the morning.

(*a*) off (*b*) up
(*c*) down (*d*) of

145. Akash worked ______ the marketing problem in a few minutes.

(*a*) at (*b*) out
(*c*) in (*d*) up

Directions: In question nos. **146** to **148**, some parts of the sentences have errors and some are correct. Find out which part of a sentence has an error and mark the appropriate correct option. If a sentence is free from error, mark the "No error" option.

146. Either bats are good enough.

Either bats are (*a*)/good (*b*)/enough (*c*)/ No error. (*d*)

147. Five aircrafts of the Indian Air Force flew over the city of Srinagar.

Five aircrafts of the Indian Air Force (*a*)/ flew over (*b*)/ the city of Srinagar (*c*)/. No error (*d*)

148. Of the three, coffee, tea and chocolate, the latter is his favourite.

Of the three, (*a*)/ coffee, tea and chocolate, (*b*)/ the latter is his favourite. (*c*)/ No error (*d*)

Directions: In question nos. **149 to 151,** sentences given with blanks are to be filled in with appropriate word(s). Four alternatives are suggested for each question. Choose the correct alternative out of the four and indicate it.

149. Sentinels were posted ______ the camp should be taken by surprise.

(*a*) lesser (*b*) less
(*c*) least (*d*) lest

150. The minister was a man of great ______.

(*a*) opulent (*b*) wealth
(*c*) malicious (*d*) wealthy

151. The porters belonged to a tribe ______ in the hills.

(*a*) dwelling (*b*) fishing
(*c*) angling (*d*) whimpering

Directions: In question nos. **152** to **154**, four alternatives are given for the Idiom/Phrase underlined in the sentence. Choose the alternative which best expresses the meaning of the Idiom/Phrase and give the answer.

152. I'd like to be a <u>fly on the wall</u> when the management discusses my project.

(*a*) A person who is an eloquent speaker.
(*b*) A person who discretely watches a situation without being noticed is called a fly on the wall.
(*c*) A person who dominates the meeting.
(*d*) A person who writes the minutes of the meeting.

153. I can't stand <u>backseat drivers</u> like my mother-in-law!

(*a*) Refers to a coupe.
(*b*) Refer to a formula I driver.
(*c*) A passenger in a car who gives unwanted advice to the driver is called a backseat driver.
(*d*) A navigator in a racing car.

154. Andrew's got <u>itchy feet</u> again because he says he's going to teach in China for a few years.

(*a*) A person who suffers a skin disease on the feet.
(*b*) A person who loves to sail in a boat.
(*c*) A person who plays soccer.
(*d*) A person who has itchy feet is someone who finds it difficult to stay in one place and likes to travel and discover new places

Directions: In question nos. **155** to **157**, out of the four alternatives, choose the one which can be substituted for the given words/sentences and indicate it.

155. A nursery where children are cared for while their parents are at work

(*a*) Mortuary (*b*) Creche
(*c*) Cemetery (*d*) Pit-stop

156. A poem of lamentation

(*a*) Eulogy (*b*) Paean
(*c*) Elegy (*d*) Hymn

157. A man employed to look after the horses of people staying at an inn.

(*a*) Ostler (*b*) Janitor

(*c*) Warden (*d*) Concierge

Directions: In question nos. **158** to **162**,a sentence/part of the sentence is underlined. Beneath each sentence four different ways of phrasing the underlined part are given. Choose the grammatically correct option. In case no improvement is needed, your answer is option (*d*)

158. Fifteen years ago today, Sachin was sitting in the plane going to Australia.

(*a*) sat (*b*) have been sitting

(*c*) is sitting (*d*) No improvement

159. They drove out of the country by car where they had been banished.

(*a*) They drove out of the country where they had been banished by car.

(*b*) They drove where they had been banished out of the country by car.

(*c*) They drove by car out of the country where they had been banished.

(*d*) No improvement

160. If he would have done his homework, he would have saved himself the embarrassment of being punished in class.

(*a*) would be doing (*b*) had done

(*c*) would do (*d*) No improvement

161. When cases pending for prosecution sanction for less than three months are taken into consideration, the number balloons up, revealing the real challenge in the banking sector.

(*a*) achieving into (*b*) aimed at

(*c*) taken upon (*d*) No improvement

162. It is heartening to note that there was an internal notification that clearly asked reporters to avoid putting serious allegations in the mouth of unnamed officials and to carry at due diligence before the newspaper publishes such reports.

(*a*) concentrate (*b*) carry out

(*c*) focus (*d*) No improvement

Directions: In question nos. **163** to **167**, you have two passage with 5 questions. Read the passage carefully and choose the best answer to each question out of the four alternatives.

Wars, fought during the course of history, have been so destructive that they have left their marks on the world, and impacted the way things are today. The main idea of a war has either been to expand power, independence, or to join a war to help allies. World War I and World War II have both similarities and differences but both wars will be remembered as infamy.

Alliances of strong countries like Hungary, Germany, and Austria caused World War I because they decided to declare war on other countries. Allies of all the countries decided to join in, which made the war more damaging and made it worse. World War II was caused by Germany because Germany was unhappy with the Treaty of Versailles. So they decided to take matters into their own hands and invaded Poland which triggered the start of the war. In World War I, most of the battles were fought in trenches, and for the first time machine guns, tanks, flamethrowers, and mustard gas were introduced. But in World War II they used more advanced machinery. One of the World War II's famous weapon is the atom bomb. This was first introduced in World War II and was used as a threat during the Cold War. These wars have left so much destruction but in WWI there was far less damage than WWII. WWI caused approximately 24 million deaths, while WWII caused about 60 million deaths, which is about 2.5 times more deaths than WWI. At the end of WWI, the League of Nations was established. And at the end of WW2, there still was no peace because people were being executed from committing war related crimes.

163. Which, according to the passage, were the main reasons that led to wars during the course of history?

(*a*) Expanding power, helping allies, Treaty of Versailles

(*b*) Treaty of Versailles, helping allies, independence

(*c*) Helping allies, independence, expanding power

(*d*) Helping allies, independence, atom bomb

164. Find the word that is same in meaning to 'infamy'.

(*a*) Infinity (*b*) Infamous

(*c*) Shadiness (*d*) Blameworthiness

165. World War I was different from World War II because

(*a*) WWI had more allies.

(*b*) of the degree of the damage caused.

(*c*) of the crimes that were caused during the war.

(*d*) None of the above

166. World War II was triggered because

(*a*) Allies of strong countries declared war on other countries.

(*b*) it was a way to test the advanced machineries.

(*c*) Germany threatened to use atom bomb during the cold war.

(*d*) Germany took control of Poland against its wishes.

167. Which of the following weapons were used in WWI for the first time?

(*a*) Trenches (*b*) Nuclear Bombs

(*c*) Flame throwers (*d*) Rifles

168. In the following question, out of the four alternatives, choose the one which best expresses the meaning of the given word and mark it.

Prosaic

(*a*) Banal (*b*) Exciting

(*c*) Allegory (*d*) Alleviation

169. In the following question, choose the word opposite in meaning to the given word and mark it.

Trite

(*a*) Bovine (*b*) Feral

(*c*) Appealing (*d*) Pedestrian

170. In the question, four words are given in each question, out of which only one word is correctly spelt. Find the correctly spelt word and mark your answer.

(*a*) Enumeration (*b*) Emuneration

(*c*) Enumereation (*d*) Enumerration

Directions: In question nos. **171** to **173**, four alternatives are given for the idioms/phrases given below. Choose the alternative which best expresses the meaning of the given idiom/phrase.

171. He's been <u>chasing his tail</u> all week collecting data but the report is still not ready.

(*a*) boot licking

(*b*) working very hard

(*c*) rushing around ineffectually

(*d*) pushing oneself to work hard

172. Good lawyers need to be able to <u>think on their feet</u> when pleading a case.

(*a*) to be able to think and speak creatively

(*b*) to be able to think and speak confidently

(*c*) to be able to think and speak honestly

(*d*) to be able to think and speak extemporaneously

173. Siddhant's comment about his friend's unethical activities was quite <u>close to home</u>.

(*a*) a lucky guess

(*b*) a relevant comment that makes one uncomfortable

(*c*) talking about something without coming to the point

(*d*) speaking a lie without hesitating

Directions for questions 174 to 176: Out of the four alternatives given below each word/phrase, choose the word which can be substituted for the given word/phrase.

174. I <u>was under the wrong impression</u> that the course was for complete beginners.

(*a*) fathomed (*b*) comprehended

(*c*) misapprehended (*d*) apprehended

175. Mahatma Gandhi was <u>the one who advocated the removal of inequalities among people</u>.

(*a*) an egalitarian (*b*) a federalist

(*c*) an absolutist (*d*) a reformist

176. <u>A person who writes historical records</u>

(*a*) Archaeologist (*b*) Annalist

(*c*) Radiologist (*d*) Genealogist

Directions: In question nos. **177** to **180**, a sentence/a part of the sentence is underlined. Below are given alternatives to the underlined part which may improve the sentence. Choose the correct alternative. In case no improvement is needed, choose "**No improvement**". Mark your answer.

177. The gravitational wave detected on September 14, 2015 was due to the <u>sketching</u> of two black holes of mass 36 and 29 times the mass of our Sun.

(*a*) illustration (*b*) pedagogy

(*c*) merger (*d*) No improvement

178. The Supreme Court guideline to banks to give prosecution sanction within four months against officers accused of collusion in criminal conspiracy to cheat them <u>have been</u> violated.

(*a*) has been (*b*) were been

(*c*) are (*d*) No improvement

179. If your <u>child is thinking about a gap year, they</u> can get good advice from this book.

(*a*) child is thinking about a gap year, he

(*b*) children is thinking about a gap year, they

(*c*) child think about a gap year, he

(*d*) No improvement

180. <u>Less than 20 children suffer from the disease each year</u>.

(*a*) Less than 20 children suffer from the diseases each year

(*b*) Fewer than 20 children suffer from the disease each year.

(*c*) Fewer than 20 children suffers from the disease each year.

(*d*) No improvement

OMR SHEET

1. (A) (B) (C) (D)
2. (A) (B) (C) (D)
3. (A) (B) (C) (D)
4. (A) (B) (C) (D)
5. (A) (B) (C) (D)
6. (A) (B) (C) (D)
7. (A) (B) (C) (D)
8. (A) (B) (C) (D)
9. (A) (B) (C) (D)
10. (A) (B) (C) (D)
11. (A) (B) (C) (D)
12. (A) (B) (C) (D)
13. (A) (B) (C) (D)
14. (A) (B) (C) (D)
15. (A) (B) (C) (D)
16. (A) (B) (C) (D)
17. (A) (B) (C) (D)
18. (A) (B) (C) (D)
19. (A) (B) (C) (D)
20. (A) (B) (C) (D)
21. (A) (B) (C) (D)
22. (A) (B) (C) (D)
23. (A) (B) (C) (D)
24. (A) (B) (C) (D)
25. (A) (B) (C) (D)
26. (A) (B) (C) (D)
27. (A) (B) (C) (D)
28. (A) (B) (C) (D)
29. (A) (B) (C) (D)
30. (A) (B) (C) (D)
31. (A) (B) (C) (D)
32. (A) (B) (C) (D)
33. (A) (B) (C) (D)
34. (A) (B) (C) (D)
35. (A) (B) (C) (D)
36. (A) (B) (C) (D)
37. (A) (B) (C) (D)
38. (A) (B) (C) (D)
39. (A) (B) (C) (D)
40. (A) (B) (C) (D)
41. (A) (B) (C) (D)
42. (A) (B) (C) (D)
43. (A) (B) (C) (D)
44. (A) (B) (C) (D)
45. (A) (B) (C) (D)
46. (A) (B) (C) (D)
47. (A) (B) (C) (D)
48. (A) (B) (C) (D)
49. (A) (B) (C) (D)
50. (A) (B) (C) (D)
51. (A) (B) (C) (D)
52. (A) (B) (C) (D)
53. (A) (B) (C) (D)
54. (A) (B) (C) (D)
55. (A) (B) (C) (D)
56. (A) (B) (C) (D)
57. (A) (B) (C) (D)
58. (A) (B) (C) (D)
59. (A) (B) (C) (D)
60. (A) (B) (C) (D)
61. (A) (B) (C) (D)
62. (A) (B) (C) (D)
63. (A) (B) (C) (D)
64. (A) (B) (C) (D)
65. (A) (B) (C) (D)
66. (A) (B) (C) (D)
67. (A) (B) (C) (D)
68. (A) (B) (C) (D)
69. (A) (B) (C) (D)
70. (A) (B) (C) (D)
71. (A) (B) (C) (D)
72. (A) (B) (C) (D)
73. (A) (B) (C) (D)
74. (A) (B) (C) (D)
75. (A) (B) (C) (D)
76. (A) (B) (C) (D)
77. (A) (B) (C) (D)
78. (A) (B) (C) (D)
79. (A) (B) (C) (D)
80. (A) (B) (C) (D)
81. (A) (B) (C) (D)
82. (A) (B) (C) (D)
83. (A) (B) (C) (D)
84. (A) (B) (C) (D)
85. (A) (B) (C) (D)
86. (A) (B) (C) (D)
87. (A) (B) (C) (D)
88. (A) (B) (C) (D)
89. (A) (B) (C) (D)
90. (A) (B) (C) (D)
91. (A) (B) (C) (D)
92. (A) (B) (C) (D)
93. (A) (B) (C) (D)
94. (A) (B) (C) (D)
95. (A) (B) (C) (D)
96. (A) (B) (C) (D)
97. (A) (B) (C) (D)
98. (A) (B) (C) (D)
99. (A) (B) (C) (D)
100. (A) (B) (C) (D)

OMR SHEET

101. Ⓐ Ⓑ Ⓒ Ⓓ	121. Ⓐ Ⓑ Ⓒ Ⓓ	141. Ⓐ Ⓑ Ⓒ Ⓓ	161. Ⓐ Ⓑ Ⓒ Ⓓ
102. Ⓐ Ⓑ Ⓒ Ⓓ	122. Ⓐ Ⓑ Ⓒ Ⓓ	142. Ⓐ Ⓑ Ⓒ Ⓓ	162. Ⓐ Ⓑ Ⓒ Ⓓ
103. Ⓐ Ⓑ Ⓒ Ⓓ	123. Ⓐ Ⓑ Ⓒ Ⓓ	143. Ⓐ Ⓑ Ⓒ Ⓓ	163. Ⓐ Ⓑ Ⓒ Ⓓ
104. Ⓐ Ⓑ Ⓒ Ⓓ	124. Ⓐ Ⓑ Ⓒ Ⓓ	144. Ⓐ Ⓑ Ⓒ Ⓓ	164. Ⓐ Ⓑ Ⓒ Ⓓ
105. Ⓐ Ⓑ Ⓒ Ⓓ	125. Ⓐ Ⓑ Ⓒ Ⓓ	145. Ⓐ Ⓑ Ⓒ Ⓓ	165. Ⓐ Ⓑ Ⓒ Ⓓ
106. Ⓐ Ⓑ Ⓒ Ⓓ	126. Ⓐ Ⓑ Ⓒ Ⓓ	146. Ⓐ Ⓑ Ⓒ Ⓓ	166. Ⓐ Ⓑ Ⓒ Ⓓ
107. Ⓐ Ⓑ Ⓒ Ⓓ	127. Ⓐ Ⓑ Ⓒ Ⓓ	147. Ⓐ Ⓑ Ⓒ Ⓓ	167. Ⓐ Ⓑ Ⓒ Ⓓ
108. Ⓐ Ⓑ Ⓒ Ⓓ	128. Ⓐ Ⓑ Ⓒ Ⓓ	148. Ⓐ Ⓑ Ⓒ Ⓓ	168. Ⓐ Ⓑ Ⓒ Ⓓ
109. Ⓐ Ⓑ Ⓒ Ⓓ	129. Ⓐ Ⓑ Ⓒ Ⓓ	149. Ⓐ Ⓑ Ⓒ Ⓓ	169. Ⓐ Ⓑ Ⓒ Ⓓ
110. Ⓐ Ⓑ Ⓒ Ⓓ	130. Ⓐ Ⓑ Ⓒ Ⓓ	150. Ⓐ Ⓑ Ⓒ Ⓓ	170. Ⓐ Ⓑ Ⓒ Ⓓ
111. Ⓐ Ⓑ Ⓒ Ⓓ	131. Ⓐ Ⓑ Ⓒ Ⓓ	151. Ⓐ Ⓑ Ⓒ Ⓓ	171. Ⓐ Ⓑ Ⓒ Ⓓ
112. Ⓐ Ⓑ Ⓒ Ⓓ	132. Ⓐ Ⓑ Ⓒ Ⓓ	152. Ⓐ Ⓑ Ⓒ Ⓓ	172. Ⓐ Ⓑ Ⓒ Ⓓ
113. Ⓐ Ⓑ Ⓒ Ⓓ	133. Ⓐ Ⓑ Ⓒ Ⓓ	153. Ⓐ Ⓑ Ⓒ Ⓓ	173. Ⓐ Ⓑ Ⓒ Ⓓ
114. Ⓐ Ⓑ Ⓒ Ⓓ	134. Ⓐ Ⓑ Ⓒ Ⓓ	154. Ⓐ Ⓑ Ⓒ Ⓓ	174. Ⓐ Ⓑ Ⓒ Ⓓ
115. Ⓐ Ⓑ Ⓒ Ⓓ	135. Ⓐ Ⓑ Ⓒ Ⓓ	155. Ⓐ Ⓑ Ⓒ Ⓓ	175. Ⓐ Ⓑ Ⓒ Ⓓ
116. Ⓐ Ⓑ Ⓒ Ⓓ	136. Ⓐ Ⓑ Ⓒ Ⓓ	156. Ⓐ Ⓑ Ⓒ Ⓓ	176. Ⓐ Ⓑ Ⓒ Ⓓ
117. Ⓐ Ⓑ Ⓒ Ⓓ	137. Ⓐ Ⓑ Ⓒ Ⓓ	157. Ⓐ Ⓑ Ⓒ Ⓓ	177. Ⓐ Ⓑ Ⓒ Ⓓ
118. Ⓐ Ⓑ Ⓒ Ⓓ	138. Ⓐ Ⓑ Ⓒ Ⓓ	158. Ⓐ Ⓑ Ⓒ Ⓓ	178. Ⓐ Ⓑ Ⓒ Ⓓ
119. Ⓐ Ⓑ Ⓒ Ⓓ	139. Ⓐ Ⓑ Ⓒ Ⓓ	159. Ⓐ Ⓑ Ⓒ Ⓓ	179. Ⓐ Ⓑ Ⓒ Ⓓ
120. Ⓐ Ⓑ Ⓒ Ⓓ	140. Ⓐ Ⓑ Ⓒ Ⓓ	160. Ⓐ Ⓑ Ⓒ Ⓓ	180. Ⓐ Ⓑ Ⓒ Ⓓ

ANSWERS

1. (a)	2. (c)	3. (a)	4. (b)	5. (c)	6. (d)	7. (a)	8. (a)	9. (a)	10. (b)
11. (a)	12. (c)	13. (a)	14. (b)	15. (c)	16. (d)	17. (c)	18. (a)	19. (c)	20. (b)
21. (b)	22. (b)	23. (c)	24. (a)	25. (c)	26. (c)	27. (c)	28. (c)	29. (b)	30. (c)
31. (b)	32. (d)	33. (a)	34. (d)	35. (b)	36. (b)	37. (b)	38. (c)	39. (a)	40. (d)
41. (a)	42. (c)	43. (b)	44. (c)	45. (a)	46. (c)	47. (c)	48. (a)	49. (a)	50. (c)
51. (a)	52. (c)	53. (c)	54. (c)	55. (b)	56. (d)	57. (d)	58. (a)	59. (a)	60. (b)
61. (b)	62. (a)	63. (b)	64. (a)	65. (c)	66. (d)	67. (b)	68. (c)	69. (c)	70. (c)
71. (c)	72. (a)	73. (a)	74. (a)	75. (d)	76. (a)	77. (c)	78. (b)	79. (a)	80. (c)
81. (c)	82. (b)	83. (d)	84. (b)	85. (b)	86. (d)	87. (c)	88. (b)	89. (b)	90. (c)
91. (a)	92. (b)	93. (c)	94. (a)	95. (a)	96. (a)	97. (d)	98. (c)	99. (b)	100. (a)
101. (c)	102. (a)	103. (c)	104. (a)	105. (c)	106. (d)	107. (b)	108. (d)	109. (a)	110. (c)
111. (a)	112. (b)	113. (d)	114. (b)	115. (c)	116. (c)	117. (a)	118. (d)	119. (d)	120. (a)
121. (d)	122. (b)	123. (b)	124. (b)	125. (a)	126. (b)	127. (c)	128. (d)	129. (a)	130. (c)
131. (a)	132. (a)	133. (a)	134. (b)	135. (c)	136. (d)	137. (c)	138. (b)	139. (a)	140. (c)
141. (d)	142. (a)	143. (c)	144. (a)	145. (b)	146. (a)	147. (a)	148. (c)	149. (d)	150. (b)
151. (a)	152. (b)	153. (c)	154. (d)	155. (b)	156. (c)	157. (a)	158. (d)	159. (c)	160. (b)
161. (d)	162. (b)	163. (c)	164. (b)	165. (b)	166. (d)	167. (c)	168. (a)	169. (c)	170. (a)
171. (c)	172. (d)	173. (b)	174. (c)	175. (a)	176. (b)	177. (c)	178. (a)	179. (a)	180. (b)

EXPLANATIONS

1. $V_1^2 = V_2^2 + V_3^2 + 2V_2V_3\cos\phi$

$(220)^2 = (122)^2 +$

$(136)^2 + 2 \times 122 \times 136 \times \cos\phi$

$\cos\phi = 0.045$

2. Given

$$R_L = 5\Omega$$

$$\therefore \quad \cos\phi = \frac{R_L}{|Z|}$$

$$|Z| = \frac{5}{0.45}$$

$$|Z| = 11.11$$

Power consumed by the load

$$P_L = \left(\frac{V_3}{|Z|}\right)^2 R_L = \left(\frac{136}{11.11}\right)^2 \times 5$$

$$= 749.1 = 750 \text{ W}$$

3. T = Torque on the coil = NBAI

Where,

N = No. of turns = 100

B = flux density

$= 200 \text{ mT} = 200 \times 10^{-3} \text{ T}$

A = area of the coil = length × depth

$= (10 \times 10^{-3}) \times (20 \times 10^{-3})$

$= 200 \times 10^{-6} \text{ m}^2$

I = Current through the coil

= 50 mA

$= 50 \times 10^{-3}$

T = NBAI

$= 100 \times (200 \times 10^{-3}) \times (200 \times 10^{-6}) \times (50 \times 10^{-3})$

$= 200\ \mu\text{N-m}$

4. Torque produced

$$T = \frac{1}{2} i^2 \frac{dL}{d\theta}$$

Where, i = 10 A and

T = 240 N-m

$= 240 \times 10^{-6}$ N-m

$$T = \frac{1}{2} i^2 \frac{dL}{d\theta}$$

$$240 \times 10^{-6} = \frac{1}{2} \times 10^2 \frac{dL}{d\theta}$$

Rate of change of self-inductance

$$= \frac{dL}{d\theta} = 4.8 \times 10^{-6} \text{ H/radian}$$

5. $$G = \frac{\Delta R / R}{\Delta l / l}$$

$$\Rightarrow \quad \frac{\Delta R}{R} = 2 \times 10^{-5}$$

$$\Rightarrow \quad \Delta R = 120 \times 2 \times 10^{-5}\,\Omega$$

$$\Rightarrow \quad \Delta R = 2.4 \times 10^{-3}\,\Omega$$

6. The current coil will be zero, because in PMMC type instruments control spring is connected in series with the moving coil.

7. $$\text{Sensitivity} = \frac{1}{I_{fsd}} = \frac{1}{100 \times 10^{-6}}$$
$$= 10\ \text{k}\Omega/\text{V for each}$$

8. Heaviside Campbell bridge method is commonly used for finding mutual inductance.

9. Complex power = VI*

$$= 10\,\angle 15^\circ \times 2\,\angle 45^\circ$$

$$= 20\,\angle 60^\circ = \left(10 + j10\sqrt{3}\right)$$

$$\Rightarrow \quad (10 + j17.32) = P + jQ$$

$$\Rightarrow \quad P = 10\ \text{W}$$

$$\Rightarrow \quad Q = 17.32\ \text{VAR}$$

10. Two wattmeter method can also take care of un-balance system

16.

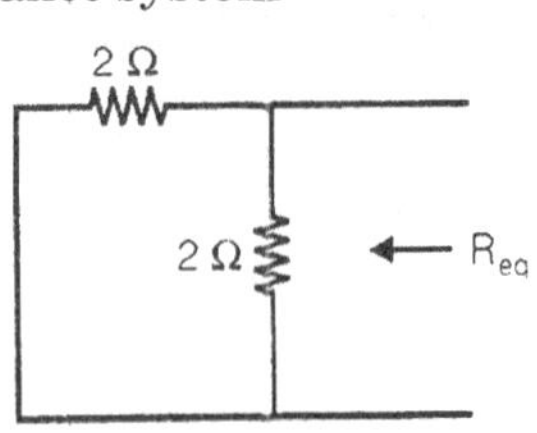

$$R_{eq} = (2 \,||\, 2) = 1\Omega$$

Time constant = $R_{eq} \times C = 1 \times 1 = 1\text{s}$

17. $Z_0^2 = Z_{OC}.Z_{SC}$

18. In case of dielectric medium (good dielectric) $\epsilon_r >> 1$ and $\sigma \to 0$ so displacement current density.

$|J_D| >> |J_C|$ conduction current density

$$\omega \in >> \sigma \text{ for good dielectric} \Rightarrow \frac{\sigma}{\omega \in} << 1$$

19. $$T_R = \frac{Z_L - Z_0}{Z_L + Z_0} = \frac{900 - 600}{900 + 600} = 0.2$$

20. Skin depth $\delta = \dfrac{1}{\sqrt{\pi f \mu \sigma}}$

26. $\nabla \times E = 0$ for conservative electric field.

27. Compare $W_M = \int_v \overline{A}.\overline{J}dv$ with electrical energy stored.

$$W_E = \frac{1}{2}\int_v \rho_v.Vdv$$

28. Net flux coming out form a closed surface i.e. $\oint \vec{D}.\vec{ds} = Q_{inclosed}$ total charge enclosed within the surface this is Gauss's law for electric field.

29. We know that propagation constant

$$\gamma = \sqrt{j\omega\mu(\sigma + j\omega \in)}$$

for a lossless medium $\sigma = 0$

$$\gamma = j\omega\sqrt{\mu \in}$$

Comparing with

$\gamma = \alpha + j\beta$, α - attenuation constant

$\beta = \omega\sqrt{\mu \in}$ β – phase constant

$$\beta = \omega\sqrt{\mu_0 \mu_r \in_0 \in_r}$$

$$\beta \propto \sqrt{\in_r}$$

30. $$\eta = \eta_0\sqrt{\frac{\mu_r}{\in_r}} \Rightarrow 60\pi = 120\pi\sqrt{\frac{1}{\in_r}} \Rightarrow \in_r = 4$$

31. For distortion less transmission line

RC = LG and

$$\alpha = R\sqrt{\frac{C}{L}} = R\sqrt{\frac{G}{R}} = \sqrt{RG}$$

32. Propagation constant in case of transmission line is given by

$$\gamma = \sqrt{R + j\omega L(G + j\omega C)}$$

Where

R = Resistance of T_X line per unit length

L = inductance of T_X line per unit length

G = Conductance of T_X line per-unit length

C = capacitance of T_X line per unit length

33. In an intrinsic semiconductor all energy levels below a certain value W_F are filled all those above W_F being empty; W_F is called the fermi level of the electrons.

34. Net capacitance C is given by for series connection.

$$\frac{1}{C} = \frac{1}{C_1} + \frac{1}{C_2} + \frac{1}{C_3} + \frac{1}{C_4} + \frac{1}{C_5} + \frac{1}{C_6}$$

$$\because C_1 > C_2 > C_3 > C_4 > C_5 > C_6$$

$$\therefore \frac{1}{C_1} < \frac{1}{C_2} < \frac{1}{C_3} < \frac{1}{C_4} < \frac{1}{C_5} < \frac{1}{C_6}$$

$$\therefore \frac{1}{C} \simeq \frac{1}{C_6}$$

$$C \cong C_6$$

35. Laplace's equation is all embracing for applying as it does where volume charge density is zero, it states that every conceivable configuration of electrodes or conductors produces a field for which $\nabla^2 V = 0$.

41. Diode D_2 is RB
$I = 0$

42. Given circuit
In the circuit $V_1 = 3.5V$
Current through Zener

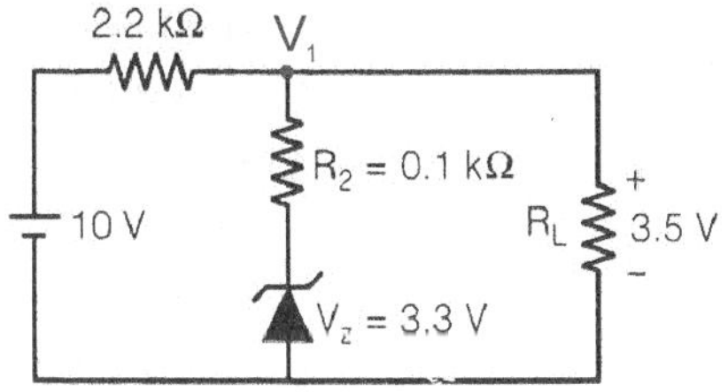

$$I_z = \frac{V_1 - V_Z}{R_Z} = \frac{3.5 - 3.3}{0.1 \times 10^3}$$

$I_z = 2mA$

43. The ripple frequency for half-wave rectifier is f.

44. $I_{rms} = \frac{I_m}{2}$ For HWR

Where $I_m = \frac{V_m}{R_f + R_L} = \frac{325}{100 + 1000} = 295.4$

$$I_{rms} = \frac{295.4}{2} = 147.7 \text{ mA}$$

45. The PIV rating of full wave rectifier with centre tap is $2V_m = 2 \times 10 = 200V$

46. $V = \frac{2V_m}{\pi} = 6.36V$

$$R_0 = 100 + \frac{30}{2} = 115\Omega$$

47. An ideal regulated power supply should have 0% regulation.

48. Current flowing through $100\Omega = \frac{10 - 2}{1k\Omega} = 8mA$

51.

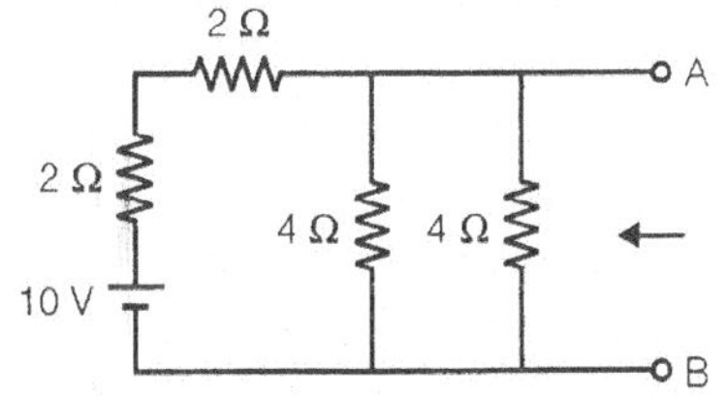

$$Z_{AB} = (4) || (4) || (4) = \frac{2 \times 4}{6} = \frac{4}{3}\Omega$$

62. Potentiometer is a instrument used for comparing emf of two cells and for calibrating ammeter voltmeter and wattmeter.

63. Decoder converts binary/BCD to alphanumeric.

64. Storing can be done only in memory and flip-flop is a memory element.

66. $Z_0 = \sqrt{Z_{o.c} \times Z_{s.c}} = \sqrt{36 \times 64} = 48\Omega$

67.
$$Z_0 = 50$$
$$Z_L = 100$$
$$\rho = \frac{Z_L - Z_0}{Z_L + Z_0} = \frac{50}{100}$$
$$\rho = \frac{1}{3}$$

69.
$$V_{max} = 4$$
$$V_{min} = 2$$
$$V_{SWR} = \frac{V_{max}}{V_{min}} = \frac{4}{2} = 2$$

70. Reflection coefficient,

$$\Gamma_L = \frac{Z_R - Z_0}{Z_R + Z_0} = \frac{\frac{Z_0}{2} - Z_0}{\frac{Z_0}{2} + Z_0} = -\frac{1}{3}$$

71.
$$\phi = \tan^{-1}\left[\frac{\sqrt{3}(2-1)}{2+1}\right]$$
$$= \tan^{-1}\left[\sqrt{3} \times \frac{1}{3}\right] = 30°$$
$$\text{p.f.} = \cos 30° = \frac{\sqrt{3}}{2} = 0.866$$

72.
$$P\Delta = 3P_Y$$
$$W_1 = 3W_2$$

73. Meters constnat

$$k = \frac{\text{No. of revolutions}}{\text{kW-hr}}$$
$$= \frac{5}{\frac{225}{1000} \times \frac{100}{3600}}$$
$$= 800 \text{ Rev/kW-hr}$$

74. $W \propto \cos(30 + \phi)$ or $\cos(30 - \phi)$

if $\phi = 60°$

$W = \cos(30° + 60°) = 0$

so, $\text{p.f.} = \cos(60°) = \frac{1}{2}$

76.

B	K	L	P
↓+2	↓+3	↓+4	↓+5
D	N	P	U

Similarly,

S	O	R	T
↓+2	↓+3	↓+4	↓+5
U	R	V	Y

77. R D N
↓ –2 ↓ –2 ↓ –2
P B L

Similarly,

N O S
↓ –2 ↓ –2 ↓ –2
L M Q

78. N I K E
↓ –4 ↓ –4 ↓ –4 ↓ –4
J E G A

Similarly,

S A L T
↓ –4 ↓ –4 ↓ –4 ↓ –4
O W H P

79. 225 = 15 × 15

Similarly, 24 × 15 = 360.

80. 4 7 5 ⟶ 4 + 7 + 5 = 16 ⎤ +2
6 3 9 ⟶ 6 + 3 + 9 = 18 ⎦

Similarly,

3 8 2 ⟶ 3 + 8 + 2 = 13 ⎤ +2
and 5 8 2 ⟶ 5 + 8 + 2 = 15 ⎦

81. The given analogy is in the form of adjective: noun. Option (c) is the correct answer since 'Bureaucratic' is an adjective while 'bureaucracy' is a noun. Options (a) and (d) are in the reverse order. Option (b) is in the form of verb : adjective.

82. As per the order in the dictionary, the correct sequence is (v), (i), (ii), (iv), (iii). Hence, option (b) is the correct answer.

83. S —+3→ V —+3→ Y
E —+8→ M —+8→ U
K —+6→ Q —+6→ W
I —+4→ M —+6→ S

Hence, IMS is odd-one.

84. Options (a), (c) and (d) are electronic media of communication, while option (b) is the print medium of communication. Hence, option (b) is the correct answer.

85. S —+2→ U —+3→ X —+4→ B
N —+2→ P —+3→ S —+3→ V
L —+2→ N —+3→ Q —+4→ U
D —+2→ F —+3→ I —+4→ M

Hence, NPSV is odd-one.

86. $217 = 6^3 + 1$

$126 = 5^3 + 1$

$344 = 7^3 + 1$

$729 = 9^3$

Hence, 729 is odd-one.

87. The correct sequence is (iii), (i), (iv), (ii). First you sketch a portrait and then you colour it. Once it is complete, it is framed and hanged on a wall. Hence, option (c) is the correct answer.

88. Rajasthan has the largest area, followed by Uttar Pradesh, Tamil Nadu, Manipur and Delhi. Hence, option (b) is the correct answer.

89. Here, 20 = 5 × 3 + 5; 62 = 14 × 4 + 6

Similarly, ? = 21 × 2 + 7 = 49.

90. M L K S
↓+2 ↓+2 ↓+2 ↓+2
O N M U
↓+2 ↓+2 ↓+2 ↓+2
Q P O W
↓+2 ↓+2 ↓+2 ↓+2
S R Q Y

91. Article 108:Joint sitting of both Houses in certain cases

92. Social accounting (also known as *social accounting and auditing, social and environmental accounting, corporate social reporting, corporate social responsibility reporting, non-financial reporting* or *accounting*) is the process of communicating the social and environmental effects of organizations' economic actions to particular interest groups within society and to society at large.

Modern forms of social accounting first produced widespread interest in the 1970s. Its concepts received serious consideration from professional and academic accounting bodies, e.g. the Accounting Standards Board's predecessor, the American Accounting Association and the American Institute of Certified Public Accountants. Business-representative bodies, e.g. the Confederation of British Industry, likewise approached the issue.

93. Forced savings is an economic situation in which consumers spend less than their disposable income, not because they want to save but because the goods they seek are not available. In a open economy, this situation would normally result in increase in prices and inflow of more goods.

94. In economics, labor demand refers to the number of hours of hiring that an employer is willing to do based on the various exogenous (externally

determined) variables it is faced with, such as the wage rate, the unit cost of capital, the market-determined selling price of its output, etc. The function specifying the quantity of labor that would be demanded at any of various possible values of these exogenous variables is called the labor demand function.

95. The Goods and Services Tax (GST) is a Value Added Tax (VAT) to be implemented in India, the decision on which is pending. It will replace all indirect taxes levied on goods and services by the Indian Central and State governments. It is aimed at being comprehensive for most goods and services.

96. A special purpose entity (SPE; or, in Europe and India, special purpose vehicle/SPV, or, in some cases in each EU jurisdiction – FVC *financial vehicle corporation*) is a legal entity (usually a limited company of some type or, sometimes, a limited partnership) created to fulfil narrow, specific or temporary objectives. Companies to isolate the firm from financial risk typically use SPEs. They are also commonly used to hide debt (inflating profits), hide ownership, and obscure relationships between different entities, which are in fact related to each other.

97. Bal Gangadhar Tilak born as Keshav Gangadhar Tilak (23 July 1856 – 1 August 1920), was an Indian nationalist, journalist, teacher, social reformer, lawyer and an independence activist. He was the first popular leader of the Indian Independence Movement. The British colonial authorities called him "Father of the Indian unrest." He was also conferred with the honorary title of "Lokmanya", which literally means "Accepted by the people (as their leader)". Tilak was one of the first and strongest advocates of "Swaraj" (self-rule) and a strong radical in Indian consciousness. His famous quote, *"Swaraj is my birthright, and I shall have it!"* is well-remembered in India even today.

99. Article 326 in The Constitution Of India 1949

326. Elections to the House of the People and to the Legislative Assemblies of States to be on the basis of adult suffrage The elections to the House of the People and to the Legislative Assembly of every State shall be on the basis of adult suffrage; but is to say, every person who is a citizen of India and who is not less than twenty one years of age on such date as may be fixed in that behalf by or under any law made by the appropriate legislature and is not otherwise disqualified under this constitution or any law made by the appropriate Legislature on the ground of non residence, unsoundness of mind, crime or corrupt or illegal practice, shall be entitled to be registered as a voter at any such election.

100. Niccolò di Bernardo dei Machiavelli (3 May 1469 – 21 June 1527) was an Italian historian, politician, diplomat, philosopher, humanist, and writer based in Florence during the Renaissance. He was for many years an official in the Florentine Republic, with responsibilities in diplomatic and military affairs. He was a founder of modern political science, and more specifically political ethics.

101. The Right of Children to Free and Compulsory Education Act or Right to Education Act (RTE), is an Indian legislation enacted by the Parliament of Indiaon 4 August 2009, which describes the modalities of the importance of free and compulsory education for children between 6 and 14 in India under Article 21a of the Indian Constitution. India became one of 135 countries to make education a fundamental right of every child when the act came into force on 1 April 2010.

102. The founder of the kanva dynasty was Vasudeva. The Kanva or Kanvayana Dynasty replaced the Sunga dynasty in Magadha, and ruled in the Eastern part of India from 75 BCE to 30 BCE.

The last ruler of the Sunga dynasty, Devabhuti, was overthrown by Vasudeva of the Kanva dynasty in 75 BC. The Kanva ruler allowed the kings of the Sunga dynasty to continue to rule in obscurity in a corner of their former dominions. Magadha was ruled by four Kanva rulers. Their dynasty was brought to an end by the ruler of the Satavahana Dynasty or 'Andhra bhritya' dynasty in Amaravati, Andhra Pradesh.

103. Provisions of the Regulating Act

- The Act limited Company dividends to 6% until it repaid a GB£1.5M loan and restricted the Court of Directors to four-year terms.
- It prohibited the servants of company from engaging in any private trade or accepting presents or bribes from the natives.
- The Act elevated Governor of Bengal, Warren Hastings to Governor-General of Bengal and subsumed the presidencies of Madras and Bombay under Bengal's control.
- The Act named four additional men to serve with the Governor-General on the Calcutta Council: Lt-Gen John Clavering, George Monson, Richard Barwell, and Philip Francis.

Barwell was the only one with previous experience in India. These councillors were commonly known as the "Council of Four".

- A supreme court was established at Fort William at Calcutta. British judges were to be sent to India to administer the British legal system that was used there.

104. Human immunodeficiency virus infection / acquired immunodeficiency syndrome (HIV/AIDS) is a diseases of the human immune system caused by infection with human immunodeficiency virus (HIV). During the initial infection, a person may experience a brief period of influenza-like illness. This is typically followed by a prolonged period without symptoms.

105. The Moplah rebellion also known as Malabar Rebellion was an armed uprising in 1921 against British authority and Hindus in the Malabar region of Southern India by Mappila Muslims and the culmination of a series of Mappila revolts that recurred throughout the 19th century and early 20th century.

106. Let the work done by A, B and C in one day be 6 units, 3 units and 2 units respectively.

$\Rightarrow$ Total work = 15 × (6 + 2) = 120 units

Hence, required time = $\frac{2 \times 120}{3+2}$ = 48 days.

107. According to the question;

MP × 0.85 × 0.75 × 0.8 = 816

$\Rightarrow$ MP = Rs. 1600

Hence, required selling price = 1600 × 0.65

= Rs. 1,040.

108. Let the values of x, y and z be 2k, 3k and 5k respectively.

$\Rightarrow 30\ k^3 = 15360$ or k = 8

Thus, x = 16, y = 24 and z = 40

Hence, putting the values of y and z in z = by + 4, we get b = 1.5.

109. Let the cost price be Rs. 90 and the marked price be Rs. 110

$\Rightarrow$ Selling price = 110 × 0.9 = Rs. 99

Hence, there would be a profit of Rs. 9 or 10% in the entire transaction.

110. Let the initial price, quantity and expenditure be P, Q and E respectively, such that P × Q = E.

$$\Rightarrow \frac{7}{4}P \times Q_2 = \frac{7}{5}E$$

$$\Rightarrow \frac{7}{4}P \times Q_2 = \frac{7}{5}(P \times Q)$$

Thus, new quantity or $Q_2 = \frac{4}{5}Q$

Hence, there would be a decrease of 20% in the quantity.

111. Distance covered by the car in first 10 hours = 10 × 25 = 250 km

Thus, it will cover remaining 200 km in $\frac{200}{25 \times 0.4}$ = 20 hrs.

Hence, average speed (in m/s)

$$= \frac{450}{30} \times \frac{5}{18} = \frac{25}{6}.$$

112. Let the amount invested be Rs. x.

$$\Rightarrow x : x \times \left\{\left(\frac{21}{20}\right)^n - 1\right\} = 8000 : 1261$$

Hence, n = 3 and required time = 1.5 years.

113. $(x - 2)(x - p) = x^2 - ax + 6$

$\Rightarrow x^2 - x(2 + p) + 2p = x^2 - ax + 6$

$\Rightarrow$ p = 3 and a = 5

Hence, $p^a = 243$.

114. $x - \frac{1}{x} = 2 \Rightarrow x^2 + \frac{1}{x^2} = 6$

Cubing both sides we get,

$$x^6 + \frac{1}{x^6} + 3 \times 6 = 216$$

Hence, $x^6 + \frac{1}{x^6} = 198$

115. $\angle C - \angle A = 70°$ ...(i)

$\angle B + \angle C = 140°$ ...(ii)

$\angle A + 2\angle B = 100°$ ...(iii)

Adding equation (i) and (iii) we get,

$\angle C + 2\ \angle B = 170°$ ...(iv)

Solving equations (ii) and (iv) we get,

$\angle B = 30°$ and $\angle C = 110°$ such that $\angle A = 40°$

Hence, $2\angle B - \angle A = 2 \times 30 - 40 = 20°$.

116.

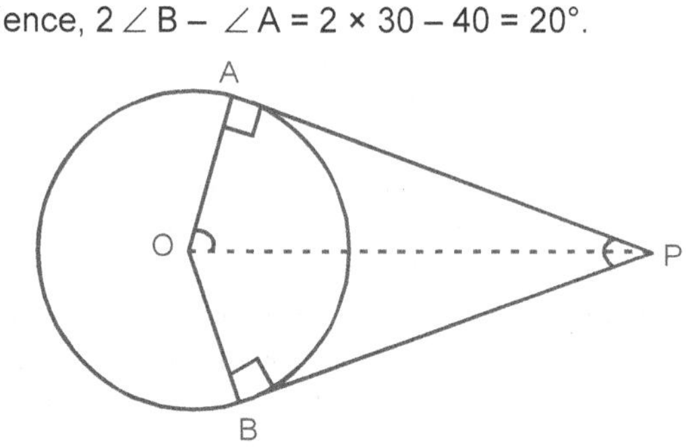

In ΔOAP,

$\angle AOP + \angle APO = 90°$

$\Rightarrow 7x + 11x = 90°$

$\Rightarrow x = 5$

Hence, $\angle BPA = 2 \times 11x = 2 \times 11 \times 5 = 110°$.

117. $\tan(5x - 10°) = \cot(20° - 5y)$

$\Rightarrow 5x - 10° = 20° - 5y = 45°$

Thus, $x = 11°$ and $y = - 5°$

Hence, $6x + 8y = 26°$.

118. When the sum of two natural numbers is given then their product will be maximum when both are equal.

Hence, required product = $12 \times 12 = 144$.

119. Let the required average be 'x'.

$\Rightarrow 60 \times 66 = 20 \times 59 + 0.4 \times 40 \times 68 + 0.6 \times 40 \times x$

Hence, $x = 70.5$.

120. $x^2 + y^2 - 4xy = (x + y)^2 - 6xy$

Now, $x + y = 2\sqrt{5}$ and $x \times y = 5 - \frac{1}{5}$

Hence, $(x + y)^2 - 6xy = 20 + 30 - \frac{6}{5} = 50 - \frac{6}{5}$.

127. "Running over" means "overflowing."

128. "Set me up" means to "establish myself."

129. "Struck down" with means "affected by."

130. This is an instance of parallelism. Since, 'worst' is used, hence the other adjective should also be in the superlative form.

131. 'Unraveled' is the best option because the scientists have used a computer model to figure out the mystery.

132. Refer to the last sentence of the first paragraph for the answer.

133. Refer to the second sentence of the first paragraph for the answer.

134. Refer to the last sentence of the first paragraph for the answer.

135. Refer to the first sentence of the second paragraph for the answer.

136. Refer to the first sentence of the second paragraph for the answer.

140. The adjective 'preferred' itself suggests comparison. Therefore, qualifying it with 'more' is redundant and so inappropriate.

141. There is no error in the given statement.

142. The correct order of placing adjectives is - dimension - shape - colour. Therefore, the correct statement would be, "A big square blue box lying..."

143. 'Passed through' means underwent.

144. 'Set off' implies started.

145. 'Worked out' refers to solved.

146. 'Either' takes a singular verb.

147. The plural of aircraft is aircraft.

148. 'Latter' is wrongly used here. 'Latter' is used only when there are two subjects. 'Last' should have been used here.

158. The given sentence is correct in itself and hence, option (d) is the answer. The statement refers to an action that is going on at some time in the past. In such cases, the past continuous tense is used.

159. The given statement is incorrect because it is not clear whether the place from where they had been banished was a country or a car. The grammatically correct sentence is given in option (c), where the modifier 'where they had been banished' is correctly placed beside 'country' making the place, where they were banished from, clear.

160. When we refer to a hypothetical, usually counterfactual, event or circumstance placed in the past (also known as conditional perfect), with the if clause (such a sentence usually has the if...then structure) we use past perfect tense. For example: If I had known about the fest,[then] I would have told him. Hence, option (B) is the correct answer.

161. The sentence needs no improvement.

162. 'Carry out' means to execute.

163. Refer to the third sentence of the passage. It says that the main idea of a war has either been to expand power, independence, or joining a war to help allies. Hence, option (C) is the correct answer.

164. 'Infamy' refers to the condition of being known for having done an evil or terrible act. 'Infamous' means well known for evil act. Hence, option (B) is the correct answer.

165. Option (A) is incorrect because we cannot infer from the passage which war had more allies. Option (B) says that the degree of damage caused made World War I different from World War II, which can be inferred from the third last sentence of the passage. Hence, option (B) is the answer. Option (C) is out of the scope of the passage.

166. What really triggered the war was Germany invading Poland. Refer to the following lines of the passage: 'So they decided ... invaded Poland which triggered the start of the war'. Hence, option (D) is the correct answer.

167. The passage states that most of the war was fought in trenches. Flame throwers, tanks, etc. were used for the first time. Hence, option (C) is the correct answer. Nuclear bomb was first used in WWII. The passage makes no mention of the use of riffle in the wars.

171. 'Chasing one's tail' means to run around ineffectually. Hence, option (c) is the correct answer.

172. 'To think on one's feet' means to be able to think and speak effectively without any preparation. Hence, option (d) is the correct answer.

173. If a remark is 'close to home', it means that it is accurate and it makes you feel uncomfortable and embarrassed.

Hence, option (b) is the correct answer.

174. 'Misapprehend' means to get a wrong idea about something. Hence, option (c) is the correct answer. 'Apprehend' means to notice and understand something. 'Fathom' means to understand the reason of something.

175. 'Egalitarian' is a person who aims for equal wealth, status, etc., for all people. Hence, option (a) is the correct answer. 'Federalist' is a supporter of the federal system of government. 'Absolutist' is a person who believes that a ruler or government should have total power at all times. 'Reformist' is someone who wants to change political or social situations.

176. An 'annalist' is a person who writes annals or historical records.

Hence, option (b) is the correct answer. An 'archaeologist' is a person who deals with past human life. A 'radiologist' is a doctor trained in radiology. A 'genealogist' is someone who traces or studies the descent of persons or families.

177. 'Merger' makes sense in the context of the sentence.

178. The subject is 'Supreme Court guideline' and it is singular.

179. The given sentence is incorrect since 'child' is singular while 'they' refers to more than one. In option (c), 'child' is singular. Therefore, 'think', which is plural, should be replaced by 'thinks'. Option (b) is incorrect since 'children', which is plural, should be followed by 'are' and not 'is'. Only option (a) is correct since it takes one of the two prepositions (he or she) that go with 'child'.

180. 'Less' is used for uncountable nouns. For example: Less money, less time, etc. 'Fewer' is used for countable nouns. Since 'children' is a countable noun, we will use 'fewer' instead of 'less'. Option (c) is incorrect since 'suffers' is singular, while 'children' is plural. Hence, option (b) is the correct answer.

PRACTICE PAPER-5

ELECTRONICS ENGINEERING

Time Allowed : 2:15 Hours **Maximum Marks : 180**

Read the following instructions carefully before you begin to answer the questions. This Booklet contains questions in English.

INSTRUCTIONS

1. This Booklet contains 180 questions in total comprising of the following five parts:

Part – I:	Domain Knowledge	[1 – 75 Questions]
Part – II:	General Intelligence and Reasoning	[76 – 90 Questions]
Part – III:	General Awareness	[91 – 105 Questions]
Part – IV:	Quantitative Aptitude	[106 – 120 Questions]
Part – V:	General English	[121 – 180 Questions]

2. All questions are compulsory and carry 1 mark each.

3. Before you start to answer the questions you must check this Booklet to ensure that it contains all the pages and see that no page is missing or repeated.

4. For every wrong answer 1/3 marks will be deducted.

5. You are supplied with the OMR-Sheet separately.

6. Answers must be indicated by completely blackening one of the corresponding ovals on OMR-Sheet against the relevant question number, by black/blue ball pen only. Answers which are not indicated by black/blue ball pen will not be awarded any mark.

7. The OMR-Sheet must be handed over to the Invigilator before you leave the Examination Hall.

8. Failure to comply with any of the above instructions will render a candidate liable to such action/penalty as may be deemed fit.

PRACTICE PAPER – 5

DOMAIN KNOWLEDGE

1. The current in the given circuit with a dependent voltage source is

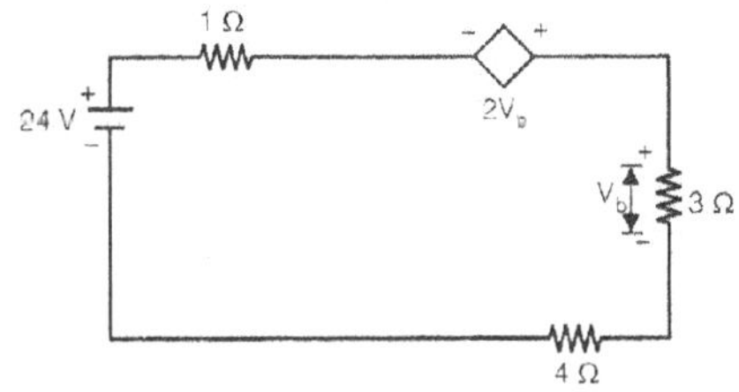

(*a*) 10 A (*b*) 12 A
(*c*) 14 A (*d*) 16 A

2. The Thevenin's equivalent resistance R_{th} for the given network is

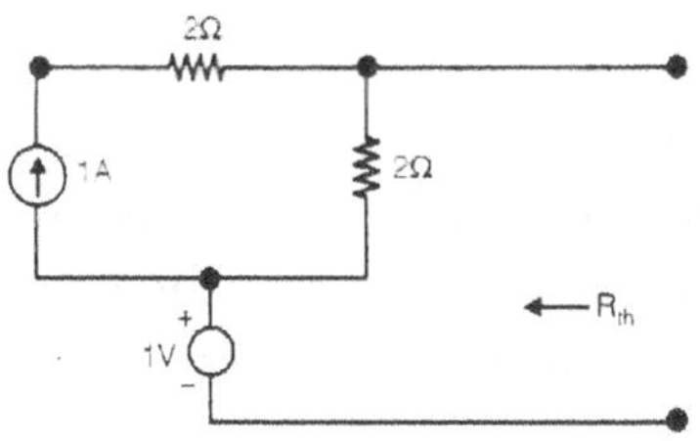

(*a*) 1 Ω (*b*) 2Ω
(*c*) 4 Ω (*d*) infinity

3. the total impedance of Z(jω) of the circuit shown below is

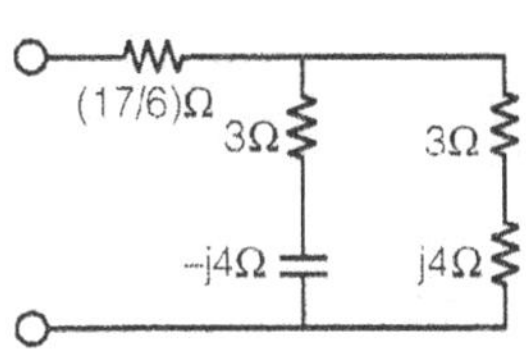

(*a*) (6+j0) Ω (*b*) (7+j0) Ω
(*c*) (0+j8) Ω (*d*) (6+j8) Ω

4. A voltage source having an internal impedance of 8 + j6 ohms supplies power to a resistive load. What should be the load resistance for maximum power transferred to it?

(*a*) 8 ohms (*b*) 6 ohms
(*c*) 10 ohms (d) $\sqrt{10}$ ohms

5. Consider the following circuit:

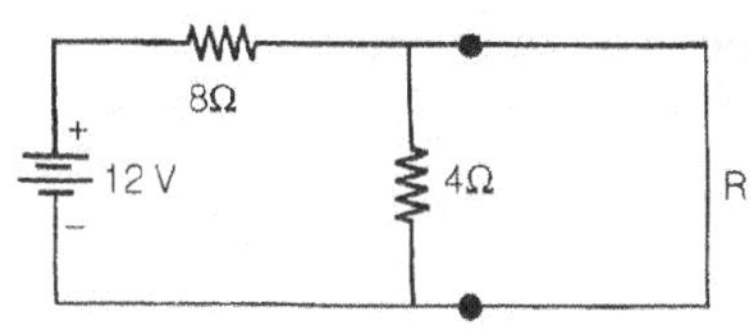

What should be the value of resistance R, in the above circuit if it has to absorb the maximum power from the source?

(*a*) 8/3 ohms (*b*) 3/8 ohms
(*c*) 4 ohms (*d*) 8 ohms

6. A balanced delta connected load has an impedance of 9 ∠ 30°ohms per phase. What is the impedance per phase of its equivalent star?

(*a*) 27 ∠ 30°Ω (*b*) 27 ∠ 90°Ω
(*c*) 3 ∠ 30°Ω (*d*) 3 ∠ 20°Ω

7. In an a.c. series RLC circuit, the voltage across R and L is 20 V, voltage across L and C is 9 V and voltage across RLC is 15 V. What is the voltage across C?

(*a*) 7V (*b*) 12V
(*c*) 16V (*d*) 21V

8. For the circuit shown below, what is the voltage across the current source I_S?

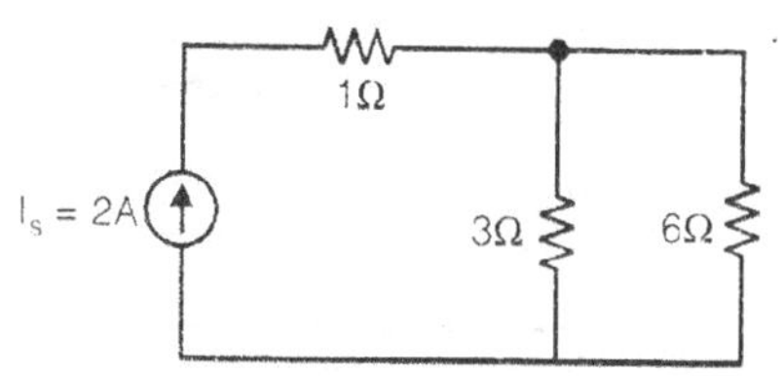

(*a*) 0 (*b*) 2V
(*c*) 3V (*d*) 6V

9. For the circuit shown below, the value of R is adjusted, so as to make the current in R_L equal to zero. What is the value of R ?

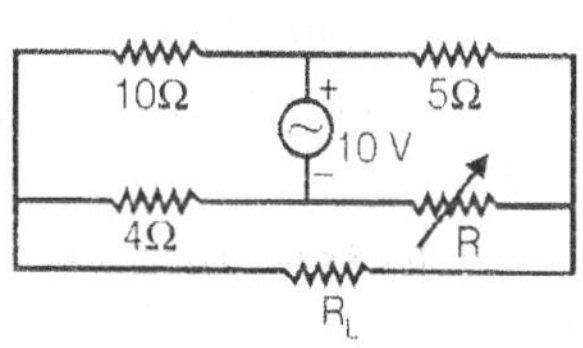

(*a*) 1Ω (*b*) 2Ω
(*c*) 3Ω (*d*) 4Ω

10. A 3-phase delta-connected sym-metrical load consumes P watt of power from a balanced supply. If the same load is connected in star to the same supply, then what is the power consumption?

(*a*) P/3 (*b*) P
(*c*) $\sqrt{3}$P (*d*) 3P

11. For the a.c circuit given below, what is the value of I?

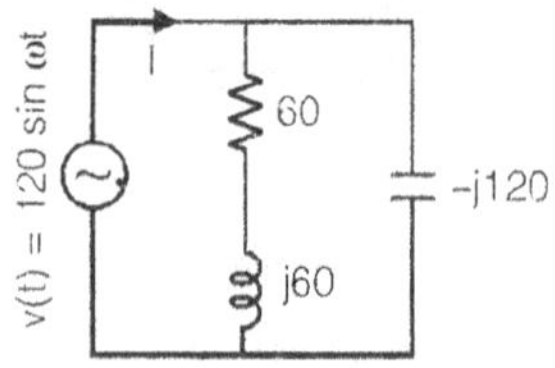

(*a*) 1 + j 1 (*b*) 1 + j0

(*c*) 0 -j 1 (*d*) 0 + j0

12. Material used for fabrication of Tunnel diode is

(*a*) Ge or GaAs (*b*) Si and Ga As

(*c*) Si and InSb (*d*) Ge and InSb

13. In a series R-L-C circuit, the maximum voltage across the capacitor occurs at a frequency

(*a*) double the resonant frequency

(*b*) equal to resonant frequency

(*c*) $\sqrt{2}$ times the resonant frequency

(*d*) below the resonant frequency

14. For the network shown below, if the current $i(t) = \sqrt{2}\sin(\omega t\text{-}30^{\circ})$, then what is the value of R?

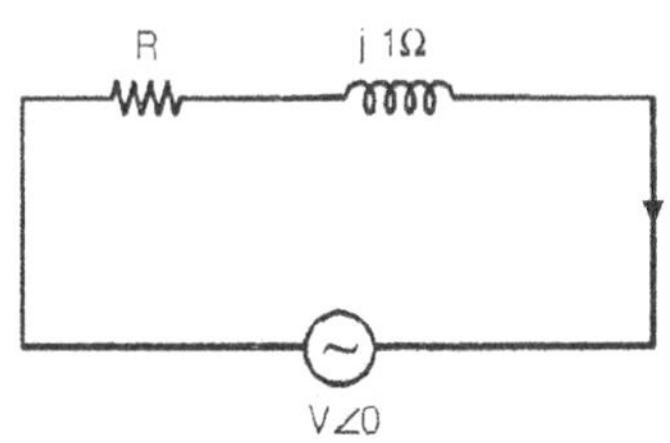

(*a*) 1Ω (*b*) 3Ω

(*c*) $\sqrt{3}\,\Omega$ (*d*) $3\sqrt{3}\,\Omega$

15. According to maximum power transfer theorem, when is the maximum power absorbed by one network from another network?

(*a*) The impedance of one of the network is half that of the outer

(*b*) The impedance of one is the complex conjugate of the other

(*c*) The impedance of one is equal to that of the other

(*d*) Only the resistive parts of both are equal

16. An ideal op-amp is an ideal

(*a*) voltage controlled current source

(*b*) voltage controlled voltage source

(*c*) current controlled current source

(*d*) current controlled voltage source

17. The action of a JFET in its equivalent circuit can be best represented as a

(*a*) Current Controlled Current Source

(*b*) current controlled voltage source

(*c*) Voltage Controlled Voltage Source

(*d*) voltage controlled current source

18. Voltage series feedback (also called series-shunt feedback) results in

(*a*) increase in both input and output impedances

(*b*) decrease in both input and output impedances

(*c*) increase in input impedance and decrease in Output impedance

(*d*) decrease in input impedance and increase in output impedance

19. In a BJT amplifier with the introduction of feedback, the input impedance is reduced, output impedance is increased, bandwidth is increased and distortion is reduced. The feedback is

(*a*) Voltage series (*b*) Current series

(*c*) Voltage shunt (*d*) Current shunt

20. If the Q of a single stage tuned amplifier is doubled then its bandwidth

(*a*) Remains same (*b*) Becomes half

(*c*) Becomes double (*d*) Becomes fourtimes

21. Positive feedback, in an amplifier circuit will....... the gain of amplifier.

(*a*) increase (*b*) decrease

(*c*) not alter (*d*) none

22. The condition which is required to get sustained oscillation is called

(*a*) Barkhausen criteria

(*b*) $(A\beta)=1$

(*c*) Phase should be zero

(*d*) All of above

23. The phase shift introduced by Wien bridge is

(*a*) 90° (*b*) 180°

(*c*) 0° (*d*) 150°

24. We use a crystal oscillator because

(*a*) It gives high output voltage

(*b*) It works at high efficiency

(*c*) The frequency of oscillations remains substantially constant

(*d*) It requires very low dc supply voltage

25. Oscillator is a

(*a*) ac to dc converter

(*b*) dc to ac converter

(*c*) ac to ac converter

(*d*) none

26. The transfer function of a system shown in figure is given by

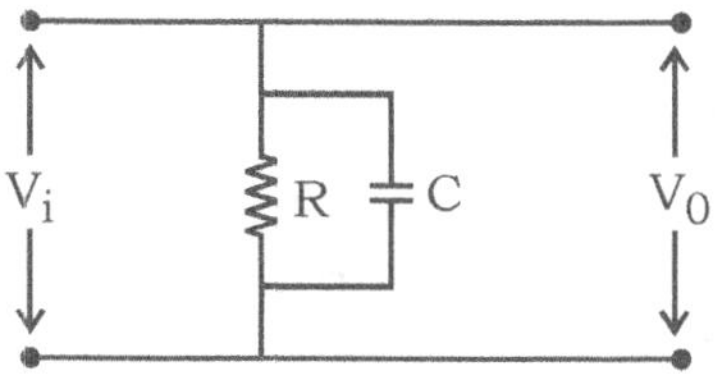

(a) $\frac{1}{SCR+1}$ (b) $\frac{1}{SCR}$

(c) 1 (d) $\frac{1}{S(SRC+1)}$

27. Which of the following converter provides DC to DC conversion?

(a) Inverter (b) Controlled Rectifier

(c) Chopper (d) AC voltage controller

28. Snubber cricuit is used for

(a) Firing the thyristor

(b) $\frac{di}{dt}$ protection

(c) $\frac{dv}{dt}$

(d) Thermal protection

29. To make current source inverter from voltage source inverter, which of the following element is connected with voltage source?

(a) Resistor in series

(b) Inductor in series

(c) Capacitor in parallel

(d) Capacitor in series

30. A signal given by $x(t) = \sum_{n=0} \cos(nt)$, is expressed is terms of fourier series components. The value of dc component of fourier series is

(a) 1 (b) 2π

(c) π (d) $\frac{1}{2}$

31. The Laplace transform of a signal x(t) is x(s) and given as

$$x(s) = \int_{-\infty}^{\infty} x(t)e^{-st}\,dt$$

The laplace variable s is described as

(a) ω (b) $j\omega$

(c) $\sigma - j\omega$ (d) $\sigma + j\omega$

32. The fundamental period of a signal $x(t) = \cos(t) + \sin\left(\sqrt{2}\,t\right)$ is

(a) 2π (b) $\sqrt{2}\,\pi$

(c) $\sqrt{2}$ (d) None of these

33. The value of the following integral $\int_{1}^{2}(3t^2+1)\delta(t)\,dt$ is where $\delta(t)$ is unit impulse signal

(a) –22 (b) 22

(c) 0 (d) 1

34. For an odd signal, its value at t = 0 is

(a) –1 (b) 0

(c) 1 (d) can not determine

35. The open loop gain of an amplifier is 200. If negative feedback with $\beta = 0.2$ is used, the closed loop gain will be

(a) 200 (b) 40.12

(c) 4.878 (d) 2.2

36. In figure as the load resistance is changed

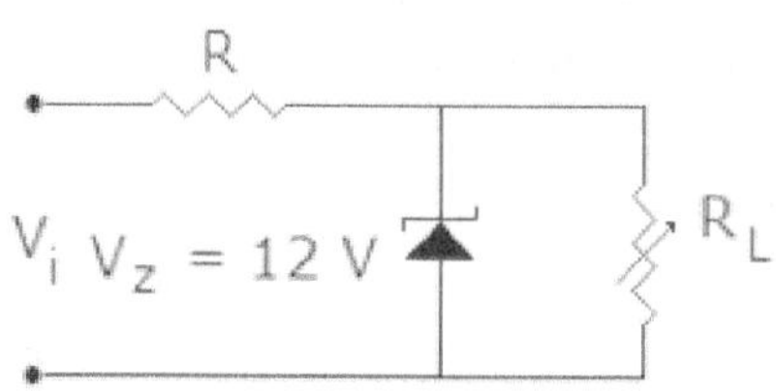

(a) the zener current remain constant

(b) the load current remains constant

(c) the output voltage remains constant

(d) the output voltage and current through R remain constant

37. In figure, voltage across R_2 = + 10 V. If V_{BE} = 0.7 V and V_E = 0.7 V and R_E = 10 kΩ, current through R_E is

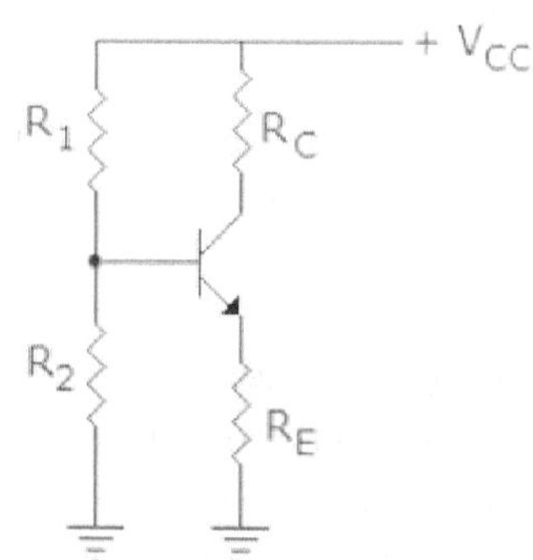

(a) 1 mA (b) 0.93 mA

(c) 2 mA (d) 1.86 mA

38. An RC oscillator uses

(*a*) one RC combination

(*b*) two RC combinations

(*c*) at least three RC combinations

(*d*) either (a) or (b)

39. AC signals are given to both inverting and non-inverting terminals of an op-amp. When will the output maximum

(*a*) when the phase difference between the inputs is 0°

(*b*) when the phase difference between the inputs in 90°

(*c*) when the phase difference between the inputs is 180°

(*d*) when the phase difference between the inputs is 0 or 180°

40. Two coils having self inductances of 10 mH and 40 mH are mutually coupled. The mutual inductance is

(*a*) 40 mH (*b*) 20 mH

(*c*) 10 mH (*d*) 5 mH

41. An op-amp has zero gain for common mode inputs. Then CMRR =

(*a*) zero (*b*) infinite

(*c*) high (*d*) low

42. Find resistance R_B to bring transistor to threshold of saturation $V_{CB} = 0$, $V_{BE} = 0.7$ V, $\alpha = 0.96$

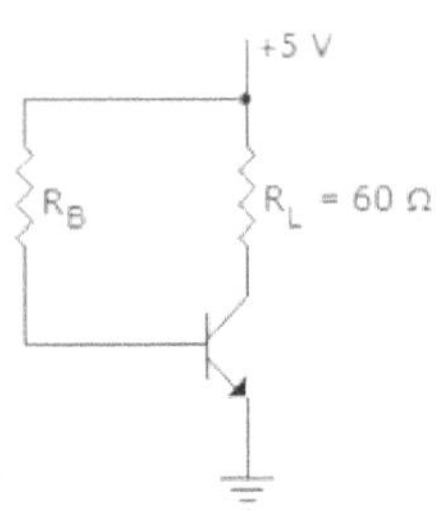

(*a*) 1.5 kΩ (*b*) 15 kΩ

(*c*) 10 kΩ (*d*) 1 kΩ

43. The output voltage waveform of a CE amplifier is fed to a dc coupled CRO. The trace on the screen will be

(*a*) dc output voltage

(*b*) ac output voltage

(*c*) sum of dc and ac output voltage

(*d*) either (*a*) or (*b*)

44. A 120 V, 30 Hz source feeds a half wave rectifier circuit through a 4 : 1 step down transformer, the average output voltage is

(*a*) 30 V (*b*) 13.5 V

(*c*) 10 V (*d*) 7.07 V

45. Calculate the output resistance of a differential amplifier circuit with $R_C = 1$ kΩ

(*a*) 1 kΩ (*b*) 2 kΩ

(*c*) 3 kΩ (*d*) 4 kΩ

46. If $e = 110 \sin(\omega t + \pi/3)$ and $i = 5 \sin(\omega t - \pi/3)$, the impedance of the circuit is

(*a*) 22 Ω (*b*) 16 Ω

(*c*) 30.8 Ω (*d*) 11 Ω

47. Ohm's law is valid for

(*a*) dc voltage and current

(*b*) ac voltage and current

(*c*) all voltages and current

(*d*) dc and ac voltages and currents

48. In a RL circuit, the angle, θ by which the current lags the voltage is

(*a*) $0 > \theta > -90°$ (*b*) 90°

(*c*) 0° (*d*) $0 < \theta < 90°$

49. A 60 W, 23 V bulb is used for 45 minutes. The electrical energy consumed in commercial units is

(*a*) 2700 units (*b*) 0.45 units

(*c*) 4.5 units (*d*) 0.045 units

50. Two capacitors of 1 F and 2 F are connected in series across a 10 V dc battery. After the capacitors have been charged, the charges across the two capacitors will be

(*a*) 10 C each (*b*) 20 C each

(*c*) 10 C and 20 C (*d*) 20 C and 10 C

51. A series RLC circuit has Q = 5. A voltage of 10 V rms is applied to it. At resonance, the voltage across C is

(*a*) 10 V rms

(*b*) 10 V peak

(*c*) 50 V rms

(*d*) 50 V peak

52. Which of the following features are offered by a bipolar junction transistor amplifier in Darlington connection?

1. High voltage gain
2. High input impedance
3. High current gain

Select the correct answer using the code given below:

(*a*) 1 and 2 only

(*b*) 2 and 3 only

(*c*) 1 and 3 only

(*d*) 1,2 and 3

53. A coil of resistance R and inductance L is connected in series with 5 F capacitor. The source frequency is equal to resonance frequency and circuit current is 0.5A Another 5 F capacitor is connected in parallel with the above capacitor. The circuit current will be

(*a*) 0.5 A (*b*) more than 0.5 A

(*c*) less than 0.5 A (*d*) 0.5 A or less

54. A capacitor having capacitance C is raised to voltage V. The energy stored will be

(*a*) CV^2 (*b*) V^2/C

(*c*) $0.5\ V^2C$ (*d*) $0.5\ C^2V$

55. The electric energy required to raise the temperature of a given amount of water is 1000 kWh. If heat losses are 25%, the total heating energy required is ________ kWh.

(*a*) 1500 (*b*) 1250

(*c*) 1333 (*d*) 1000

56. In saturation region JFET acts like a

(*a*) bipolar device (*b*) short circuit

(*c*) switch (*d*) resistance

57. In an op-amp a common mode signal is applied to

(*a*) inverting terminal

(*b*) non-inverting terminal

(*c*) both terminals

(*d*) one or both the terminals

58. Which of the following op-amp circuits has zero resistance in the feedback path?

(*a*) Inverting amplifier

(*b*) Summing amplifier

(*c*) Source follower

(*d*) Integrating amplifier

59. A bipolar transistor is a ________ controlled devices whereas a FET is a ________ controlled devices.

(*a*) current, voltage (*b*) current, current

(*c*) voltage, current (*d*) voltage, voltage

60. In a N-P-N transistor of the emitter junction is reverse biased and collector junction is also reversed biased, the transistor will operate in

(*a*) active region (*b*) saturation region

(*c*) cut-off region (*d*) inverted region

61. The hexadecimal number 'A0' has the decimal value

(*a*) 80 (*b*) 256

(*c*) 100 (*d*) 160

62. For the logic circuit of the given figure, the minimized expression is

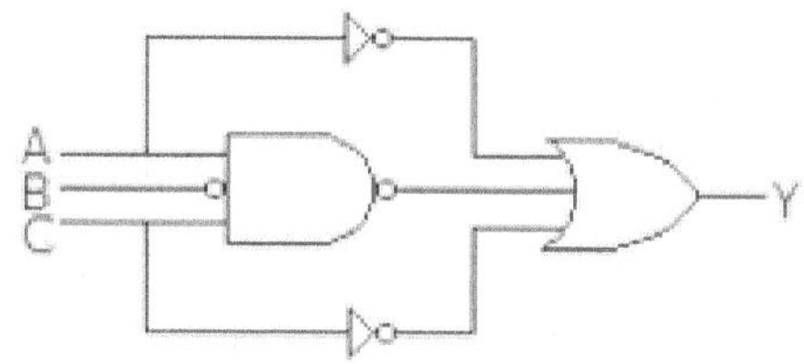

(*a*) $Y = \overline{A\overline{B}C}$ (*b*) Y = A + B + C

(*c*) Y = A + B (*d*) Y = ABC

63. Which condition determines the causality of the LTI system in terms of its impulse response?

(*a*) Only if the value of an impulse response is zero for all negative values of time

(*b*) Only if the value of an impulse response is unity for all negative values of time

(*c*) Only if the value of an impulse response is infinity for all negative values of time

(*d*) Only if the value of an impulse response is negative for all negative values of time

64. Under which conditions does an initially relaxed system become unstable?

(*a*) only if bounded input generates unbounded output

(*b*) only if bounded input generates bounded output

(*c*) only if unbounded input generates unbounded output

(*d*) only if unbounded input generates bounded output

65. Which of the following is minimum error code?

(*a*) Octal code (*b*) Grey code

(*c*) Binary code (*d*) Excess 3 code

66. The hexadecimal number $(3E8)_{16}$ is equal to decimal number

(*a*) 1000 (*b*) 982

(*c*) 768 (*d*) 323

67. An electric circuit with 15 branches and 8 nodes will have independent

(*a*) 4 loop equations (*b*) 8 loop equations

(*c*) 15 loop equations (*d*) 10 loop equations

68. Superposition theorem is not applicable for

(*a*) voltage calculation

(*b*) bilateral elements

(*c*) power calculations

(*d*) passive elements

69. An FM wave uses a 25V, 500 Hz modulating frequency and has a modulation index of 50. The deviation is

(*a*) 500 Hz (*b*) 100 Hz
(*c*) 1250 Hz (*d*) 25000 Hz

70. For the circuit given in figure below the power delivered by the 2 volt source is given by

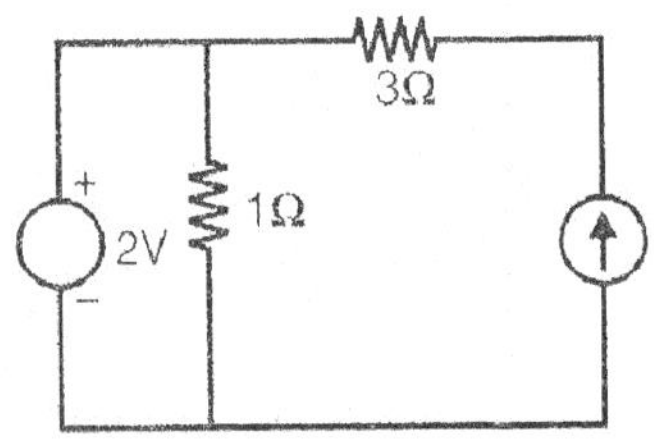

(*a*) 4 W (*b*) 2 W
(*c*) -2 W (*d*) -4 W

71. At resonant frequency a RLC series circuit draws maximum current due to the reason that

(*a*) the difference between capacitance reactance and inductance reactance is zero
(*b*) the impedance is more than the resistance
(*c*) the voltage across the capacitor equals the applied voltage
(*d*) the power factor is less than unity

72. In a network made up of linear resistors and ideal voltage sources, values of all resistors are doubled. Then the voltage across each resistor is

(*a*) Doubled
(*b*) Halved
(*c*) Decreases four times
(*d*) Not changed

73. In the given circuit, the current source is sinusoidal with frequency equal to the resonant frequency of the circuit What is the value of current through the resistor?

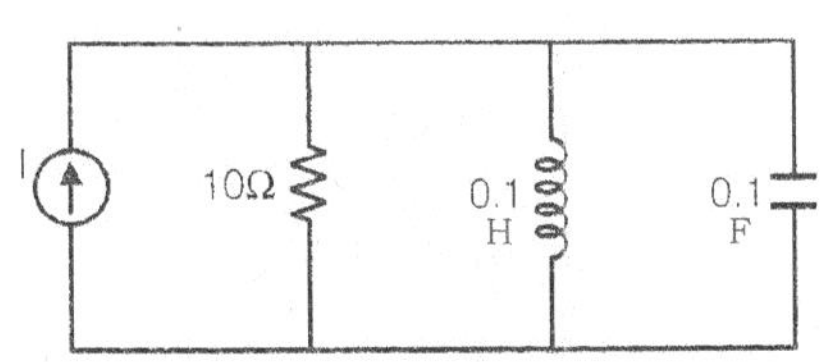

(*a*) 0 (*b*) 0.1I
(*c*) I (*d*) 10 I

74. One of the main functions of RF stage amplifier in Super Heterodyne Radio Receiver is to

(*a*) provide improve tracking:
(*b*) permit better adjacent channel rejection.
(*c*) increase tuning range of the receiver
(*d*) improve rejection of image frequency.

75. In a half-wave rectifier. If an ac supply is 50 Hz, then what is the ac ripple at output?

(a) 50 Hz (b) 45 Hz
(c) 60 Hz (d) 150 Hz

GENERAL INTELLIGENCE AND REASONING

76. If – = Less than, × = Not equal to, | = Greater than, $ = Not less than, # = Not greater than, @ = Equal to and A $ B @ C is true, then which of the following is correct?

(*a*) C @ B # A (*b*) C @ B | A
(*c*) A $ C – B (*d*) B – A # C

77. Identify one response which would be a correct inference from the given premises stated according to the following symbols:

'A' stands for less than
'D' stands for equal to
'B' stands for addition
'F' stands for subtraction
'G' stands for division
'E' stands for multiplication
'I' stands for greater than
Premises (3M I 4N) and (2N D 5K)

(*a*) 4M I 20K (*b*) 2M A K
(*c*) 2K I N (*d*) M I 2K

Directions: In question nos. **78** to **82**, select the missing number from the given responses.

78. 21 27 14 58 33 71
6 9 ?

(*a*) 12 (*b*) 13
(*c*) 15 (*d*) 18

79. 2, 5, 10, 25, 50, 125, ?

(*a*) 240 (*b*) 440
(*c*) 250 (*d*) 380

80. 2, 5, 12, 27, 58, ?

(*a*) 109 (*b*) 113
(*c*) 121 (*d*) 129

81. 48 21 105 52 ? 29
30 56 45

(*a*) 122 (*b*) 136
(*c*) 84 (*d*) 71

82. 317 312 303
623 618 609
834 829 ?

(*a*) 822 (*b*) 808
(*c*) 812 (*d*) 820

Directions: In question nos. **83** to **86**, select the related number / word / letters from the given alternatives.

83. Frog : Amphibian :: Koala : ?
(*a*) Aerial (*b*) Terrestrial
(*c*) Aquatic (*d*) Arboreal

84. OAGE : UGMK :: READ : ?
(*a*) WJGJ (*b*) XKGJ
(*c*) WJHK (*d*) XKHK

85. 17 : 11 :: 73 : ?
(*a*) 58 (*b*) 63
(*c*) 69 (*d*) 80

86. Periodic : Constant :: Sporadic : ?
(*a*) Witty (*b*) Irregular
(*c*) Instantaneous (*d*) Suave

Directions: In question nos. **87** and **88**, which one set of letters when sequentially placed at the gaps in the given letter series shall complete it?

87. a _ _ a a a b b _ b b a a _ b b
(*a*) bbaa (*b*) bbab
(*c*) aabb (*d*) abab

88. AYB_, AY_Z, A_BZ, _ YBZ
(*a*) ZBYA (*b*) YABZ
(*c*) YBZA (*d*) ABYZ

89. Find out the wrong number in the sequence
106, 104, 101, 94, 86, 76, 64.
(*a*) 104 (*b*) 76
(*c*) 101 (*d*) 86

Directions: In question nos. **90**, select the one which is different from the other three responses.

90. (*a*) MUTE = 1321205 (*b*) PRAY = 1618125
(*c*) NEWS = 1452319 (*d*) BOLD = 2151204

GENERAL AWARENESS

91. Which one of the following Vedas is not part of Vedatrayi?
(*a*) Rig Veda (*b*) Sama Veda
(*c*) Yajur Veda (*d*) Atharva Veda

92. How many atoms of chlorine are there in a molecule of Magnesium Chloride?
(*a*) 1 (*b*) 2
(*c*) 3 (*d*) 4

93. The trade union act was passed in which year?
(*a*) 1920
(*b*) 1924
(*c*) 1926
(*d*) 1928

94. Game Theory was propounded by
(*a*) John Von Neumamn
(*b*) Alfred Marshall
(*c*) Adam Smith
(*d*) J.M. Keynes

95. Bovine Spongiform Encephalopathy is the scientific name of which of the following diseases?
(*a*) Mad cow disease (*b*) Leprosy
(*c*) Cholera (*d*) Small Pox

96. $C_6H_{12}O_6$ is the chemical formula for
(*a*) Methane gas (*b*) Glucose
(*c*) Ethanol (*d*) Citric Acid

97. In fluid mechanics, Mach Number is one if the speed of an object is equal to the
(*a*) speed of sound
(*b*) speed of light
(*c*) rotational speed of the earth
(*d*) speed at which the earth revolves around the sun

98. Which of the following is used as an additive booster in gasoline?
(*a*) Lysergic acid diethylamide
(*b*) Carbon Monoxide
(*c*) Chlorophyll
(*d*) Tetraethyl Lead

99. Which of the following plants belong to genus Allium?
(*a*) Tomato (*b*) Onion
(*c*) Chilli (*d*) Coriander

100. At room temperature, fluorescence is exhibited by which of the following elements?
(*a*) Silicon (*b*) Carbon
(*c*) Hydrogen (*d*) Phosphorus

101. Which of the following elements is not present in "laughing gas"?
(*a*) Hydrogen (*b*) Nitrogen
(*c*) Oxygen (*d*) All of the above

102. Which of the following elements is not a part of that group in the modern periodic table to which other three belong?
(*a*) Sodium (*b*) Potassium
(*c*) Lithium (*d*) Hydrogen

103. Which of the following elements is kept in kerosene to prevent it from reacting with air vigorously and catching fire?
(*a*) Chromium (*b*) Tungsten
(*c*) Sodium (*d*) Helium

104. Which of the following animals is a ruminent?

(a) Dog (b) Cow

(c) Cat (d) Mouse

105. Calcium in our body is absorbed with the help of

(a) Vitamin A (b) Vitamin C

(c) Vitamin D (d) Vitamin E

QUANTITATIVE APTITUDE

106. A can do a work in $12\frac{1}{2}$ days and B can do the same work in $16\frac{2}{3}$ days. If they together work on it for 4 days, then the fraction of the work that is left is

(a) $\frac{11}{25}$ (b) $\frac{27}{50}$

(c) $\frac{19}{50}$ (d) $\frac{12}{25}$

107. If the area of a circle is A cm^2, radius of the circle is r cm and perimeter of the semicircle is P cm, then

(a) $rP = A + 2r^2$ (b) $P = A + 2r$

(c) $AP = \frac{r^2}{4}$ (d) $\frac{A}{r} = P$

108. A shopkeeper gains 27.5% after allowing a discount of 15% on the marked price of an article. Find his profit percent if the articles are sold after allowing a 20% discount on the marked price.

(a) 20% (b) 100%

(c) 27% (d) 63%

109. A invested Rs. 14,500 in a project. After 2 months, B joined A with a capital of Rs. 15,000 while after 3 months A withdrew Rs. 3,500. At the start of the 5th month, B invested Rs. 5,000 more and at the same time C joins the business with a capital of Rs. 25,000. The share of C exceeds that of B, out of a total profit of Rs. 1,49,100 after one year by-

(a) Rs. 2,800 (b) Rs. 3,500

(c) Rs. 4,750 (d) Rs. 6,300

110. A dealer sold a scooter at a profit of 10%. Had he bought the scooter at 15% less price and sold it at Rs. 75 more, he would have gained 30%. The cost price of the scooter was

(a) Rs. 15,600

(b) Rs. 16,000

(c) Rs. 14,200

(d) Rs. 15,000

111. Kiran spends 60% of her income. Her income is increased by 30% and she decreased her expenditure by 5%. Her savings are increased by

(a) 72.5% (b) 21.6%

(c) 82.5% (d) 17.5%

112. A boat can travel with a speed of 17 km/hr in still water. If the speed of stream is 5 km/hr, then the time taken by boat to go 57 km downstream and then come back is

(a) $7\frac{15}{44}$ hrs (b) $4\frac{3}{37}$ hrs

(c) 8 hrs (d) 12 hrs

113. If $x = \frac{1}{\sqrt{3}+\sqrt{2}}$, then $x^2 + 2\sqrt{2}x + 1$ is equal to

(a) 0 (b) $\sqrt{2} - 1$

(c) $\sqrt{2} + 1$ (d) 2

114. If x = 222, y = 223, z = 225, then the value of $x^3 + y^3 + z^3 - 3xyz$ is

(a) 6690 (b) 5690

(c) 3690 (d) 4690

115. If two medians YA and ZB of a triangle XYZ, interest each other at O and if OY = ZO, $\angle YOZ = 60°$ and YZ = 10 cm, then the ratio of the area of the triangle YOZ to the area of triangle XYZ is

(a) 2 : 1 (b) 3 : 1

(c) 1 : 2 (d) 1 : 3

116. XYZ is a cyclic triangle and the bisectors of $\angle XYZ$, $\angle YXZ$ and $\angle XZY$ meet the circle at M, N and O respectively.The sum of the angles $\angle MNO$ and $\angle NMO$ is

(a) $90° + \frac{Z}{2}$ (b) $90° - \frac{X}{2}$

(c) $90° - \frac{Z}{2}$ (d) $90° + \frac{Y}{2}$

117. If $0° \le A \le 90°$, the simplified form of the given expression sinAcosA(tanA – cot A)

$$-\left(\frac{1}{\operatorname{cosec}A \sec A} - 1\right)\left(\frac{1}{\cot A} - \frac{1}{\tan A}\right) \text{ is}$$

(a) $\cos^2 A$

(b) –2tan2A

(c) – 2cot2A

(d) $1 - 2\cos^2 A$

118. Eighteen years ago, the ratio of A's age to B's age was 21 : 26. The ratio of their present ages is 6 : 7. What is the present age of A?

(*a*) 70 years (*b*) 50 years

(*c*) 40 years (*d*) 60 years

119. Out of the five numbers the average of the first four is 13.5 and that of the last four is 18.5. If the first number is 8, then the average of the first and last numbers is

(*a*) 23 (*b*) 15

(*c*) 24 (*d*) 18

120. If $\frac{x^{12}+1}{x^6}=9$, then the value of $\frac{x^{36}+1}{x^{18}}$ is

(*a*) 535 (*b*) 615

(*c*) 640 (*d*) 702

GENERAL ENGLISH

Directions: In question nos. **121** to **123**, each of the following sentences contains a blank to be filled in with an appropriate word. Choose the correct alternative out of the four given options.

121. Night came ____. A full moon ____ high over the trees.

(*a*) by, risen (*b*) on, rose

(*c*) in, rise (*d*) by, raised

122. 'Feel free to contact me ______ my email address', said Flintstone.

(*a*) upon (*b*) on

(*c*) at (*d*) by

123. Even the president's own bipartisan commission can't agree ______ what to do.

(*a*) with (*b*) on

(*c*) to (*d*) at

Directions: In question nos. **124** to **126**, some parts of the sentences have errors and some are correct. Find out which part of a sentence has an error and mark the answer corresponding to the appropriate letter (A, B, C). If a sentence is free from error, mark your answer as (*d*).

124. Between you and me,(*a*)/ it's Rahul's terrible habit that (*b*)/ both Somiya and me find annoying. (*c*)/ No error (*d*)

125. I'm seeing Namit (*a*)/ at 5 and then I am (*b*)/ have dinner with Dikshant. (*c*)/ No error (*d*)

126. If I were working (*a*)/ in Bandra, I will (*b*)/ commute by the local train. (*c*)/ No error (*d*)

Directions: In question nos. **127** to **129**, four alternatives are given for the idioms/phrases given below. Choose the alternative which best expresses the meaning of the given idiom/phrase.

127. Until now, Aladdin couldn't figure out how to <u>get his foot in the door</u>, but now he has the perfect contact.

(*a*) enter an organisation as an initial step towards success

(*b*) enter into a competition that offers huge prize money

(*c*) step out of a difficult situation which involves money

(*d*) use the available opportunity for one's own good

128. Miss Jasmine <u>kicked the bucket</u> last week.

(*a*) had a fit (*b*) died

(*c*) retired (*d*) fired someone

129. Dr Jekyll, the chief of critical care at the hospital, said that the medical procedure was <u>in the home stretch.</u>

(*a*) time consuming. (*b*) started.

(*c*) mid-way. (*d*) nearly completed.

130. In the following question, out of the four alternatives, choose the one which best expresses the meaning of the given word and mark it.

Obliterate

(*a*) Extradite (*b*) Mason

(*c*) Meadow (*d*) Extirpate

131. In the following question, choose the word opposite in meaning to the given word and mark it.

Insolent

(*a*) Brazen (*b*) Impolite

(*c*) Arrogant (*d*) Refined

132. In following question, four words are given in each question, out of which only one word is correctly spelt. Find the correctly spelt word and mark your answer in the answer sheet.

Find the correctly spelt word:

(*a*) Hermeneutic (*b*) Hermenetic

(*c*) Hermenutic (*d*) Hermenytic

Directions: In question nos. **133** to **135**, Out of the four alternatives, choose the one which can be substituted for the given words/sentences.

133. Science of rearing of animals for farming

(*a*) Aviculture (*b*) Horticulture

(*c*) Agriculture (*d*) Apiculture

134. An agent who negotiates contracts of purchase and sale

(*a*) Salesman (*b*) Broker

(*c*) Bargainer (*d*) Moderator

135. A person who believes something is very important and should be followed all the time

(*a*) Stickler (*b*) Perseverer

(*c*) Slacker (*d*) Moral

Directions: In question nos. **136** to **140**, you have a passage with 5 questions. Read the passage carefully and choose the best answer to each question out of the four alternatives given.

Passage – I

The positive nutritional effects led the Olmec to believe chocolate drink possessed mystic qualities, so it was generally reserved for important figures at sacred ceremonies. The Olmec passed the chocolate drink on to the Maya civilization, who found that chocolate was a mood enhancer and energy booster. They passed it on to perhaps the beverage's most famous historical forefathers, the Aztecs. Legendary Aztec leader Montezuma II was known to demand cacao beans from conquered peoples and supposedly drank five goblets of hot chocolate every day in a display of power and opulence. Besides his own enjoyment of the beverage, he only allowed those who contributed military service to drink chocolate. When Hernan Cortes and his soldiers encountered the Aztecs, one of his men wrote about Montezuma's consumption of the curious cacao-made drink and how the Spanish themselves were also served the beverage "all frothed up." Ultimately, Cortes conquered the Aztecs, and brought the popular drink to Spain, from which it spread throughout Europe and, eventually, the world.

136. Which of the following civilizations might not have enjoyed a chocolate drink?

(*a*) Maya civilization (*b*) Aztec
(*c*) Olmec (*d*) Cortes

137. Who among the following must not have been allowed to drink chocolate during the rule of Montezuma II?

(*a*) Soldiers
(*b*) Important figures
(*c*) The Aztec leader of the time
(*d*) Visitors

138. Find an option that is same in meaning to the word given in bold.

Opulence

(*a*) Authority (*b*) Wealth
(*c*) Political control (*d*) Greatness

139. Which of the following is not a characteristic of the chocolate drink?

(*a*) Energy booster (*b*) Mood enhancer
(*c*) Nutritional (*d*) Intoxicating

140. What made Montezuma II drink five goblets of chocolate each day?

(*a*) He wanted to show how great he was.
(*b*) He wanted to assert his authority.
(*c*) He wanted to suppress his people.
(*d*) He felt energized after taking the drink.

Directions: In question nos. **141** to **145**, a sentence / part of the sentence is underlined. Beneath each sentence four different ways of phrasing the underlined part are given. Choose the grammatically correct option. In case no improvement is needed, your answer should be option (*d*).

141. The children <u>were out playing</u> games as I speak.

(*a*) are out playing (*b*) had been out playing
(*c*) will be out playing (*d*) No improvement

142. Not even a single Indian woman <u>wears saree</u> to work in USA.

(*a*) wear saree (*b*) wears a saree
(*c*) wear a saree (*d*) No improvement

143. Merely informing people of a fact does not mean that you will convince them <u>for</u> its truth.

(*a*) to (*b*) of
(*c*) about (*d*) No improvement

144. Adolf Dassler didn't view the Berlin Games as <u>a automobile</u> for Nazi propaganda but as a chance to launch his humble athletic shoe business.

(*a*) a bucket of bolts (*b*) an automobile
(*c*) a vehicle (*d*) No improvement

145. In summers, people tend to buy <u>less</u> jackets.

(*a*) fewer (*b*) little
(*c*) lesser (*d*) No improvement

146. Choose the one which best expresses the meaning of the given word.

Exude

(*a*) Vent (*b*) Seep
(*c*) Pour (*d*) Gush

147. Choose the word opposite in meaning to the given word.

Passive

(*a*) Resigned (*b*) Resistant
(*c*) Yielding (*d*) Forbearing

Directions: In question no. **148**, four words are given in each question, out of which only one word is spelt correctly. Find the correctly spelt word and mark your answer.

148. (*a*) Assasin (*b*) Assassin
(*c*) Asasin (*d*) Asassin

Directions: In question nos. **149** to **151**, out of the four alternatives, choose the one which can be substituted for the given words/sentences.

149. A word that carries within it a synonym of itself

(*a*) Joey word (*b*) Kangaroo word
(*c*) Metaphor (*d*) Acronym

150. A study of rocks is called
(*a*) Trichology (*b*) Ornithology
(*c*) Gemology (*d*) Petrology

151. A reservoir
(*a*) Sump (*b*) Dump
(*c*) Lump (*d*) Clump

Directions: In question nos. **152** to **154**, four alternatives are given for the idiom/phrase underlined in the sentence. Choose the alternative which best expresses the meaning of the given idiom/phrase.

152. Namit's boss has been taking him to task for his irresponsibility towards the work he was assigned.
(*a*) to be angry at someone
(*b*) to give additional work to someone
(*c*) to criticize somebody
(*d*) to be indifferent towards someone

153. While other businesses are shutting down due to recession, pharmaceutical companies are sitting pretty with high profits.
(*a*) to be in an excellent situation
(*b*) to be in an advantageous position
(*c*) to be at the top
(*d*) to be in an unexpected state

154. The rescue team's arrival was like manna from heaven for the hostages.
(*a*) a person who needs help
(*b*) something that seems problematic at first but turns out to be good
(*c*) a way out
(*d*) unexpected help or aid

Directions: In question nos. **155** to **157**, sentences are given with blanks to be filled with an appropriate word(s). Four alternatives are suggested for each question. Choose the correct alternative 5out of the four.

155. Someone's been putting it _____ that you plan to shut down the organisation.
(*a*) away (*b*) up
(*c*) aside (*d*) about

156. The people of both the nations are united _____ borders by religion and history.
(*a*) through (*b*) by
(*c*) in (*d*) across

157. Your brother hasn't yet _____from the anaesthetic that the doctor gave him.
(*a*) come about (*b*) come out
(*c*) come at (*d*) come round

Directions: In question nos. **158** to **160**, read each of the following sentences to find out whether there is any grammatical error in it. The error, if any, will be in one part of the sentence. The number of that part is the answer. If there is 'No error' the answer is (*d*).

158. Archie is one of (*a*)/ the cleverest boys who (*b*)/ has passed through the school. (*c*)/ No error (*d*)

159. The play started by (*a*)/ the time Tenali Raman arrived (*b*)/ at the biggest theater in the world. (*c*)/ No error (*d*)

160. Billu's taste in (*a*)/ comic books is (*b*)/ different than Pinky's (*c*)/ No error (*d*)

Directions: In question nos. **161** to **165**, part of the sentence/whole sentence is underlined. Below are given alternatives that improve the sentence. Choose the correct option. In case no improvement is desired, your answer is D.

161. Rahul cannot be given this secret job as he will fly-by-night.
(*a*) Fly-to-night (*b*) fly-in-night
(*c*) fly-the-night (*d*) no improvement

162. You will not succeed unless you are optimistic.
(*a*) successful unless you are
(*b*) success unless you don't be
(*c*) succeed unless you don't be an
(*d*) no improvement

163. He puts out an air of dignity
(*a*) put out (*b*) puts on
(*c*) puts off (*d*) no improvement

164. Do you have little bread left?
(*a*) some (*b*) many
(*c*) any (*d*) no improvement

165. A Hindi that Seta speaks is broken.
(*a*) The Hindi (*b*) A Hindi
(*c*) An Hindi (*d*) no improvement

Directions: In question nos. **166** to **170**, you have a passage with 5 questions in it. Read the passage carefully and choose the best answer to each question.

In September 1620, a small ship called the Mayflower left Plymouth, England, carrying 102 passengers — an assortment of religious separatists seeking a new home where they could freely practice their faith and other individuals lured by the promise of prosperity and land ownership in the New World. After a treacherous and uncomfortable crossing that lasted 66 days, they dropped anchor near the tip of Cape Cod, far north of their intended destination at the mouth of the Hudson River. One month later, the Mayflower crossed Massachusetts Bay, where the Pilgrims, as they are now commonly known, began the work of establishing a village at Plymouth.

Throughout that first brutal winter, most of the colonists remained on board the ship, where they suffered from exposure, scurvy and outbreaks of contagious disease. Only half of the Mayflower's original passengers and crew lived to see their first New England spring. In March, the remaining settlers moved ashore, where they received an astonishing visit from an Abenaki Indian who greeted them in English. Several days later, he returned with another Native American, Squanto, who shared their pain. Squanto taught the Pilgrims, weakened by malnutrition and illness, how to cultivate corn, extract sap from maple trees, catch fish in the rivers and avoid poisonous plants. He also helped the settlers forge an alliance with the Wampanoag, a local tribe, which would endure for more than 50 years and tragically remains one of the sole examples of harmony between European colonists and Native Americans.

166. Find a word in the passage which means 'a collection of things or people'.

(*a*) Bunch (*b*) Bundle
(*c*) Assortment (*d*) Pack

167. Find a word in the passage which is the opposite of 'unimpressive'.

(*a*) Boring (*b*) Unsurprising
(*c*) Ordinary (*d*) Astonishing

168. Which of the following is a disease that the colonists on board suffered from?

(*a*) Scurvy (*b*) Rickets
(*c*) Beriberi (*d*) Tuberculosis

169. The phrase 'forge an alliance' means

(*a*) A forced kinship
(*b*) Developing a hoaxed liaison
(*c*) Creating a successful relationship
(*d*) Havingbitter feelings

170. Where did the ship stop at the end of 66 days of uncomfortable sailing?

(*a*) Plymouth (*b*) England
(*c*) Massachusetts Bay (*d*) Cape Cod

171. In the following question, a word is followed by four options. Select the option that best expresses the meaning of the given word.

Sleuth

(*a*) Butcher (*b*) Executioner
(*c*) Invigilator (*d*) Investigator

172. In the following question, a word is followed by four options. Select the option that is opposite in meaning to the given word.

Apposite

(*a*) Suitable (*b*) Harmonious
(*c*) Immaterial (*d*) Disagreeable

173. In the question, four words are given in each question, out of which only one word is spelt correctly. Find the correctly spelt word and mark your answer in the answer sheet.

(*a*) Confete (*b*) Confeti
(*c*) Confetti (*d*) Confette

Directions: In question nos. **174** to **176**, out of the four alternatives, choose the one which can be substituted for the given words/sentences.

174. A person who is hard to manage or control

(*a*) Factious (*b*) Ruly
(*c*) Fractious (*d*) Tractable

175. Uncontrollable craving for alcohol

(*a*) Pyromania (*b*) Dipsomania
(*c*) Hypomania (*d*) Megalomania

176. Cultivation of ornamental flowers

(*a*) Orthodontic (*b*) Agriculture
(*c*) Floriculture (*d*) Anthropology

Directions: In question nos. **177** to **179**, four alternatives are given for the idiom/phrase underlined in the sentence. Choose the alternative which best expresses the meaning of the given idiom/phrase.

177. <u>Eat one's heart out</u>

(*a*) To feel unhappy about something
(*b*) To feel merry about something
(*c*) To say disheartening words about someone
(*d*) To quarrel with someone

178. We talked about everything <u>under the sun</u>.

(*a*) in an open area. (*b*) in existence.
(*c*) during the day. (*d*) about the planets.

179. She had to <u>go cap in hand</u> to the committee to get a grant for the proposal.

(*a*) to go somewhere excitedly to ask for a favour
(*b*) to go somewhere hastily to ask for a favour
(*c*) to nervously ask for a favour
(*d*) to ask for a favour which makes you feel ashamed

Direction: In question no. **180**, sentences are given with blanks to be filled with an appropriate word(s). Four alternatives are suggested for each question. Choose the correct alternative out of the four.

180. No amount of populist pep talk can ______ the fact that Ulysses is a demanding book.

(*a*) obscure
(*b*) camouflage
(*c*) impersonate
(*d*) undrape

OMR SHEET

1.	Ⓐ Ⓑ Ⓒ Ⓓ	26.	Ⓐ Ⓑ Ⓒ Ⓓ	51.	Ⓐ Ⓑ Ⓒ Ⓓ	76.	Ⓐ Ⓑ Ⓒ Ⓓ
2.	Ⓐ Ⓑ Ⓒ Ⓓ	27.	Ⓐ Ⓑ Ⓒ Ⓓ	52.	Ⓐ Ⓑ Ⓒ Ⓓ	77.	Ⓐ Ⓑ Ⓒ Ⓓ
3.	Ⓐ Ⓑ Ⓒ Ⓓ	28.	Ⓐ Ⓑ Ⓒ Ⓓ	53.	Ⓐ Ⓑ Ⓒ Ⓓ	78.	Ⓐ Ⓑ Ⓒ Ⓓ
4.	Ⓐ Ⓑ Ⓒ Ⓓ	29.	Ⓐ Ⓑ Ⓒ Ⓓ	54.	Ⓐ Ⓑ Ⓒ Ⓓ	79.	Ⓐ Ⓑ Ⓒ Ⓓ
5.	Ⓐ Ⓑ Ⓒ Ⓓ	30.	Ⓐ Ⓑ Ⓒ Ⓓ	55.	Ⓐ Ⓑ Ⓒ Ⓓ	80.	Ⓐ Ⓑ Ⓒ Ⓓ
6.	Ⓐ Ⓑ Ⓒ Ⓓ	31.	Ⓐ Ⓑ Ⓒ Ⓓ	56.	Ⓐ Ⓑ Ⓒ Ⓓ	81.	Ⓐ Ⓑ Ⓒ Ⓓ
7.	Ⓐ Ⓑ Ⓒ Ⓓ	32.	Ⓐ Ⓑ Ⓒ Ⓓ	57.	Ⓐ Ⓑ Ⓒ Ⓓ	82.	Ⓐ Ⓑ Ⓒ Ⓓ
8.	Ⓐ Ⓑ Ⓒ Ⓓ	33.	Ⓐ Ⓑ Ⓒ Ⓓ	58.	Ⓐ Ⓑ Ⓒ Ⓓ	83.	Ⓐ Ⓑ Ⓒ Ⓓ
9.	Ⓐ Ⓑ Ⓒ Ⓓ	34.	Ⓐ Ⓑ Ⓒ Ⓓ	59.	Ⓐ Ⓑ Ⓒ Ⓓ	84.	Ⓐ Ⓑ Ⓒ Ⓓ
10.	Ⓐ Ⓑ Ⓒ Ⓓ	35.	Ⓐ Ⓑ Ⓒ Ⓓ	60.	Ⓐ Ⓑ Ⓒ Ⓓ	85.	Ⓐ Ⓑ Ⓒ Ⓓ
11.	Ⓐ Ⓑ Ⓒ Ⓓ	36.	Ⓐ Ⓑ Ⓒ Ⓓ	61.	Ⓐ Ⓑ Ⓒ Ⓓ	86.	Ⓐ Ⓑ Ⓒ Ⓓ
12.	Ⓐ Ⓑ Ⓒ Ⓓ	37.	Ⓐ Ⓑ Ⓒ Ⓓ	62.	Ⓐ Ⓑ Ⓒ Ⓓ	87.	Ⓐ Ⓑ Ⓒ Ⓓ
13.	Ⓐ Ⓑ Ⓒ Ⓓ	38.	Ⓐ Ⓑ Ⓒ Ⓓ	63.	Ⓐ Ⓑ Ⓒ Ⓓ	88.	Ⓐ Ⓑ Ⓒ Ⓓ
14.	Ⓐ Ⓑ Ⓒ Ⓓ	39.	Ⓐ Ⓑ Ⓒ Ⓓ	64.	Ⓐ Ⓑ Ⓒ Ⓓ	89.	Ⓐ Ⓑ Ⓒ Ⓓ
15.	Ⓐ Ⓑ Ⓒ Ⓓ	40.	Ⓐ Ⓑ Ⓒ Ⓓ	65.	Ⓐ Ⓑ Ⓒ Ⓓ	90.	Ⓐ Ⓑ Ⓒ Ⓓ
16.	Ⓐ Ⓑ Ⓒ Ⓓ	41.	Ⓐ Ⓑ Ⓒ Ⓓ	66.	Ⓐ Ⓑ Ⓒ Ⓓ	91.	Ⓐ Ⓑ Ⓒ Ⓓ
17.	Ⓐ Ⓑ Ⓒ Ⓓ	42.	Ⓐ Ⓑ Ⓒ Ⓓ	67.	Ⓐ Ⓑ Ⓒ Ⓓ	92.	Ⓐ Ⓑ Ⓒ Ⓓ
18.	Ⓐ Ⓑ Ⓒ Ⓓ	43.	Ⓐ Ⓑ Ⓒ Ⓓ	68.	Ⓐ Ⓑ Ⓒ Ⓓ	93.	Ⓐ Ⓑ Ⓒ Ⓓ
19.	Ⓐ Ⓑ Ⓒ Ⓓ	44.	Ⓐ Ⓑ Ⓒ Ⓓ	69.	Ⓐ Ⓑ Ⓒ Ⓓ	94.	Ⓐ Ⓑ Ⓒ Ⓓ
20.	Ⓐ Ⓑ Ⓒ Ⓓ	45.	Ⓐ Ⓑ Ⓒ Ⓓ	70.	Ⓐ Ⓑ Ⓒ Ⓓ	95.	Ⓐ Ⓑ Ⓒ Ⓓ
21.	Ⓐ Ⓑ Ⓒ Ⓓ	46.	Ⓐ Ⓑ Ⓒ Ⓓ	71.	Ⓐ Ⓑ Ⓒ Ⓓ	96.	Ⓐ Ⓑ Ⓒ Ⓓ
22.	Ⓐ Ⓑ Ⓒ Ⓓ	47.	Ⓐ Ⓑ Ⓒ Ⓓ	72.	Ⓐ Ⓑ Ⓒ Ⓓ	97.	Ⓐ Ⓑ Ⓒ Ⓓ
23.	Ⓐ Ⓑ Ⓒ Ⓓ	48.	Ⓐ Ⓑ Ⓒ Ⓓ	73.	Ⓐ Ⓑ Ⓒ Ⓓ	98.	Ⓐ Ⓑ Ⓒ Ⓓ
24.	Ⓐ Ⓑ Ⓒ Ⓓ	49.	Ⓐ Ⓑ Ⓒ Ⓓ	74.	Ⓐ Ⓑ Ⓒ Ⓓ	99.	Ⓐ Ⓑ Ⓒ Ⓓ
25.	Ⓐ Ⓑ Ⓒ Ⓓ	50.	Ⓐ Ⓑ Ⓒ Ⓓ	75.	Ⓐ Ⓑ Ⓒ Ⓓ	100.	Ⓐ Ⓑ Ⓒ Ⓓ

OMR SHEET

101. Ⓐ Ⓑ Ⓒ Ⓓ	121. Ⓐ Ⓑ Ⓒ Ⓓ	141. Ⓐ Ⓑ Ⓒ Ⓓ	161. Ⓐ Ⓑ Ⓒ Ⓓ
102. Ⓐ Ⓑ Ⓒ Ⓓ	122. Ⓐ Ⓑ Ⓒ Ⓓ	142. Ⓐ Ⓑ Ⓒ Ⓓ	162. Ⓐ Ⓑ Ⓒ Ⓓ
103. Ⓐ Ⓑ Ⓒ Ⓓ	123. Ⓐ Ⓑ Ⓒ Ⓓ	143. Ⓐ Ⓑ Ⓒ Ⓓ	163. Ⓐ Ⓑ Ⓒ Ⓓ
104. Ⓐ Ⓑ Ⓒ Ⓓ	124. Ⓐ Ⓑ Ⓒ Ⓓ	144. Ⓐ Ⓑ Ⓒ Ⓓ	164. Ⓐ Ⓑ Ⓒ Ⓓ
105. Ⓐ Ⓑ Ⓒ Ⓓ	125. Ⓐ Ⓑ Ⓒ Ⓓ	145. Ⓐ Ⓑ Ⓒ Ⓓ	165. Ⓐ Ⓑ Ⓒ Ⓓ
106. Ⓐ Ⓑ Ⓒ Ⓓ	126. Ⓐ Ⓑ Ⓒ Ⓓ	146. Ⓐ Ⓑ Ⓒ Ⓓ	166. Ⓐ Ⓑ Ⓒ Ⓓ
107. Ⓐ Ⓑ Ⓒ Ⓓ	127. Ⓐ Ⓑ Ⓒ Ⓓ	147. Ⓐ Ⓑ Ⓒ Ⓓ	167. Ⓐ Ⓑ Ⓒ Ⓓ
108. Ⓐ Ⓑ Ⓒ Ⓓ	128. Ⓐ Ⓑ Ⓒ Ⓓ	148. Ⓐ Ⓑ Ⓒ Ⓓ	168. Ⓐ Ⓑ Ⓒ Ⓓ
109. Ⓐ Ⓑ Ⓒ Ⓓ	129. Ⓐ Ⓑ Ⓒ Ⓓ	149. Ⓐ Ⓑ Ⓒ Ⓓ	169. Ⓐ Ⓑ Ⓒ Ⓓ
110. Ⓐ Ⓑ Ⓒ Ⓓ	130. Ⓐ Ⓑ Ⓒ Ⓓ	150. Ⓐ Ⓑ Ⓒ Ⓓ	170. Ⓐ Ⓑ Ⓒ Ⓓ
111. Ⓐ Ⓑ Ⓒ Ⓓ	131. Ⓐ Ⓑ Ⓒ Ⓓ	151. Ⓐ Ⓑ Ⓒ Ⓓ	171. Ⓐ Ⓑ Ⓒ Ⓓ
112. Ⓐ Ⓑ Ⓒ Ⓓ	132. Ⓐ Ⓑ Ⓒ Ⓓ	152. Ⓐ Ⓑ Ⓒ Ⓓ	172. Ⓐ Ⓑ Ⓒ Ⓓ
113. Ⓐ Ⓑ Ⓒ Ⓓ	133. Ⓐ Ⓑ Ⓒ Ⓓ	153. Ⓐ Ⓑ Ⓒ Ⓓ	173. Ⓐ Ⓑ Ⓒ Ⓓ
114. Ⓐ Ⓑ Ⓒ Ⓓ	134. Ⓐ Ⓑ Ⓒ Ⓓ	154. Ⓐ Ⓑ Ⓒ Ⓓ	174. Ⓐ Ⓑ Ⓒ Ⓓ
115. Ⓐ Ⓑ Ⓒ Ⓓ	135. Ⓐ Ⓑ Ⓒ Ⓓ	155. Ⓐ Ⓑ Ⓒ Ⓓ	175. Ⓐ Ⓑ Ⓒ Ⓓ
116. Ⓐ Ⓑ Ⓒ Ⓓ	136. Ⓐ Ⓑ Ⓒ Ⓓ	156. Ⓐ Ⓑ Ⓒ Ⓓ	176. Ⓐ Ⓑ Ⓒ Ⓓ
117. Ⓐ Ⓑ Ⓒ Ⓓ	137. Ⓐ Ⓑ Ⓒ Ⓓ	157. Ⓐ Ⓑ Ⓒ Ⓓ	177. Ⓐ Ⓑ Ⓒ Ⓓ
118. Ⓐ Ⓑ Ⓒ Ⓓ	138. Ⓐ Ⓑ Ⓒ Ⓓ	158. Ⓐ Ⓑ Ⓒ Ⓓ	178. Ⓐ Ⓑ Ⓒ Ⓓ
119. Ⓐ Ⓑ Ⓒ Ⓓ	139. Ⓐ Ⓑ Ⓒ Ⓓ	159. Ⓐ Ⓑ Ⓒ Ⓓ	179. Ⓐ Ⓑ Ⓒ Ⓓ
120. Ⓐ Ⓑ Ⓒ Ⓓ	140. Ⓐ Ⓑ Ⓒ Ⓓ	160. Ⓐ Ⓑ Ⓒ Ⓓ	180. Ⓐ Ⓑ Ⓒ Ⓓ

ANSWERS

1. (*b*)	2. (*b*)	3. (*b*)	4. (*c*)	5. (*a*)	6. (*c*)	7. (*a*)	8. (*d*)	9. (*b*)	10. (*b*)
11. (*b*)	12. (*a*)	13. (*b*)	14. (*c*)	15. (*b*)	16. (*b*)	17. (*d*)	18. (*c*)	19. (*d*)	20. (*b*)
21. (*a*)	22. (*d*)	23. (*c*)	24. (*c*)	25. (*b*)	26. (*c*)	27. (*c*)	28. (*c*)	29. (*b*)	30. (*a*)
31. (*d*)	32. (*d*)	33. (*c*)	34. (*b*)	35. (*c*)	36. (*d*)	37. (*b*)	38. (*c*)	39. (*c*)	40. (*b*)
41. (*b*)	42. (*a*)	43. (*c*)	44. (*b*)	45. (*a*)	46. (*a*)	47. (*c*)	48. (*d*)	49. (*d*)	50. (*b*)
51. (*c*)	52. (*b*)	53. (*c*)	54. (*c*)	55. (*b*)	56. (*d*)	57. (*c*)	58. (*c*)	59. (*a*)	60. (*c*)
61. (*d*)	62. (*a*)	63. (*a*)	64. (*a*)	65. (*b*)	66. (*a*)	67. (*b*)	68. (*c*)	69. (*d*)	70. (*b*)
71. (*a*)	72. (*d*)	73. (*c*)	74. (*d*)	75. (*a*)	76. (*a*)	77. (*d*)	78. (*b*)	79. (*c*)	80. (*c*)
81. (*d*)	82. (*d*)	83. (*d*)	84. (*b*)	85. (*c*)	86. (*b*)	87. (*a*)	88. (*a*)	89. (*c*)	90. (*d*)
91. (*d*)	92. (*b*)	93. (*c*)	94. (*a*)	95. (*a*)	96. (*b*)	97. (*a*)	98. (*d*)	99. (*b*)	100. (*d*)
101. (*a*)	102. (*d*)	103. (*c*)	104. (*b*)	105. (*c*)	106. (*a*)	107. (*a*)	108. (*a*)	109. (*a*)	110. (*d*)
111. (*c*)	112. (*a*)	113. (*d*)	114. (*d*)	115. (*d*)	116. (*a*)	117. (*c*)	118. (*d*)	119. (*d*)	120. (*d*)
121. (*b*)	122. (*c*)	123. (*b*)	124. (*d*)	125. (*c*)	126. (*b*)	127. (*a*)	128. (*b*)	129. (*d*)	130. (*d*)
131. (*d*)	132. (*b*)	133. (*c*)	134. (*b*)	135. (*a*)	136. (*d*)	137. (*b*)	138. (*b*)	139. (*d*)	140. (*b*)
141. (*a*)	142. (*b*)	143. (*b*)	144. (*c*)	145. (*a*)	146. (*b*)	147. (*b*)	148. (*b*)	149. (*b*)	150. (*d*)
151. (*a*)	152. (*c*)	153. (*a*)	154. (*d*)	155. (*d*)	156. (*d*)	157. (*d*)	158. (*c*)	159. (*a*)	160. (*c*)
161. (*d*)	162. (*d*)	163. (*b*)	164. (*c*)	165. (*a*)	166. (*c*)	167. (*d*)	168. (*a*)	169. (*c*)	170. (*d*)
171. (*d*)	172. (*c*)	173. (*c*)	174. (*c*)	175. (*b*)	176. (*c*)	177. (*a*)	178. (*b*)	179. (*d*)	180. (*b*)

EXPLANATIONS

1. Applying KVL in the loop

$24 - 1I + 2V_b - V_b - 4I = 0$

Where $V_b = 3I$

$24 - 5I + V_b = 0$

$24 - 5I + 3I = 0$

$I = 12A$

2. For finding R_{th} voltage source is short circuited and current source is open circuited.

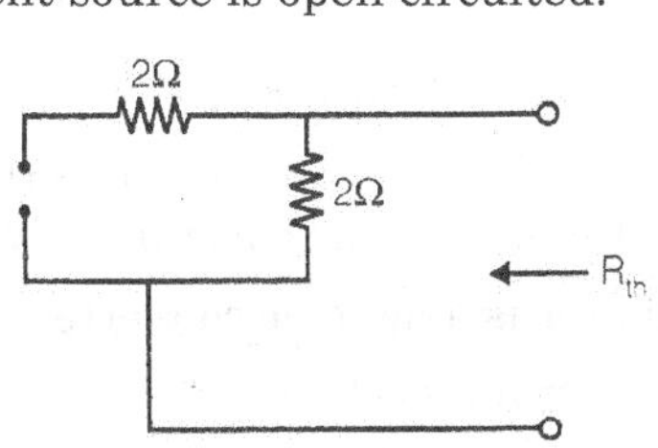

$R_{th} = 2\Omega$

3. $Z(j\omega) = +[(3 + j4) \,||\, (3 - j4)]\dfrac{17}{6}$

$$= \left[\frac{(3 + j4)(3 - j4)}{3 + j4 + 3 - j4}\right] + \frac{17}{6} = \left[\frac{9 + 16}{6}\right] + \frac{17}{6}$$

$Z(j\omega) = (7 + j0)\Omega$

4.

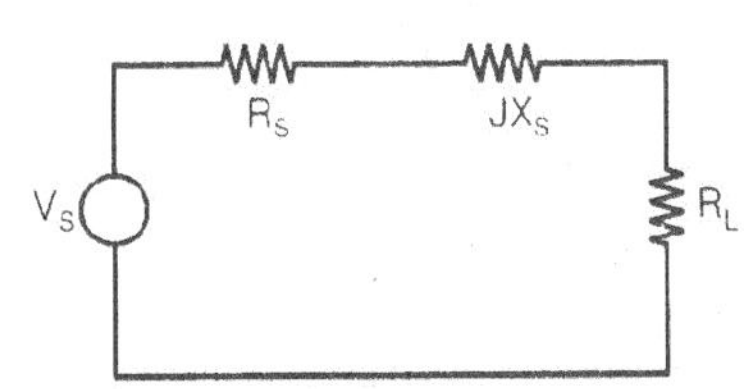

For max power transfer

$$R_L = \sqrt{R_s^2 + X_s^2} = \sqrt{8^2 + 6^2} = 10\Omega$$

5. For maximum power from source

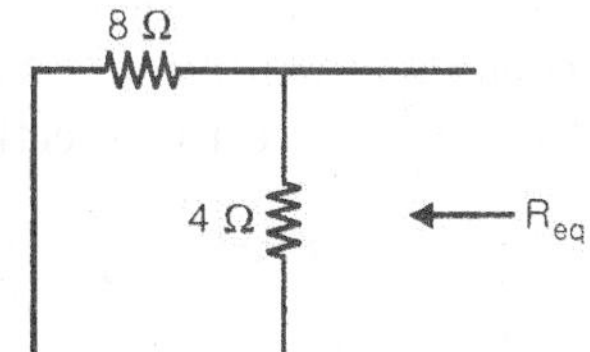

$$R_{eq} = 8 \parallel 4 = \frac{32}{12} = \frac{8}{3}\Omega$$

$$R = R_{eq} = \frac{8}{3}\Omega$$

6.

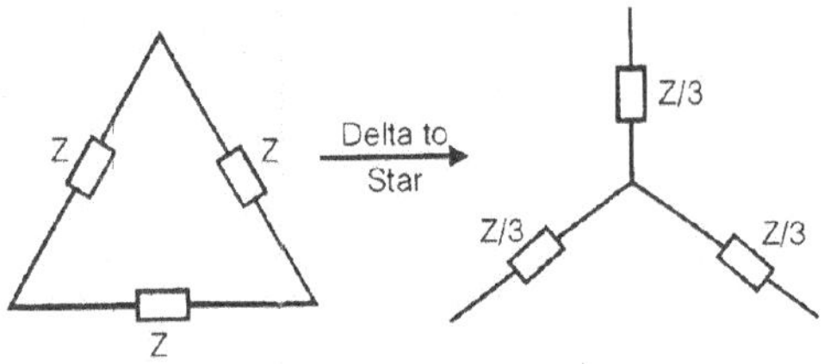

$$Z_{eq} = \frac{Z}{3} = \frac{9\angle 30^\circ}{3} = 3\angle 30^\circ$$

7.

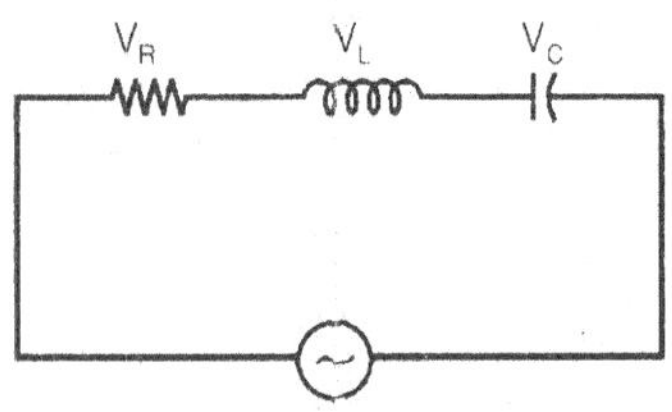

$$V_S = \sqrt{R_R^2 + (V_L - V_C)^2}$$

$V_L - V_C = 9V$ and $V_S = 15$

$$V_R = \sqrt{V_S^2 - (V_L - V_C)^2} = \sqrt{15^2 - 9^2} = 12V$$

$$\sqrt{R_R^2 + V_L^2} = 20$$

$$V_L = \sqrt{20^2 - V_L^2} = \sqrt{20^2 - 12^2} = 16V$$

$V_L - V_C = 9V$

$\Rightarrow V_C = V_L - 9 = 16\text{-}9 = 7$ V

8.

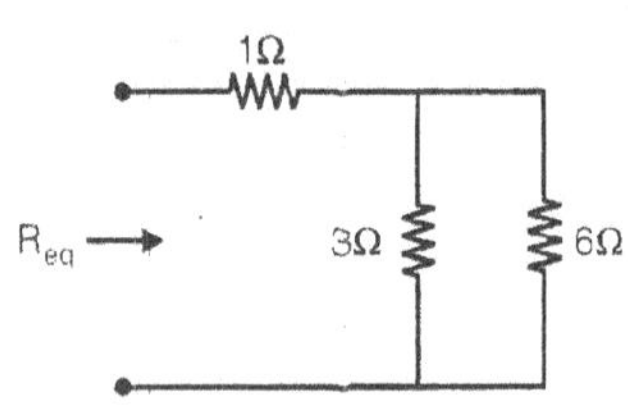

$$R_{eq} = 1 + (3 \parallel 6) = 3\Omega$$

Voltage across current source

$I_S R_{eq} = 2 \times 3 = 6V$

9. Apply Wheatstone's bridge balance equation,

$$\frac{4}{R} = \frac{10}{5} \Rightarrow R = 2\Omega$$

10. Power consumed will remain same whether we connect in delta or star connection.

11. $Y = \frac{1}{60 + j60} + \frac{1}{-j120} = 8.33 \times 10^{-3} \angle 0^\circ$

$I = VY = 120\angle 0 \times 8.33 \times 10^{-3}\angle 0 = 1 + j0$

13. In series R-L-C circuit, the capacitor voltage is Q times the source voltage at resonant frequency. Q-factor first increases with frequency and after resonant frequency it decreases with frequency.

14. Angle between voltage across R & inductor will be 30º as given.

So, $\tan 30 = \frac{X_L}{R}; R = \frac{X_L}{\tan 30} = \sqrt{3}.X_L$

as $X_L = 1\Omega$

$R = \sqrt{3}\Omega$

15.

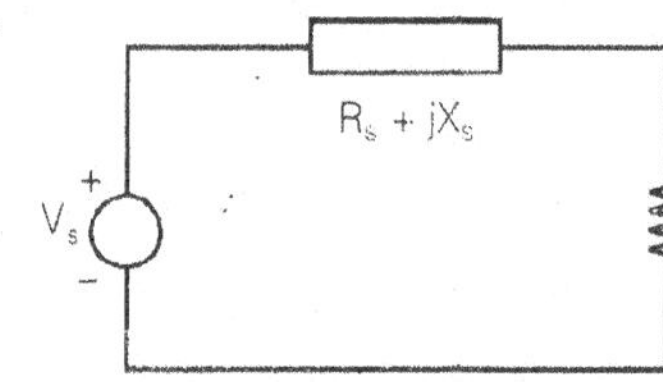

for such network

$$R_L = \sqrt{R_S^2 + X_S^2}$$

For maximum power transfer

16. An ideal op-amp is a voltage controlled voltage source.

i.e. $V_0 \propto V_d$

17. JFET is a voltage controlled current source

$i_d \propto V_{GS}$

18. Voltage series feedback results in increase in input impedance and decrease in output impedance.

19. It is a current shunt.

20. $Q \propto \frac{1}{BW}$

21. By using positive feedback, an amplifier circuit will increase the gain of amplifier.

22. The condition which is required to get sustained oscillation is called Barkhausen criteria.

$|A\beta| = 1$ (or) phase should be zero.

23. The phase shift introduced by Wien bridge is 0º

24. The use of a crystal oscillator is the frequency of oscillations remains substantially constant.

25. Oscillator is a dc to ac converter.

26. For the given system

$V_i = V_0$

Taking laplace transform

$V_i(s) = V_0(s)$

$$\frac{V_0(s)}{V_i(s)} = 1$$

30. $x(t) = \sum_{n=0} \cos(nt) = \cos(0) = 1$

32. $x(t) = x_1(t) + x_2(t)$

$X_1(t) = \cos(rt)$ is periodic with

$$T_1 = \frac{2\pi}{1} = 2\pi$$

$X_2(t) = \sin\left(\sqrt{2}\,t\right)$ is periodic with

$$T_2 = \frac{2\pi}{\sqrt{2}} = \sqrt{2}\pi$$

Since $\frac{T_1}{T_2} = \sqrt{2}$ is an irrational number

so x(t) is non periodic.

33. We know that $\int_{-\infty}^{\infty} \delta(t)dt = \int_{0^-}^{0^+} \delta(t)dt = 1$ and x(t).

$\delta(t) = x(0).\delta(0) = x(0)$

Region of integration does not consist zero, so

$$\int_1^2 (3t^2 + 1)\delta(t)dt = 0$$

34. For an odd signal, it must pass through origin.

35. Closed loop gain $= \frac{200}{1 + 0.2 \times 200}$

36. Output voltage = 12 V,

$$I_R = \frac{V_i - 12}{R}$$

37. Current through $R_E = \frac{10 - 0.7}{10 \times 10^3} = 0.93$ mA

38. One RC combination can give a phase shift of less then 90°. Therefore 3 RC combinations are required for 180° phase shift.

39. For non-inverting terminal input, output is in phase with input. For inverting terminal input, output is 180° out of phase with input. If inputs have 180° phase difference, outputs will be in phase and additive.

40. Maximum mutual inductance $= \sqrt{L_1 L_2} = 20$ mH.

41. Since common mode signal is reduced to zero, CMRR is infinite.

42. $R_B = \frac{V_B - V_{BE}}{I_B}$

$$I_C = \frac{V_{CC} - V_{CEsat}}{R_2} = \frac{5.07}{60} \Rightarrow 0.07 \text{ A}$$

$$\beta = \frac{\alpha}{1 - \alpha} = 24$$

$$I_B = \frac{I_C}{\beta} = \frac{0.07}{24} = 3 \text{ mA}$$

$$R_B = \frac{5 - 0.7}{3\text{mA}} = 1.5 \text{ k}\Omega$$

43. Due to direct coupling between output and CRO both dc and ac output voltage will appear on screen of CRO.

44. $V_{dc} = \frac{120}{4}\left(\frac{\sqrt{2}}{\pi}\right) = 13.5$ V

46. $|Z| = \frac{V}{I} = \frac{\text{Peak value of V}}{\text{Peak value of I}}$

47. KCL and KVL are valid for voltages and currents of all wave shapes during transient as well as steady state conductions.

48. In RL circuit the current lags the voltage by an angle which is greater than 0° but less than 90°.

49. $\text{kWh} = \frac{60 \times 45}{1000 \times 60} = 0.45$

50. Since capacitors are in series, charge will be same.

$$\text{Total capacitance} = \frac{1 \times 2}{3} = \frac{2}{3}\mu\text{F}$$

$$\text{Charge} = \frac{2}{3} \times 30 = 20\mu\text{C}.$$

51. $V_C = QV = 50$ V rms.

52. In Darlington connection it will have high Zt and A, but voltage gain is unity.

53. Since C increases, the circuit will not be in resonance. Therefore, current decreases (In series resonant circuit current is maximum at resonance).

54. Energy stored = $0.5\ CV^2$

55. Total heating energy required = 1000 kWh + 25% of 1000 kWh.

61. $'\text{Ao}' = 10 \times 10^1 + 0 = 160$

62. $Y = \overline{\overline{A}BC} + \overline{A} + \overline{C} = \overline{A} + \overline{B} + \overline{C} + \overline{A} + \overline{C} = \overline{A} + \overline{C} + \overline{B}$

$$= Y = \overline{A\overline{B}C}$$

66. $3 \times 16^2 + 14 \times 16^1 + 8 = 1000.$

67. (b-n+1) links associated with fundamental loops. So b – n + 1 = 15 – 8 + 1

= 8

68. Superposition theore is not applicable for nonlinear parameters.

$\because P = I^2R$ or $\frac{V^2}{R}$

not applicable for power calculation.

69. Modulation index = m_i

$m_i = \frac{\Delta f}{m_f} \Rightarrow \Delta f = 50 \times 500 = 25000Hz$

70.

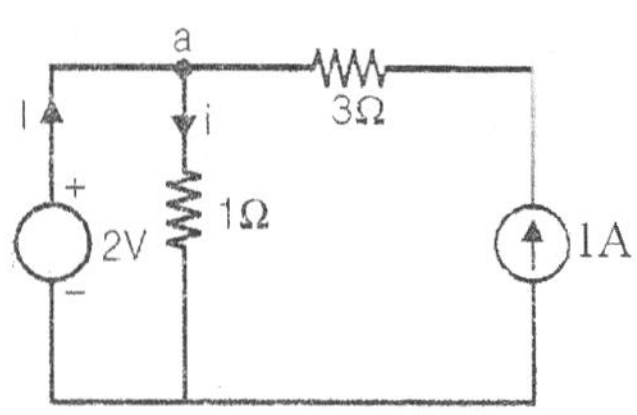

From figure

$i = \frac{2}{1} = 2$ Amp

By applying KCL at node a

I = i - 1 = 2 – 1 = 1 Amp

i.e. the current delivered by voltage source = 1 amp.

power delivered by voltage source = 2 ×1 = 2 W

71. $Z(j\omega) = R + j\left(\omega L - \frac{1}{\omega C}\right)$

At resonance, imaginary term of $Z(j\omega)$ becomes zero i.e. capacitive reactance is equal to inductive reactance and $Z(j\omega)$ has minimum value R.

72. Ideal voltage source keeps the terminal voltage constant so accordingly current will change and the voltage across each resistor is unchanged following superposition principle.

73. At resonance in parallel R-L-C circuit input impedance is equal to R.

74. The image frequency can be eliminated by sufficient attenuation on the incoming signal by the RF amplifier filter of the superheterodyne receiver.

75. The ripple frequency for half-wave rectifier is f.

76. A $ B @ C means A ≥ B = C

Option (a): C @ B # A mean C = B ≤ A

Option (b): C @ B lA mean C = B > A

Option (c): A $ C – B mean A ≥ C < B

Option (d): B – A # C mean B < A ≤ C

Hence, option (a) is correct.

77. Premises (3M I 4N) and (2N D 5K): 3M > 4N and 2N = 5K i.e.

$M > \frac{10K}{3}$

Option (a): 4M I 20K means 4M > 20K i.e. M > 5K

Option (b): 2M A K means 2M < K

Option (c): 2K I N means 2K > N

Option (d): M I 2K means M > 2K

Hence, option (d) is true.

78. $\frac{21+27}{8} = 6$

$\frac{14+58}{8} = 9$

$\frac{33+71}{8} = \mathbf{13}.$

79. 2 × **2.5** = 5

5 × **2** = 10

10 × **2.5** = 25

25 × **2** = 50

50 × **2.5** = 125

125 × **2** = **250**

80. 2 × 2 + 1 = 5

5 × 2 + 2 = 12

12 × 2 + 3 = 27

27 × 2 + 4 = 58

58 × 2 + 5 = **121**

81. 48 – 21 = 30 **– 3**

105 – 52 = 56 **– 3**

71 – 29 = 45 **– 3**

82. 317 + 306 = 623, 623 + 211 = 834

312 + 306 = 618, 618 + 211 = 829

303 + 306 = 609, 609 + 211 = **820**

83. Frogs live on land as well as water. Thus, they are called amphibians. Koalas live on trees. Thus, they are called arboreal. Hence, option (D) is the correct answer.

84. O(15) + 6 = U(21)

A(1) + 6 = G(7)

G(7) + 6 = M(13)

E(5) + 6 = K(11)

Similarly,

R(18) + 6 = X(24)

E(5) + 6 = K(11)

A(1) + 6 = G(7)

D(4) + 6 = J(10)

Hence, the answer is XKGJ.

85. 17 – **(7 – 1)** = 11

73 – **(7 – 3)** = **69**

86. The synonym of 'periodic' is 'constant'. Similarly, the synonym of 'sporadic' is 'irregular'. Hence, option (B) is the correct answer. 'Suave' is someone confident, polite and elegant.

87. **abb**a / aabb / **a**bba / a**a**bb

88. AYB**Z**/ AY**B**Z/ A**Y**BZ/ **A**YBZ

89. 106 – 2 = 104

104 – 4 = **100**

100 – 6 = 94

94 – 8 = 86

86 – 10 = 76

76 – 12 = 64

90. In coding of MUTE, PRAY and NEWS code for alphabet, whose position number is less than 10 is coded in single digit but in BOLD code for D is 04.

92. Magnesium Chloride is $MgCl_2$ which has one atom of magnesium and two atoms of chlorine.

95. Bovine spongiform encephalopathy (Mad Cow Disease) is a fatal neurodegenerative disease (encephalopathy) in cattle that causes a spongy degeneration in the brain and spinal cord.

97. Mach number is a dimensionless quantity representing the ratio of speed of an object moving through a fluid and the local speed of sound.

98. Tetraethyllead, abbreviated TEL, is an organolead compound with the formula $(CH_3CH_2)_4Pb$. It was used with gasoline as an inexpensive octane booster.

99. Allium is a monocot genus of flowering plants, informally referred to as the onion genus.

100. The glow of phosphorus itself originates from oxidation of the white (but not red) phosphorus - a process now termed chemiluminescence.

101. Laughing gas is common name of Nitrous Oxide (N_2O).

102. Hydrogen (H), while placed in column 1 of the modern periodic table, is not considered to be a in the group Alkali Metals.

103. Sodium reacts vigorously with air and burns up. Therefore, it must be stored in kerosene to avoid this reaction.

104. Ruminent animals quickly swallow the food and store it in a part of the stomach called rumen. The food gets partially digested here and is called cud.

105. Vitamin D is a group of fat-soluble secosteroids responsible for enhancing intestinal absorption of calcium, iron, magnesium, phosphate and zinc.

106. Let the total work be 50 units (LCM of $\frac{25}{2}$ and $\frac{50}{3}$), such that the work done by each of them in 1 day be 4 units and 3 units respectively.

Work done by them in 4 days = (4 + 3) × 4 = 28 units.

Hence, part of the work left = $\frac{22}{50} = \frac{11}{25}$.

107. $P = \pi r + 2r$ and $A = \pi r^2$

$\Rightarrow P = \frac{A}{r} + 2r$

$\Rightarrow rP = A + 2r^2$

108. Let M and C be the marked price and cost price respectively.

Then, 0.85 M = 1.275C ⇒ M = 1.5C

and 0.8 M = 0.8 × 1.5C = 1.20C

Hence, profit percentage = 20%.

109. Ratio of the investments of A, B and C

= (14500 × 3 + 11000 × 9) : (15000 × 2 + 20000 × 8) : 25000 × 8

= 57 : 76 : 80

Hence, required amount = $\frac{80-76}{213} \times 149100$

= Rs. 2,800.

110. Let the cost price be Rs. x.

Selling price = Rs. 1.1x

New CP = Rs. 0.85x

Thus, 0.85x × 1.3 = 1.1x + 75

⇒ x = Rs. 15,000.

111. Let the income be Rs. 100.

New income = Rs. 130.

New expenditure = 60 × 0.95 = Rs. 57

New savings = 130 – 57 = Rs. 73

Hence, percentage increase in her savings

= $\frac{73-40}{40} \times 100 = 82.5\%$.

112. Required time

$$= \frac{57}{17-5} + \frac{57}{17+5} = \frac{57}{12} + \frac{57}{22} = 7\frac{15}{44} \text{ hrs.}$$

113. Given that, $x = \frac{1}{\sqrt{3}+\sqrt{2}} = \sqrt{3}-\sqrt{2}$

$$x^2 = \left(\sqrt{3}-\sqrt{2}\right) = 5 - 2\sqrt{6}$$

$$\therefore x^2 + 2\sqrt{2}x + 1$$

$$= 5 - 2\sqrt{6} + 2\sqrt{2}(\sqrt{3}-\sqrt{2}) + 1 = 2.$$

114. $x^3 + y^3 + z^3 - 3xyz$

$$= (x+y+z)(x^2+y^2+z^2-xy-yz-zx)$$

$$= (x+y+z)\frac{(x-y)^2+(y-z)^2+(z-x)^2}{2}$$

Putting, x = 222, y = 223 and z = 225 we get the required answer as 4690.

115.

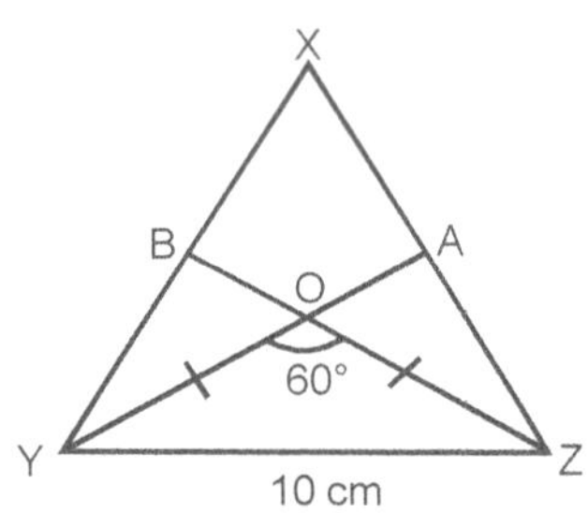

As YA and ZB are medians.

$\therefore$ O is the circumcenter of ΔXYZ.

Hence, area of ΔYOZ : Area of ΔXYZ = 1 : 3.

116.

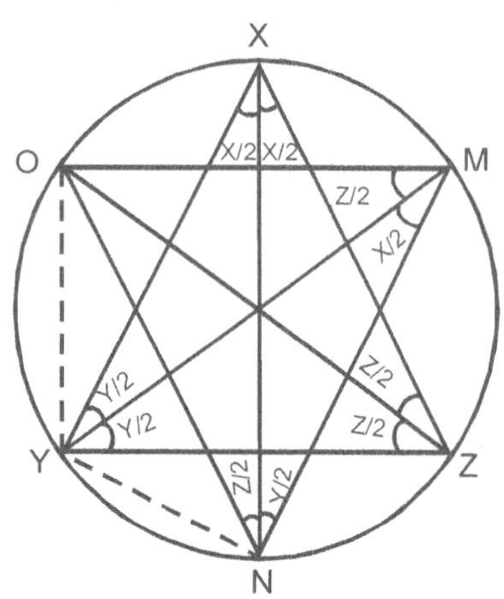

$$\angle NMO + \angle MNO = \angle\frac{X}{2} + \angle\frac{Y}{2} + \angle\frac{Z}{2} + \angle\frac{Z}{2}$$

$$\Rightarrow \angle NMO + \angle MNO = 90° + \angle\frac{Z}{2}.$$

117. sinAcosA(tanA – cot A) –

$$\left(\frac{1}{\text{cosecA secA}} - 1\right)\left(\frac{1}{\cot A} - \frac{1}{\tan A}\right)$$

= sinAcosA(tanA – cotA)

– (sinAcosA – 1) (tanA – cotA)

$$= \tan A - \cot A = \frac{\sin^2 A - \cos^2 A}{\cos A \sin A}$$

$$= \frac{-2\cos 2A}{\sin 2A} = -2\cot 2A.$$

118. Subtracting 18 from each of the options, we get multiple of 21 i.e. 60 – 18 = 42 in only option (D).

Now, in order to confirm this option, take the present age of B as 70 which gives a multiple of 26 after subtracting 18 i.e. 70 – 18 = 52 = 26 × 2.

Hence, option (D) is the right option.

119. Let the five numbers be p, q, r, s and t.

Now, p + q + r + s = 54 ... (i)

q + r + s + t = 74 ... (ii)

From (i) and (ii), we get

t – p = 20

$\Rightarrow$ t = 20 + 8 = 28

Hence, required average $= \frac{28+8}{2} = 18$.

120. Let $x^6 = a$

$$\Rightarrow a + \frac{1}{a} = 9 \text{ and } \frac{x^{36}+1}{x^{18}} = a^3 + \frac{1}{a^3}$$

Now,

$$a^3 + \frac{1}{a^3} = \left[a + \frac{1}{a}\right]\left[\left(a + \frac{1}{a}\right)^2 - 3a \times \frac{1}{a}\right]$$

Hence, the required value = 9 × (81 – 3) = 702.

121. The first sentence is in past tense. Therefore, the second blank will take a word that is in past tense. Thus, option (c) is incorrect. Option (d) is incorrect since 'raise' refers to lifting something manually. Option (a) is incorrect since 'risen' is always followed by 'had/have'. 'Rose' is simple past tense and is consistent with the structure of the sentence. 'Come on' means start to arrive or happen. Hence, option (c) is the correct answer.

122. Option (c), 'at', is the correct preposition to be filled in the blank. The preposition 'at' is used to indicate presence or occurrence, intended goal, the means, cause or manner, age or position in time, etc.

123. The correct preposition is 'on'. 'On' is used with 'agree' when we talk about an issue or a point of debate. Hence, option (b) is the correct answer. 'Agree with' is used when we accept the point of someone or something. 'Agree to' is used when we accept the demands or queries of a person. 'Agree at' is grammatically incorrect.

124. The given sentence is correct. A preposition such as 'between' should be followed by an objective pronoun (such as me, him, her and us) rather than a subjective pronoun (such as I, he, she and we). In part (c), the correct expression is 'Somiya and me' since it is the object of the sentence. Hence, part (d) is the correct answer.

125. There is a parallelism error in the sentence. The correct sentence should be "... I'm having dinner with Dikshant". Hence, part (c) is the correct option.

126. Clauses containing 'if' are like hypothesis, suppositions or desires and in such cases, we use past tense forms to talk about the present and the future. Therefore, the correct sentence will be 'If I were working in Bandra, I would commute by the local train'. Hence, the correct answer is part (b).

127. 'To get one's foot in the door' means to manage to enter an organisation that could bring one success. Hence, option (a) is the correct answer.

128. 'Kick the bucket' means to die. Hence, option (b) is the correct answer.

129. 'In the home stretch' refers to the last part of an activity, etc. Hence, option (d) is the correct answer.

133. 'Agriculture' deals with the science of growing plants and breeding animals for farming. Hence, option (C) is the correct answer. 'Aviculture' is concerned with the raising and caring of birds, especially of wild birds in captivity. 'Horticulture' deals with the study and science of flowers, fruits and vegetables. 'Apiculture' refers to keeping of bees especially on large scale.

134. An agent who negotiates contracts of purchase and sale is called a 'broker'. The contracts can be contracts of securities, commodities, real estate, etc. Hence, option (b) is the correct answer. A 'salesman' is a person whose job is to sell things. A 'bargainer' is a negotiator of the terms of a trade. A 'moderator' is someone who leads a discussion in a group and tells each person when to speak.

135. A person who believes that a particular type of behaviour is important is called a 'stickler'. Hence, option (a) is the correct answer. 'Perseverer' is a persistent person. Someone who is lazy and avoids work is called a 'slacker'. Person who has certain social values is said to be 'moral'.

136. 'Cortes' was not a civilization. He was a person whose army conquered the Aztecs. Hence, option (d) is the correct answer.

137. The passage says that when the chocolate was passed to the Aztecs, their leader Montezuma II, drank five goblets of hot chocolate and only those who contributed towards military service were allowed to drink it. Therefore, Options (a) and (c) are incorrect. The passage further states that the drink was served to Hernan Cortes and his soldiers, the Spaniyards, when they encountered the Aztecs. This means that visitors were also served the drink. Thus, we can say that option (b), important figures is the correct answer.

138. 'Opulence' means wealth, riches, capital or fortune. Hence, option (b) is the correct answer.

139. Option (c) can be inferred from the first sentence of the passage. The passage further states that the Mayans found that the chocolate drink was a mood enhancer and energy booster. Nothing is said about the drink being intoxicating. Hence, option (d) is the correct answer.

140. Look at the line "Legendary Aztec leader Montezuma II ... display of power and opulence." It says that Montezuma II drank five cups of chocolate drink each day because he wanted to display his power and wealth. Hence, option (b) is the correct answer.

141. Option (a) is correct since present continuous tense is used for an action that is going on at the time of speaking. The given sentence is incorrect as it is in past continuous tense, option (b) is in past perfect continuous tense and option (c) is in future continuous tense.

142. Single countable nouns always take an article before them. Therefore, the given sentence and option (a) are incorrect. 'Woman' is singular. Therefore, the verb should also be singular, i.e., it should be 'wears'. Hence, option (b) is the correct answer.

143. Convince takes either 'to' or 'of'. 'To' is used in the sense of convincing someone to do something. For example: I tried to convince him to go to the market with me. 'Of' is used in the sense of convincing somebody of something. For example: You will need to convince them of your enthusiasm for your job. The given sentence is in the latter form and hence, option (b) is the correct answer.

144. Option (c) is the correct answer. 'Vehicle for something' is used to express feelings or ideas as a way of achieving something. 'Bucket of bolts' refers to a piece of machinery that is not worth more than its scrap value.

145. 'Fewer' is used to refer to people or things in plural. 'Jackets' is plural. Hence, option (a) is the correct answer. 'Less' is used when something that cannot be counted is being referred to. 'Little' means small, which does not logically fit in the meaning of the sentence.

146. 'Exude' means to flow out slowly, so does seep. Hence, option (b) is the correct answer. 'Gush', which means to flow out qickly, is the antonym of 'exude'.

147. 'Passive' is used to describe someone who allows things to happen or accepts what other people decide for him. 'Resistant' is used to describe someone who is opposed to something. Hence, option (b) is the correct answer. Something or someone is called 'yielding' when it is not hard or rigid. 'Forbearing' is someone who avoids saying or doing something.

148. The correct spelling is 'Assassin'. Hence, option (b) is the correct answer.

149. A 'kangaroo word' is a word that contains letters of another word. The two words are synonyms of each other. 1 'joey word' is a word that is contained in a kangaroo word. For example: 'alone' is a kangaroo word while 'one', which is formed by using the letters of the word 'alone', is a joey word. Hence, option (b) is the answer. 1 'metaphor' is a figure of speech that describes a subject by asserting that it is, on some point of comparison, the same as another otherwise unrelated object. 'Acronym' is a word formed from the first letters of the words that make up the name of something. For example: SARS is an acronym of severe acute respiratory syndrome.

150. 'Petrology' is defined as the study of rocks. Hence, option (d) is the correct answer. 'Trichology' deals with the scientific study of the health of hair and scalp. 'Ornithology' refers to the study of birds. 'Gemology' is the science of gems.

151. 'Sump' is a low part or area where liquid collects when it drains from something. Hence, option (a) is the correct answer. 'Dump' means to put down in a careless or untidy way. 'Lump' refers to a hard or solid piece that does not have a particular shape. 'Clump' is the sound made by somebody putting their feet down very heavily.

152. 'Take someone to task' means to criticize somebody strongly for something they have done. Hence, option (c) is the correct answer.

153. 'Sitting pretty' means to be in a good situation, especially when others are not. Hence, option (a) is the correct answer.

154. 'Manna from heaven' means an unexpected help or comfort. Hence, option (d) is the correct answer.

155. 'Put (something) about' means to give people some news or information that may be false. Hence, option (d) is the correct answer. 'Put (something) away' means to keep something back in its place. 'Put up' means to suggest an idea for other people to discuss. For example: put up an idea, put up a case, etc. 'Put aside' means to put away or to stop using

156. The sentence means that the people of both the nations are separated by borders but united by religion and history. 'Across' means from one side to the other side of something. Hence, option (d) is the correct answer. Options (a), (b) and (c) will render the sentence redundant. 'Through' means by means of. 'By' is usually used after a passive verb, to show who or what does, creates or causes something. 'In' means at a point within an area or space.

157. 'Come round/around' means to become conscious again. Hence, option (d) is the correct answer. 'Come about something' means to arrive somewhere in order to get something.

158. When the subject of the verb is a relative pronoun ('that', in this case), then the verb agrees in number and person with the antecedent of the relative (boys). Therefore, 'has' in part (c) is grammatically incorrect, rendering part (c) the answer. The correct sentence should be 'Archie is one of the cleverest boys who have passed through the school.'

159. When two actions happened in the past, then past perfect tense is used for the one that happened before the other. In the sentence, the action (starting of the play) was completed before the second action (Tenali Raman's arrival). Thus, the correct sentence will be "The play had started by the time Tenali Raman arrived at the biggest theater in the world." Hence, part (a) is the correct answer.

160. The preposition 'than' is used with 'different' when a noun is compared with a clause. For example: Her appearance was very different than I'd expected. When two nouns are compared, like in this case, then the preposition 'from' is used with 'different'. Hence, part (c) is the correct answer.

161. "Fly-by-night" is used to describe someone who is untrustworthy, irresponsible or unreliable. Hence, the sentence does not require any change and perfectly makes sense.

162. The other options are grammatically incorrect. Option (a) would have been correct if there was a "be" before "successful."

163. "To put on an air of dignity" refers to "assuming a sense of dignity." It makes sense while the other options do not. "Put out" means extinguish, "Put off" refers to evading someone while "put in" refers to "made" or "sent in."

164. The words "some" and "any" are used for countable and uncountable nouns. In general, we could say that some means a few/a little and "any" means none in negative clauses or a few/a little in questions. In the context of the sentence, "any" is better.

165. Here, we want to refer to the kind of Hindi language that Seta speaks. Hence, we are specifying Seta's "Hindi'. Hence, "the Hindi" will be the answer.

166. The correct answer is option (c). 1 collection of things or people is called 'assortment'.

167. 'Astonishing' refers to something that causes great wonder or excitement. 'Unimpressive' means ordinary or not special in any way. Hence, option (d) is the correct answer.

168. Refer to the first sentence of the second paragraph. It says that the colonists on board suffered from exposure, scurvy and outbreaks of contagious disease. Hence, option (d) is the correct answer.

169. 'Forge an alliance' means to put a lot of effort into making a relationship successful or strong. Hence, option (c) is the correct answer.

170. Refer to the second sentence of the passage - 'After a treacherous and uncomfortable... lasted 66 days, they dropped anchor near the tip of Cape Cod...'. Hence, option (d) is the correct answer.

171. 'Sleuth' is a person who looks for information to solve crimes. An 'investigator' does the same. Hence, option (d) is the correct answer.

172. 'Apposite' means appropriate for a situation. 'Immaterial' means not important in a particular situation. Hence, option (c) is the correct answer.

173. The correct spelling is 'confetti'. It refers to the small pieces of brightly coloured paper that people throw at celebrations. Hence, option (c) is the correct answer.

174. A person who is hard to manage or control is called 'fractious'. Hence, option (c) is the correct answer. 'Factious' is someone who is a part of a larger group and has different ideas and opinions than the rest of the group. 'Ruly' is someone who is obedient. Someone who can be easily managed or controlled is called 'tractable'.

175. 'Dipsomania' refers to the uncontrollable craving for alcoholic beverages. Hence, option (b) is the correct answer. 'Pyromania' is the mental illness that makes people have a strong desire to set things on fire. 'Hypomania' is a mild form of mania marked by elation and hyperactivity. 'Megalomania' is a strong feeling that you want to have more and more power.

176. 'Floriculture' refers to the cultivation of ornamental plants. Hence, option (c) is the correct answer. 'Orthodontic' means treatment of problems relating to teeth and jaws. 'Agriculture' refers to the science and occupation of farming. 'Anthropology' refers to the study of human race.

177. The idiom means to feel very unhappy, especially because you want somebody/something you cannot have. Hence, option (a) is the correct answer.

178. 'Under the sun' means on earth. It is used to emphasise that you are talking about a large number of things that are in existence. Hence, option (b) is the correct answer.

179. 'Go cap in hand' means to ask somebody for something in a polite manner that makes you feel less important or ashamed. Hence, option (d) is the correct answer.

180. The sentence means that no amount of encouraging speech can mask the fact that Ulysses is a demanding book. 'Camouflage' means to hide something by covering it up or making it harder to see. Hence, option (b) is the correct answer. 'Impersonate' means to pretend to be another person. 'Undrape' means to unveil. 1 fact that is already known to people cannot be undraped or unveiled. Same is the case with 'obscure', which means to make something difficult to understand or know.

PRACTICE PAPER-6

ELECTRONICS ENGINEERING

Time Allowed : 2:15 Hours **Maximum Marks : 180**

Read the following instructions carefully before you begin to answer the questions. This Booklet contains questions in English.

INSTRUCTIONS

1. This Booklet contains 180 questions in total comprising of the following five parts:

Part – I:	Domain Knowledge	[1 – 75 Questions]
Part – II:	General Intelligence and Reasoning	[76 – 90 Questions]
Part – III:	General Awareness	[91 – 105 Questions]
Part – IV:	Quantitative Aptitude	[106 – 120 Questions]
Part – V:	General English	[121 – 180 Questions]

2. All questions are compulsory and carry 1 mark each.

3. Before you start to answer the questions you must check this Booklet to ensure that it contains all the pages and see that no page is missing or repeated.

4. For every wrong answer 1/3 marks will be deducted.

5. You are supplied with the OMR-Sheet separately.

6. Answers must be indicated by completely blackening one of the corresponding ovals on OMR-Sheet against the relevant question number, by black/blue ball pen only. Answers which are not indicated by black/blue ball pen will not be awarded any mark.

7. The OMR-Sheet must be handed over to the Invigilator before you leave the Examination Hall.

8. Failure to comply with any of the above instructions will render a candidate liable to such action/penalty as may be deemed fit.

PRACTICE PAPER – 6

DOMAIN KNOWLEDGE

1. Introducing a resistor in the emitter of a common emitter amplifier stabilizes the dc operating point against variations in
   (*a*) only the temperature
   (*b*) only the β
   (*c*) Both temperature and β
   (*d*) none

2. In the circuit of the figure, assume that the transistor is in the active region. It has a large β and its base- emitter voltage is 0.7 V. The value of I_C is

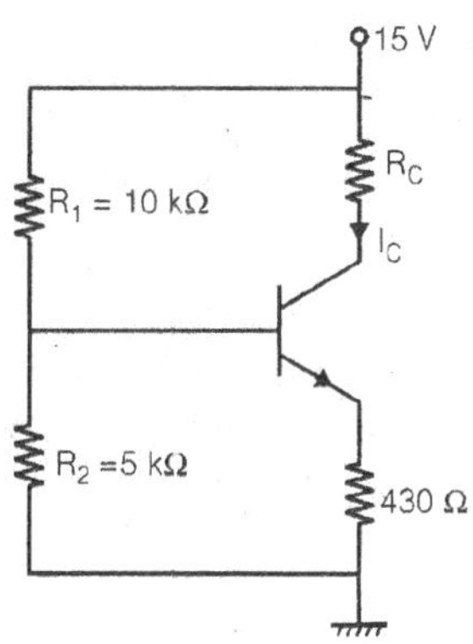

   (*a*) Indeterminate since Rc is not given
   (*b*) 1 mA
   (*c*) 5 mA
   (*d*) 10 mA

3. Assuming V_{CEsat} = 0.2 V and β = 50, the minimum base current (I_B) required to drive the transistor in the figure to saturation is

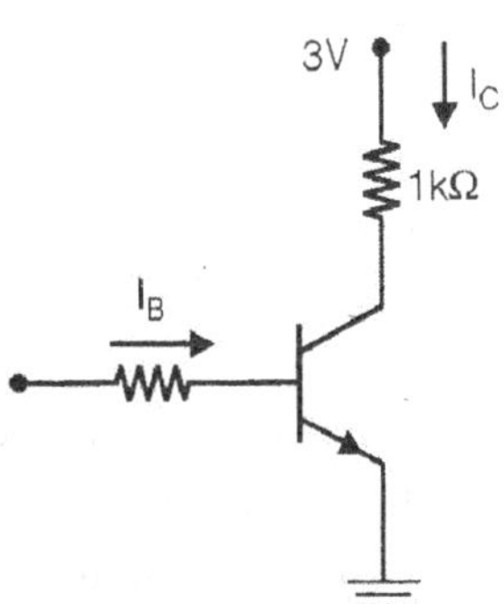

   (*a*) 56 μA (*b*) 140 μA
   (*c*) 60 μA (*d*) 3 μA

4. For an npn transistor connected as shown in the figure, V_{BE} = 0.7 V, given that reverse saturation current of the junction at room temperature 300 K is 10^{-13} A, the emitter current is

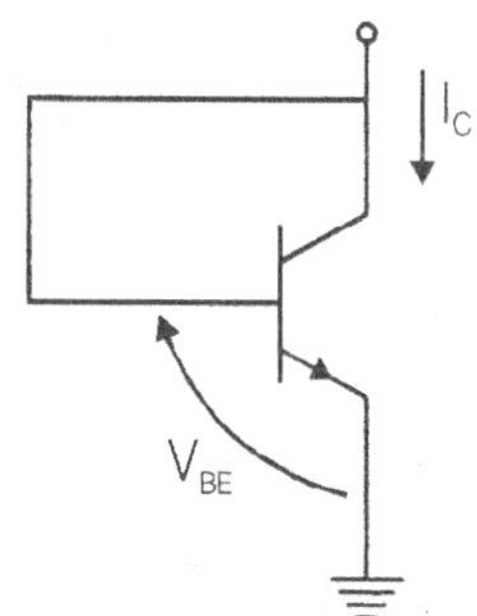

   (*a*) 30 mA (*b*) 39 mA
   (*c*) 49 mA (*d*) 20 mA

5. Thermal run away will take place if the quiescent point is such that
   (*a*) $V_{CE} > \frac{1}{2} V_{CC}$ (*b*) $V_{CE} < V_{CC}$
   (*c*) $V_{CE} < 2V_{CC}$ (*d*) $V_{CE} < \frac{1}{2} V_{CC}$

6. If the transistor is in saturation, then
   (*a*) I_C is always equal to $\beta_{dc} I_B$
   (*b*) I_C is always equal to $-\beta_{dc} I_B$
   (*c*) I_C is greater than or equal to $\beta_{dc} I_B$
   (*d*) I_C is less than or equal to $\beta_{dc} I_B$

7. The current gain of a bipolar transistor drops at high frequencies because of
   (*a*) transistor capacitances
   (*b*) high current effects in the base
   (*c*) parasitic inductive elements
   (*d*) the early effect

8. Negative feedback in an amplifier
   (*a*) reduces gain
   (*b*) increases frequency and phase distortions
   (*c*) reduces bandwidth
   (*d*) increases noise

9. Power consumed by a balanced 3-phase, 3-wire load is measured by two wattmeter method. The first wattmeter reads twice that of the second. then the load impedance angle in radians is
   (*a*) π/12 (*b*) π/8
   (*c*) π/6 (*d*) π/3

10. For a series R-L-C circuit energized with a sinusoidal voltage source of frequency 4 rad/s, the applied voltage lags the current by an angle of $\tan^{-1} 2$ degrees. Then the value of R for L = 1 H and C = 0.05 F is

(*a*) 4.0 ohm (*b*) 2.0 ohm
(*c*) 1.0 ohm (*d*) 0.5 ohm

11. the voltage (V) and current (*a*) across load are $v(t) = 100 \sin(\omega t)$

$i(t) = 10 \sin(\omega t\text{-}60^{\circ}) + 2\sin(3\omega t) + 5\sin(5\omega t)$

The average power consumed by the load is

(*a*) 350 W (*b*) 250 W
(*c*) 150 W (*d*) 200 W

12. In an ideal operational amplifier the voltage gain for the common mode signal is

(*a*) 0 (*b*) 0.5
(*c*) 2.0 (*d*) infinite

13. The magnitude of the gain in the inverting operational amplifier circuit shown below is Y, with the switch open, when the switch is closed the magnitude of the gain is

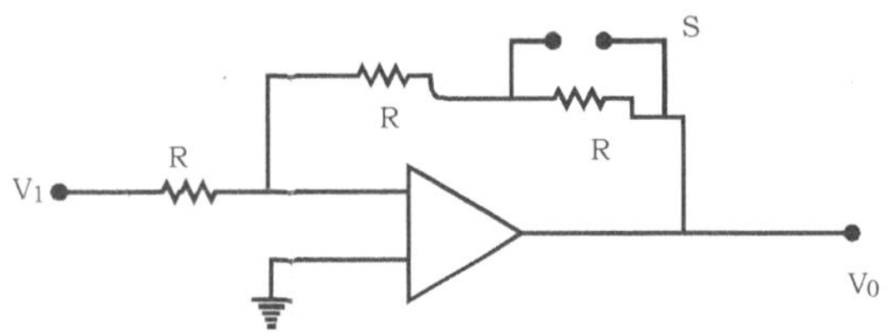

(*a*) -Y (*b*) Y / 2
(*c*) 2Y (*d*) - 2Y

14. The binary representation of 100110 is numerically equivalent to

(*a*) The decimal representation of 46
(*b*) The octal representation of 46
(*c*) The octal representation of 26
(*d*) The decimal representation of 26

15. The peak voltage rating of a diode in a bridge rectifier is X times larger than that of a full wave rectifier feeding the same de output voltage, where the value of X is

(*a*) 0.5 (*b*) 1.0
(*c*) $\sqrt{2}$ (*d*) 2.0

16. Determine V_0 for the circuit shown in figure V_2 is the zener voltage of the zener diode

(*a*) 10 V
(*b*) 12 V
(*c*) 12.5 V
(*d*) 22 V

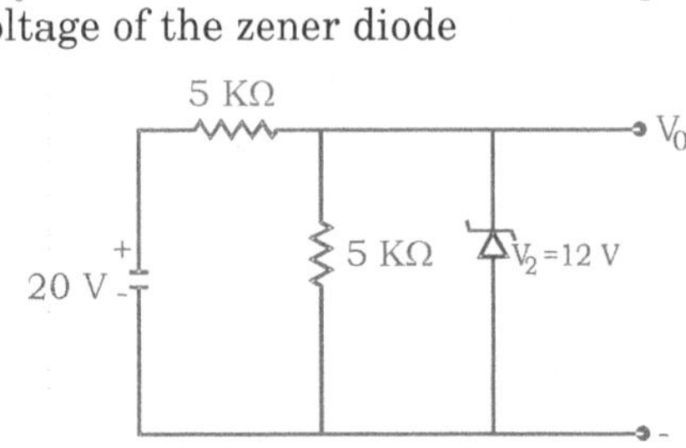

17. In the circuit shown below, the zener diode is ideal. The minimum value of R_L up to which the output voltage remains constant is

(*a*) 27 Ω
(*b*) 45 Ω
(*c*) 15 Ω
(*d*) 24 Ω

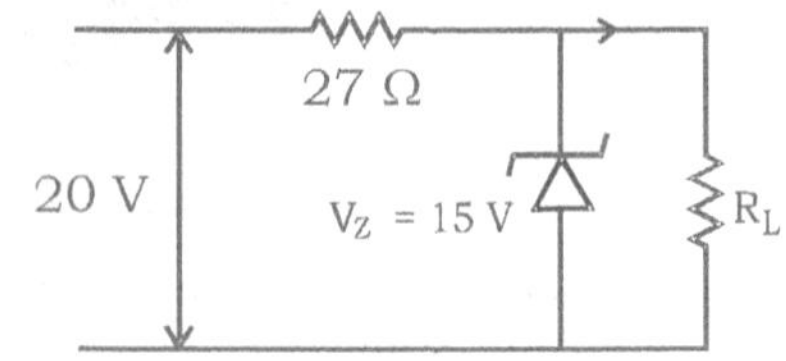

18. The Boolean expression $X = B + A.\overline{B} + A.B$ is equivalent to

(*a*) $A + B$ (*b*) $\overline{A}.B$
(*c*) $\overline{A + B}$ (*d*) $A \times B$

19. The output frequency of a decade counter that is clocked from a 50 KHz signal

(*a*) 12.5 kHz (*b*) 50 kHz
(*c*) 5 kHz (*d*) 500 kHz

20. One of the advantage of base modulation over collector modulation of a transistor class a amplifier is

(*a*) the lower modulating power required
(*b*) higher power output per transistor
(*c*) better efficiency
(*d*) better linearity

21. A single phase full bridge diode rectifier delivers a constant load current of 10A. Average and RMS values of source current are respectively

(*a*) 5 A, 10 A (*b*) 10 A, 10 A
(*c*) 5 A, 5 A (*d*) 0 A, 10 A

22. The boolean expression

$ABC + \overline{A}\overline{B}C + AB\overline{C} + +\overline{A}B\overline{C}$ is equiva-lent to

(*a*) OR GATE (*b*) AND GATE
(*c*) EX-NOR GATE (*d*) EX-OR GATE

23. An operational amplifier is basically a

(*a*) law gain AC amplifier
(*b*) high gain DC amplifier
(*c*) high gain RC coupled amplifier
(*d*) Low gain transformer coupled amplifier

24. The boolean expression $ABCD + A\overline{B}CD + ABC\overline{D} + A\overline{B}C\overline{D}$ is equivalent

(*a*) A (*b*) AC
(*c*) ABC (*d*) 1

25. The sign flag of 8085 microprocessor is set to 1, if

(*a*) the result of an arithmetic operation is 1
(*b*) there is a carry from additional or borrow from subtraction
(*c*) the most significant bit of the result of an arithmetic
(*d*) None of these

26. The value of i in the circuit is

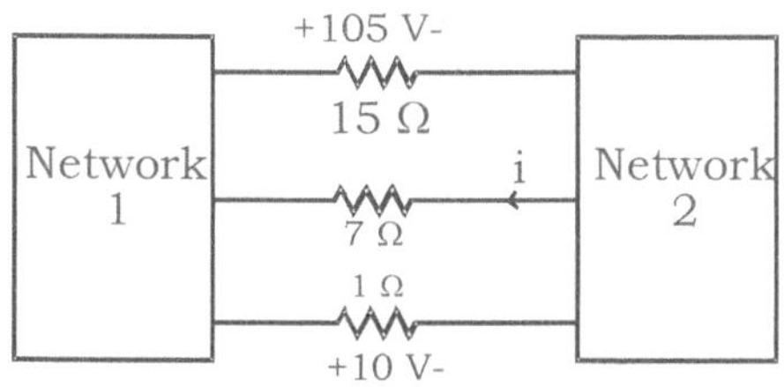

(a) 7 A (b) 0 A
(c) 17 A (d) 10 A

27. The value of current i for circuit shown in figure is

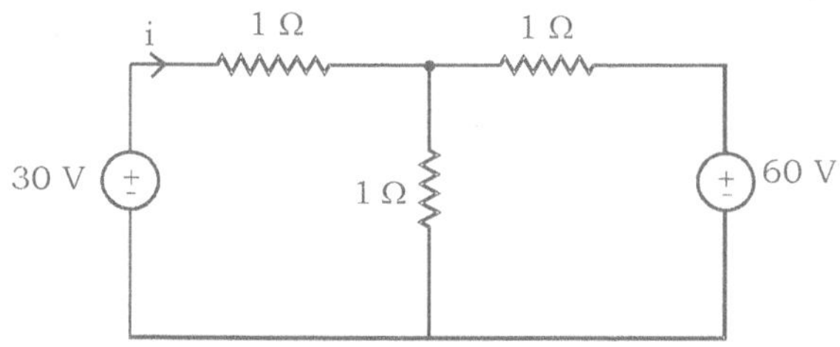

(a) 20 A (b) 0 A
(c) -20 A (d) 1A

28. A switch is connected across a 10 Ω resistor as shown in figure. When a source is connected to A-B terminals the switch S is closed for 7 seconds and open for 3 seconds. This process repeats with period to 10 seconds. What is the value of equivalent resistance offered by this arrangement

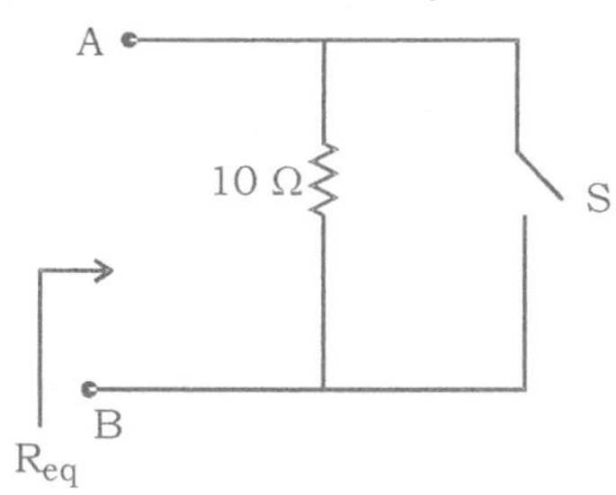

(a) 10 Ω (b) 0 Ω
(c) 3 Ω (d) 1 Ω

29. The value of load impedance to absorb maximum power through the source is

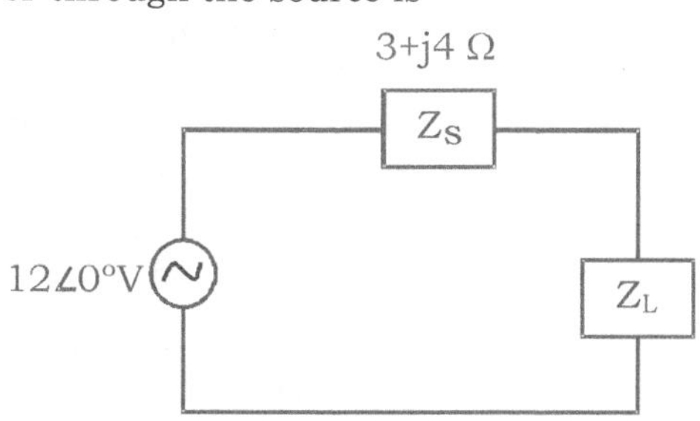

(a) 3 + j Ω
(b) 3 Ω
(c) 4 Ω
(d) 3-j4 Ω

30. For the circuit switch S cross at t = 0. The voltage across capacitor is

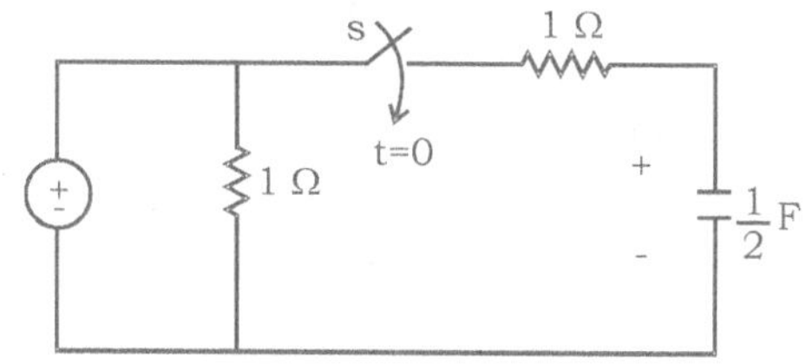

(a) $10[1-e^{-2t}]$ (b) $10e^{-2t}$
(c) $10[1+e^{-2t}]$ (d) $10e^{-t}$

31. For maximum power transfer to the load R_L the value of R_L is

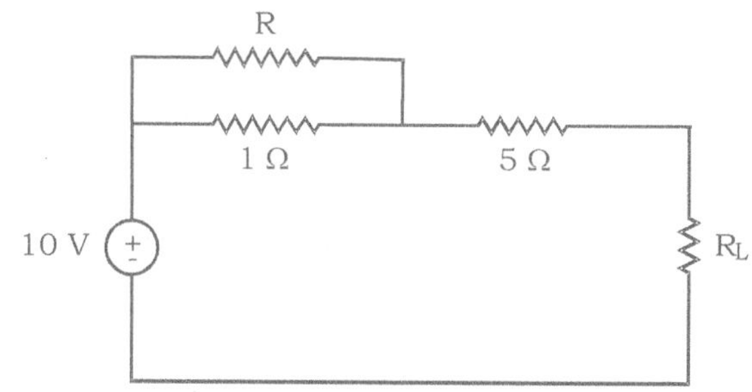

(a) 1 Ω (b) 0 Ω
(c) 5 Ω (d) 6 Ω

32. The value of dampling ratio ξ for the circuit shown in figure is

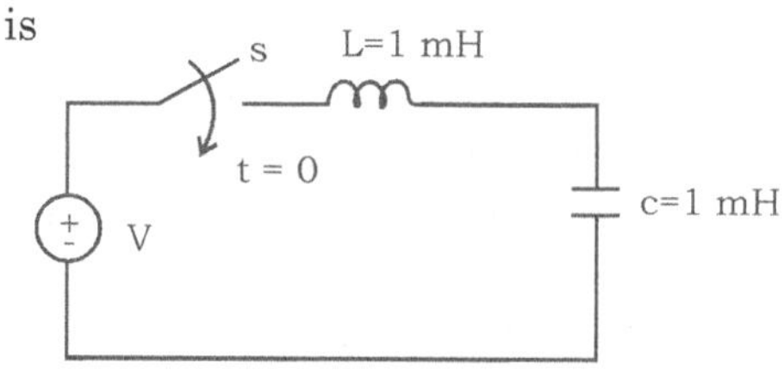

(a) ξ = 0
(b) ξ <1
(c) ξ = 1
(d) ξ >1

33. Which of the following is a linear element
(a) voltage source
(b) current source
(c) unchanged capacitor
(d) energized inductor

34. An NPN Bipolar Transistor with a current gain of 100 is biased in saturation mode. If the base current is increased by ΔI_b, then
(a) $\Delta I_C > 100\,\Delta I_b$ (b) $\Delta I_C = 100\,\Delta I_b$
(c) $\Delta I_C < 100\,\Delta I_b$ (d) $\Delta I_C = \Delta I_b$

35. Let β be the short circuit common emitter current gain of a BJT biased in normal active mode. Let I_C be the collector current. Is this β
(a) a monotonically increasing function of I_C?
(b) a monotonically decreasing function of I_C?
(c) initially an increasing function of I_C, which reaches a plateau, and then decreases with increasing I_C?
(d) independent of I_C?

36. An N-channel enhancement mode MOSFET with threshold voltage of 1 V is biased at V_{GS} = 2 V and V_{DS} = 2 V. If the drain voltage is doubled to 4 V, the drain to source current will

(*a*) double

(*b*) more than double

(*c*) increase only slightly

(*d*) become half

37. To double the drain current of an N-channel enhancement mode MOSFET biased in saturation

(*a*) Channel length should be doubled

(*b*) Channel width should be halved

(*c*) Channel length should be halved

(*d*) Oxide thickness should be doubled

38. Two identical resistive loads consume W watts each when connected in parallel across an ideal current source of I amperes. If, instead, they were connected in series with the same source, their total consumption

(*a*) would halve

(*b*) would double

(*c*) would remain the same

(*d*) would increase by a factor of 4

39. A wattmeter will read zero under the following condition

(*a*) The voltage and current are exactly in phase

(*b*) The voltage and current have the same time periods but the voltage is sinusoidal whereas the current is a square wave

(*c*) The voltage frequency is twice the current frequency

(*d*) The current is dc and the voltage is sinusoidal

40. In a three-phase delta-connected balanced load

(*a*) line current is equal to the phase current

(*b*) line current is three times the phase current

(*c*) line current is $\sqrt{3}$ times the phase current

(*d*) line current is the sum of the three phase currents

41. Which of the following is generally true? RC snubber is used to protect the SCR

(*a*) against device overvoltage

(*b*) against device overcurrent

(*c*) against forward dv/dt

(*d*) against di/dt

42. For a 50 Hz ac input, the ripple frequency in the output that a full-wave rectifier produces in equal to

(*a*) 25 Hz (*b*) 50 Hz

(*c*) 100 Hz (*d*) 200 Hz

43. As compared to a BJT amplifier, an amplifier made using a JFET is likely to have

(*a*) very high voltage gain

(*b*) very high bandwidth

(*c*) very high voltage swing

(*d*) very high input resistance

44. The voltage gain of a common emitter amplifier is

(*a*) directly proportional to collector bias current

(*b*) inversely proportional to collector bias current

(*c*) independent of collector bias current

(*d*) proportional to square of collector bias current

45. If the emitter bypass capacitor in a common emitter amplifier is removed, then

(*a*) input resistance will decrease

(*b*) lower cut-off frequency will increase

(*c*) voltage gain will decrease

(*d*) upper cut-off frequency will decrease

46. Use of a series-shunt negative feedback in an amplifier will

(*a*) increase input impedance & decrease output impedance

(*b*) decrease input impedance & decreases output impedance

(*c*) decreases input impedance & increase output impedance

(*d*) increase input impedance & increase output impedance

47. The number of different Boolean functions of 4 variables is

(*a*) 2^{16} (*b*) 16^2

(*c*) 4^2 (*d*) 16^4

48. In superheterodyne radio receiver, the RF amplifier

(*a*) removes the need for the mixer stage

(*b*) improves sensitivity of the receiver

(*c*) creates image channel interference

(*d*) removes the need for the detector stage

49. If a lossless line is terminated with a load impedance 40 + j30 Ω, then the characteristic impedance of the line for minimum possible standing wave ratio (SWR) will be

(*a*) 70 Ω (*b*) 25 Ω

(*c*) 50 Ω (*d*) 100 Ω

50. Two ceramic capacitors have the following parameters. One has capacitance C_1 and ceramic thickness d_1. The other has capacitance C_2 and ceramic thickness d_2. When the two are connected in parallel, the combination has a capacitance of

(*a*) $C = d_1 + d_2$ (*b*) $C = d_1 d_2 /(d_1+d_2)$

(*c*) $C = C_1 + C_2$ (*d*) $C = C_1C_2/(C_1+C_2)$

51. The conductivity σ as a function of 1/T, where T is the temperature, for a semiconducting material varies as shown in the figure. Using this information, state whether a resistance made from intrinsic semiconductor will have

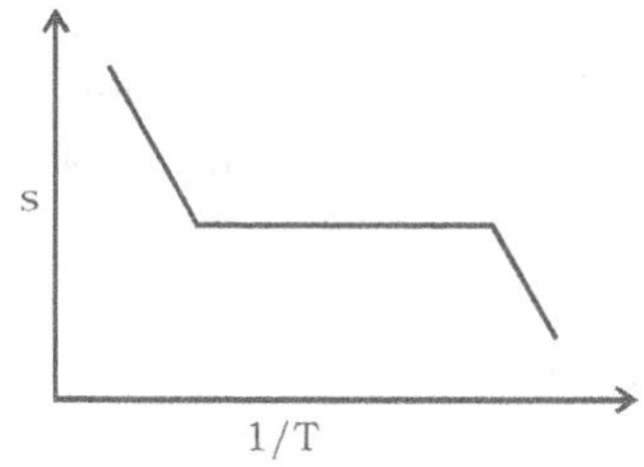

(a) Positive temperature coefficient of resistance

(b) Negative temperature coefficient of resistance

(c) Zero temperature coefficient of resistance

(d) Initially positive and later negative temperature coefficient of resistance

52. The time rate of change of a voltage applied across a 1 μF capacitor is 2 V/s. This means that the current flowing through the capacitor is

(a) 2×10^{-6} A (b) 2 A

(c) 0.5×10^{-6} A (d) 0.5 A

53. The time rate of change of a current passed through a 1mH inductor is 2 mA/s. This means that the voltage across the inductor is

(a) 0.5×10^{-6} V (b) 0.5 V

(c) 2×10^{-6} V (d) 2 V

54. Which of the following statements is true?

(a) Both refrigerators and thermocouples can be made using the Peltier effect.

(b) Both refrigerators and thermocouples can be made using the Seebeck effect.

(c) Refrigerators can be made using the Peltier effect and thermocouples can be made using the Seebeck effect.

(d) Refrigerators can be made using the Seebeck effect and thermocouples can be made using the Peltier effect.

55. As the frequency is increased, the charging MVAR in a cable

(a) decreases

(b) increases

(c) remain the same

(d) decreases or remains the scheme

56. In a three cavity klystron amplifier, the oscillations are excited in

(a) input cavity (b) output cavity

(c) intermediate cavity (d) both (a) and (b)

57. 2's complement of binary number 0101 is

(a) 1011 (b) 1111

(c) 1101 (d) 1110

58. An OR gate has 4 inputs. One input is high and the other three are low. The output

(a) is low

(b) is high

(c) is alternately high and low

(d) may be high or low depending on relative magnitude of inputs

59. Decimal number 10 is equal to binary number

(a) 1110 (b) 1010

(c) 1001 (d) 1000

60. Which of the following can be accessed only sequentially?

(a) Floppy disk (b) Hard disk

(c) Magnetic tape (d) ROM

61. Binary number 1101.101 is equivalent to decimal number?

(a) 13.5 (b) 13.75

(c) 13.625 (d) 13.875

62. Which of the following computer memories is fastest?

(a) Cache (b) Primary

(c) Mass storage (d) Off line back up

63. During processing the instructions, data, intermediate results and final results in a computer are held in

(a) CPU (b) ALU

(c) Primary memory (d) Secondary memory

64. The circuit given below is constituted by an iron-cored coil and a capacitor. At resonance, the circuit behaves like

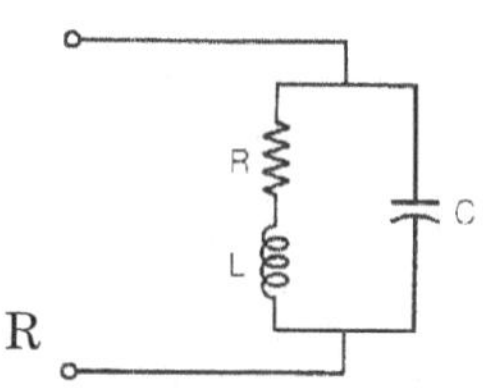

(a) an open circuit

(b) a short circuit

(c) a pure resistor of value R

(d) a pure resistor of value much higher than R

65. The total number of branches in a network is equal to b. The graph of the network has n number of branches. The minimum number of line currents is

(a) b + n (b) b

(c) b - n (d) n

66. The number of independent KVL and KCL equations for a network withn-nodes and *l* links are respectively

(a) *l* and n

(b) *l* and n - 1

(c) n-a and *l*

(d) n-and *l* - 1

67. In a series resonance circuit, at resonance selectivity Q is equal to

(a) $\frac{1}{R\sqrt{LC}}$ (b) $\frac{1}{R}\sqrt{\frac{C}{L}}$

(c) $\frac{1}{R}\sqrt{\frac{L}{C}}$ (d) $\frac{1}{R}\sqrt{LC}$

68. SR Flip flop can be converted to T-type flip-flop if?

(a) S is connected to Q

(b) R is connected to Q

(c) Both S and R are shortend

(d) S and R are connected to Q and Q' respectively

69. The main difference between JK and RS flip-flop is that?

(a) JK flip-flop does not need a clock pulse

(b) there is feedback in JK flip-flop

(c) JK flip-flop accepts both inputs as 1

(d) JK flip-flop is acronym of junction cathode multivibrator

70. Kirchhoff's law is applicable to

(a) passive networks only

(b) a.c. circuits only

(c) d.c. circuits only

(d) both a.c. as well d.c. circuits

71. The superposition theorem is not applicable to

(a) voltage only (b) bilateral element

(c) power calculation (d) passive element

72. Tuned voltage amplifiers are not used

(a) in public address systems

(b) in radio receivers

(c) where a band of frequencies is to be selected and amplified

(d) in television receivers

73. A power amplifier delivers 50 W output at 50% efficiency. The ambient temperature is 25°C. If the maximum allowable junction temperature is 150°C. Then the maximum thermal resistance θ_{jc} that can be tolerated is

(a) 2.5°C/W (b) 5°C/W

(c) 10°C/W (d) none

74. In crystal oscillator, frequency determining parameters are

(a) thickness of crystal

(b) angle of cut

(c) physical size of crystal

(d) all of the above

75. In transistor tuned oscillator, frequency of oscillations is determined by

(a) RC network (b) Tank circuit

(c) RC parallel circuit (d) None

GENERAL INTELLIGENCE AND REASONING

Directions : In the following question select the related letter/word from the given alternatives.

76. Baboon : Infant :: Beaver : ?

(a) Kitten (b) Kid

(c) Joey (d) Fawn

Directions: In question nos. **77** to **81**, a series is given, with one term missing. Choose the correct alternative from the given ones that will complete the series.

77. BGL, FKP, JOT, ___?___ .

(a) MSY (b) NSX

(c) NRX (d) MRY

78. BX, EU, HR, KO, ___?___ , QI, ___?___ .

(a) NL and TF (b) ML and TE

(c) NK and TF (d) MK and TE

79. 72, 46, 521, 612, 343, ___?___ .

(a) 678 (b) 545

(c) 215 (d) 154

80. W 5 K, T 12 M, Q 20 O, N 29 Q, ___?___ .

(a) L 45 R (b) K 39 S

(c) K 37 S (d) K 43 R

81. 17, 150, ___?___ , 1500, 1667, 15000, 16667.

(a) 1607 (b) 1067

(c) 167 (d) 177

Directions: In question nos. **82** and **83**, find the odd word/number from the given alternatives.

82. (a) French Beans (b) Gourd

(c) Pumpkin (d) Jackfruit

83. (a) Solar energy (b) Biomass

(c) Fossil fuel (d) Radiant energy

84. What is the least number that must be added to 2851 to make it a perfect square?

(a) 55 (b) 65

(c) 45 (d) 70

85. Karan is taller than Mukesh and is shorter than Anil. Rakesh is shorter than Sunil, who in turn is shorter than Anil. Who is the tallest?

(a) Anil (b) Sunil

(c) Mukesh (d) Karan

Directions : In the following question from the given alternatives select the word which cannot be formed using the letters of the given word.

86. ACCOUTREMENTS

(a) ACCOUNTS (b) TOURNAMENT

(c) MOURNS (d) TENETS

87. In the following question, the number of letters skipped in between adjacent letters in the series is consecutive odd numbers. Which of the following series observes this rule?

(*a*) B D G J P (*b*) C E I O V

(*c*) D F J P X (*d*) E G K P X

88. A man and his wife have four sons and three daughters. All four sons are married and have five children each. Find the total number of members in the family.

(*a*) 33 (*b*) 25

(*c*) 28 (*d*) 29

89. If 42 + 74 = 66 and 35 + 17 = 64, what is 57 + 21?

(*a*) 68 (*b*) 36

(*c*) 72 (*d*) 44

Directions: In question no. **90**, select the missing number from the given responses.

90.

6	15	35
10	21	55
12	18	?

(*a*) 45 (*b*) 28

(*c*) 24 (*d*) 36

GENERAL AWARENESS

91. Badminton player PV Sindhu recently won historic sliver medal in women's singles badminton event at Rio Olympics. Who won gold in the event?

(*a*) Wang Yihan (*b*) Sain Nehwal

(*c*) Carolina Marin (*d*) Dolores Marco

92. The Hindi word 'rupiya' (Indian currency) is derived from Sanskrit word 'rupya' which literally means

(*a*) wrought iron (*b*) wrought silver

(*c*) wrought aluminium (*d*) wrought copper

93. Cracking in brass is caused by

(*a*) Ammonia (*b*) Nitric acid

(*c*) Hydrogen peroxide (*d*) Sulphuric acid

94. Article 2 of the Indian constitution states

(*a*) Admission or establishment of new States

(*b*) Formation of new States and alteration of areas, boundaries or names of existing States

(*c*) a law that grants special autonomous status to Jammu and Kashmir

(*d*) deals with the failure of the Constitutional machinery of an Indian state

95. Which part of the human eye acts as an aperture?

(*a*) Retina (*b*) Pupil

(*c*) Sclera (*d*) Cornea

96. Which personality is mentioned in the Buddhist text 'Milinda Panha'?

(*a*) Nagasena (*b*) Chanakya

(*c*) Kalidas (*d*) Hemachandra

97. Which coating of mucins gives them considerable water-holding capacity and makes them resistant to proteolysis?

(*a*) Fat coating (*b*) Sugar coating

(*c*) Protein coating (*d*) Glycoprotein coating

98. Which of the following are the possible outcomes after Quantitative easing?

1. Increase in Inflationary tendency in the economy.
2. Stimulus to the growth prospects of economy.
3. Availability of more credit in the market.

Correct code:

(*a*) 1 and 2 only (*b*) 2 and 3 only

(*c*) 1 and 3 only (*d*) 1, 2 and 3

99. The scientific method of dating based on the analysis of patterns of tree rings is known as

(*a*) Dendrology (*b*) Dendrochronology

(*c*) Chronology (*d*) Acanthochronology

100. Recently which Airport awarded Best Tourist Friendly Airport ?

(*a*) Nagpur (*b*) Lucknow

(*c*) Goa (*d*) Tirupati

101. 'Making India Awesome' is a book authored by

(*a*) Chetan Bhagat (*b*) Arundhati Roy

(*c*) V.S. Naipaul (*d*) Amitav Ghosh

102. Which of the following is useful in ice-cream making?

(*a*) Agar-Agar (*b*) Algal extract

(*c*) Algal bloom (*d*) Algal excreta

103. Kushok Bakula Rimpochhe Airport is situated in

(*a*) Assam (*b*) Tripura

(*c*) Jammu and Kashmir (*d*) Mizoram

104. Dengue is caused by the

(*a*) Virus

(*b*) Bacteria

(*c*) Nematode

(*d*) Protozoan

105. Where was the Asia's first export processing zone set up?

(*a*) Mangalore

(*b*) Kandla

(*c*) Howrah

(*d*) Gurgaon

QUANTITATIVE APTITUDE

106. A person borrows ₹33,000 at 20% compound interest. How much he has to pay equally at the end of each year, to settle his loan in two years?

(*a*) ₹22,450 (*b*) ₹21,600
(*c*) ₹17,000 (*d*) ₹19,600

107. If the expression $2x^2 - 6x + 5$ is written in the form $2\left(x-\frac{3}{2}\right)^2 + 2q^2$, then the possible values of q are

(*a*) $\pm\frac{1}{2}$ (*b*) $\pm\frac{2}{3}$
(*c*) $\pm\sqrt{\frac{3}{2}}$ (*d*) $\pm\sqrt{\frac{1}{2}}$

108. A faulty clock shows correct time at 6 a.m. It lags 1sec at the end of every 10 sec. What will be the time shown by the clock at 9 p.m.?

(*a*) 7:00 p.m. (*b*) 7:30 p.m.
(*c*) 8:00 p.m. (*d*) 8:30 p.m.

109. What is the value of x, if $x = \cfrac{1}{3+\cfrac{3}{5+\cfrac{4}{5+\cfrac{7}{2}}}}$?

(*a*) $\frac{31}{110}$ (*b*) $\frac{31}{101}$
(*c*) $\frac{41}{101}$ (*d*) $\frac{3}{11}$

110. If $\frac{a}{b} = \frac{1}{4}$ and $\frac{b}{c} = \frac{2}{3}$, then $\frac{12a^2 - 8c^2}{33a^2 + c^2}$ is equal to

(*a*) –4 (*b*) –1
(*c*) 0 (*d*) $\frac{2}{17}$

111. Radius of two concentric circles C_1 and C_2 with centre O are 6 cm and 10 cm respectively. A tangent drawn to circle C_1 at point P meet the circle C_2 at points A and B. The area of the triangle AOB is

(*a*) 24 cm^2 (*b*) 36 cm^2
(*c*) 48 cm^2 (*d*) 56 cm^2

112. Amit goes 10 m North, then 15 m due West, then 14 m more due North. If he finally goes 8 m due East, then the distance between his initial and final position is

(*a*) 10 m
(*b*) 24 m
(*c*) 25 m
(*d*) 31 m

113. Two triangles ABC and PQR are such that $\angle ABC = \angle QPR$ and $\angle BCA = \angle PRQ$, then one can infer that $\Delta ABC \cong \Delta PQR$ when

(*a*) BC = PR (*b*) $\angle BAC = \angle PQR$
(*c*) AB = PQ (*d*) BC = QR

114. Two chords AB and CD of a circle with centre O intersect each other at P, $\angle DCA = 28°$ and $\angle APD = 62°$ then the $\angle CDB$ is

(*a*) 34° (*b*) 62°
(*c*) 90° (*d*) 118°

115. The difference of a number consisting of three different digits from the number formed by reversing the digits is always divisible by

(*a*) 9 (*b*) 10
(*c*) 11 (*d*) Both (*a*) and (*c*)

116. Pipe A can empty a tank in 12 minutes, pipe B can empty the tank in 18 minutes while the pipe C can empty the tank in 36 minutes. In how much time the tank will be empty if all three pipes are opened together?

(*a*) 22 minutes (*b*) 10 minutes
(*c*) 6 minutes (*d*) 5 minutes

117. Four men and ten women are assigned to complete a work. They complete one third of the work in 4 days, then 2 more men and 2 more women are also engaged to complete the work. They completed two-ninth of the work in next two days. If the remaining work is to be completed in 4 days, then how many more women must be employed?

(*a*) 60 (*b*) 75
(*c*) 90 (*d*) None of these

118. A conical vessel of base radius 9 cm and height 8 cm is filled with milk. If the milk leaks through a hole at the bottom of the conical vessel into a cylindrical jar of radius 3 cm, then find the level of the milk in the jar after the milk has leaked completly.

(*a*) 12 cm (*b*) 15 cm
(*c*) 24 cm (*d*) 27 cm

119. A shopkeeper sells an article at $12\frac{1}{2}$% loss. If he sells it for ₹51.80 more, then he gains 6%. What is the cost price of the article?

(*a*) ₹210 (*b*) ₹240
(*c*) ₹280 (*d*) ₹300

120. A shopkeeper gives 25% discount on buying an article and 15% additional discount if bought between 4 p.m. to 5 p.m. What will be the total discount offered to the customer?

(*a*) 40% (*b*) 10%
(*c*) 36.25% (*d*) 33.75%

GENERAL ENGLISH

Direction : In the following question out of the four alternatives, choose the one which best expresses the meaning of the given word and mark it in the answer sheet.

121. Exude
(*a*) Vent (*b*) Seep
(*c*) Pour (*d*) Gush

122. Imperious
(*a*) Dangerous (*b*) Precious
(*c*) Feeble (*d*) Authoritarian

Directions: In question nos. **123** and **124**, choose the word opposite in meaning to the given word and mark it in the answer sheet.

123. Passive
(*a*) Resigned (*b*) Resistant
(*c*) Yielding (*d*) Forbearing

124. Incorrigible
(*a*) Irredeemable (*b*) Unruly
(*c*) Reformable (*d*) Inveterate

Directions: In question nos. **125** to **127**, Out of the four alternatives, choose the one which can be substituted for the given words/sentences.

125. Science of rearing of animals for farming
(*a*) Aviculture (*b*) Horticulture
(*c*) Agriculture (*d*) Apiculture

126. Bhawana always achieves her targets on time and she enjoys doing difficult tasks.
(*a*) is narcoleptic (*b*) is pyromaniac
(*c*) is workaholic (*d*) is glutton for work

127. Easy to understand
(*a*) Confirmed (*b*) Evident
(*c*) Witty (*d*) Chronic

Directions: In question nos. **128** to **130**, four alternatives are given for the idioms/phrases given below. Choose the alternative which best expresses the meaning of the given idiom/phrase.

128. He's been chasing his tail all week collecting data but the report is still not ready.
(*a*) boot licking
(*b*) working very hard
(*c*) rushing around ineffectually
(*d*) pushing oneself to work hard

129. The director of the school runs a tight ship.
(*a*) organized and disciplined
(*b*) haywire and upsetting
(*c*) valueless and expensive
(*d*) firmly and effectively

130. To catch some rays
(*a*) To tan in the sun
(*b*) To succeed
(*c*) To rise with less efforts
(*d*) To get promotion

Direction : In the following question out of four words only one word is spelt correctly. Find the correctly spelt word and mark your answer in the answer sheet.

131. (*a*) Casete (*b*) Cassete
(*c*) Casette (*d*) Cassette

Directions: In question nos. **132** to **134**, some parts of the sentences have errors and some are correct. Find out which part of a sentence has an error and mark the answer corresponding to the appropriate letter (A,B,C). If a sentence is free from error, mark your answer as (*d*).

132. They're considering (*a*) / to making an appeal (*b*) / against the judgment. (*c*) / No error (*d*)

133. If I were working (*a*) / in Bandra, I will (*b*) / commute by the local train. (*c*) / No error (*d*)

134. My daughter pointed out last night (*a*) / that by next March, my nephew (*b*)/ shall be married for 10 years. (*c*) / No error (*d*)

Directions: In question nos. **135** to **137**, sentences are given with blanks to be filled with an appropriate word(s). Four alternatives are suggested for each question. Choose the correct alternative out of the four.

135. While the chicken is _____ way, you should prepare a rice mix and steam peas.
(*a*) in (*b*) on
(*c*) under (*d*) below

136. She asked her husband, "We should exercise daily, _______?"
(*a*) shouldn't we (*b*) should we
(*c*) shall we (*d*) should she

137. Tourists ______ on Niagara Falls after the giant waterfall froze. One of three waterfalls has frozen completely, while the Horseshoe Falls ______ to flow unimpeded.
(*a*) disunited, continues (*b*) disbanded, continued
(*c*) converged, continues (*d*) assembled, continued

Directions: In question nos. **138** to **140**,a sentence/part of the sentence is underlined. Beneath each sentence four different ways of phrasing the underlined part are given. Choose the grammatically correct option. In case no improvement is needed, your answer is option (*d*)

138. A lot of problems have arosen from her failure to meet the deadline.
(*a*) arose (*b*) arised
(*c*) arisen (*d*) No improvement

139. Fifteen years ago today, Sachin was sitting in the plane going to Australia.

(*a*) sat (*b*) have been sitting

(*c*) is sitting (*d*) No improvement

140. They drove out of the country by car where they had been banished.

(*a*) They drove out of the country where they had been banished by car.

(*b*) They drove where they had been banished out of the country by car.

(*c*) They drove by car out of the country where they had been banished.

(*d*) No improvement

Directions : In question no. **141** to **145**, you have given a passage with 5 questions. Read the following passage carefully and choose the best answer to each question out of the four alternatives.

Wars, fought during the course of history, have been so destructive that they have left their marks on the world, and impacted the way things are today. The main idea of a war has either been to expand power, independence, or to join a war to help allies. World War I and World War II have both similarities and differences but both wars will be remembered as infamy.

Alliances of strong countries like Hungary, Germany, and Austria caused World War I because they decided to declare war on other countries. Allies of all the countries decided to join in, which made the war more damaging and made it worse. World War II was caused by Germany because Germany was unhappy with the Treaty of Versailles. So they decided to take matters into their own hands and invaded Poland which triggered the start of the war. In World War I, most of the battles were fought in trenches, and for the first time machine guns, tanks, flamethrowers, and mustard gas were introduced. But in World War II they used more advanced machinery. One of the World War II's famous weapon is the atom bomb. This was first introduced in World War II and was used as a threat during the Cold War. These wars have left so much destruction but in WWI there was far less damage than WWII. WWI caused approximately 24 million deaths, while WWII caused about 60 million deaths, which is about 2.5 times more deaths than WWII. At the end of WWI, the League of Nations was established. And at the end of WW2, there still was no peace because people were being executed from committing war related crimes.

141. Which, according to the passage, were the main reasons that led to wars during the course of history?

(*a*) Expanding power, helping allies, Treaty of Versailles

(*b*) Treaty of Versailles, helping allies, independence

(*c*) Helping allies, independence, expanding power

(*d*) Helping allies, independence, atom bomb

142. Find the word that is same in meaning to 'infamy'.

(*a*) Infinity (*b*) Infamous

(*c*) Shadiness (*d*) Blameworthiness

143. World War I was different from World War II because

(*a*) WWI had more allies.

(*b*) of the degree of the damage caused.

(*c*) of the crimes that were caused during the war.

(*d*) None of the above

144. World War II was triggered because

(*a*) Allies of strong countries declared war on other countries.

(*b*) it was a way to test the advanced machineries.

(*c*) Germany threatened to use atom bomb during the cold war.

(*d*) Germany took control of Poland against its wishes.

145. Which of the following weapons were used in WWI for the first time?

(*a*) Trenches (*b*) Nuclear Bombs

(*c*) Flame throwers (*d*) Riffles

Directions: In question nos. **146** and **147**,out of the four alternatives, choose the one which best expresses the meaning of the given word and mark it in the answer sheet.

146. Awash

(*a*) Swamp (*b*) Arid

(*c*) Dampish (*d*) Soaked

147. Endue

(*a*) Provide (*b*) Divest

(*c*) Strip (*d*) Instill

Directions: In the following question choose the word opposite in meaning to the given word and mark it in the answer sheet.

148. Timid

(*a*) Horrified (*b*) Audacious

(*c*) Aggressive (*d*) Embarrassed

149. Precarious

(*a*) Insecure (*b*) Stable

(*c*) Ambiguous (*d*) Dangerous

Directions: In question nos. **150** to **152**, Out of the four alternatives, choose the one which can be substituted for the given words/sentences.

150. An agent who negotiates contracts of purchase and sale

(*a*) Salesman (*b*) Broker

(*c*) Bargainer (*d*) Moderator

151. A person who believes something is very important and should be followed all the time

(*a*) Stickler (*b*) Perseverer

(*c*) Slacker (*d*) Moral

152. A person who thinks that he is better than others and thinks and talks too much about himself
(*a*) Immodest (*b*) Egotist
(*c*) Presumptuous (*d*) Inordinate

Directions: In question nos. **153** to **155**, four alternatives are given for the idioms/phrases given below. Choose the alternative which best expresses the meaning of the given idiom/phrase.

153. With more clergy hooked up to computers, the demand for specialist software is growing.
(*a*) associated (*b*) dispersed
(*c*) divorced (*d*) snubbed

154. Good lawyers need to be able to think on their feet when pleading a case.
(*a*) to be able to think and speak creatively
(*b*) to be able to think and speak confidently
(*c*) to be able to think and speak honestly
(*d*) to be able to think and speak extemporaneously

155. Siddhant's comment about his friend's unethical activities was quite close to home.
(*a*) a lucky guess
(*b*) a relevant comment that makes one uncomfor-table
(*c*) talking about something without coming to the point
(*d*) speaking a lie without hesitating

Directions: In the following questionFour words are given in each question, out of which only one word is spelt correctly. Find the correctly spelt word and mark your answer in the answer sheet.

156. (*a*) Assasin (*b*) Assassin
(*c*) Asasin (*d*) Asassin

Directions: In question nos. **157** to **159**, some parts of the sentences have errors and some are correct. Find out which part of a sentence has an error and mark the answer corresponding to the appropriate letter (A,B,C). If a sentence is free from error, mark your answer as (*d*).

157. Between you and me,(*a*) / it's Rahul's terrible habit that (*b*)/ both Somiya and me find annoying. (*c*) / No error (*d*)

158. I'm seeing Namit (*a*) / at 5 and then I am (*b*) / have dinner with Dikshant. (*c*) / No error (*d*)

159. I am reading (*a*) / this book for two months (*b*) / but I've only read half of it. (*c*) / No error (*d*)

Directions: In question nos. **160** to **162**, sentences are given with blanks to be filled with an appropriate word(s). Four alternatives are suggested for each question. Choose the correct alternative out of the four.

160. Paper airplane contest awards the flimsy fliers that are a cut ______ the rest.
(*a*) through (*b*) for
(*c*) above (*d*) among

161. Eating disorders are serious, potentially life-threatening ________ that affect a person's emotional and physical health. People struggling with an eating disorder need to ________ professional help.
(*a*) conditions, seek (*b*) traits, seek
(*c*) condition, take (*d*) trait, take

162. They then heated the result in a _____, oxygen-free environment to distil out what they could in the form of oil.
(*a*) pressurizing (*b*) pressure
(*c*) pressurized (*d*) pressurizer

Directions: In question nos. **163** to **165**,a sentence/part of the sentence is underlined. Beneath each sentence four different ways of phrasing the underlined part are given. Choose the grammatically correct option. In case no improvement is needed, your answer is option (*d*)

163. The next flight is on time and will be arriving at five in the evening.
(*a*) arrived (*b*) arrives
(*c*) has been arriving (*d*) No improvement

164. He is playing since five o' clock.
(*a*) has been (*b*) would be
(*c*) will be (*d*) No improvement

165. If the dresses were cheaper, I would buy them.
(*a*) would have bought (*b*) would be buying
(*c*) will buy (*d*) No improvement

Directions : In question nos. **166** to **170** You have given passage with 5 questions. Read the passage carefully and choose the best answer to each question out of the four alternatives.

They lived in well-planned cities, made exquisite jewelry, and enjoyed the ancient world's best plumbing. But the people of the sophisticated Indus civilization—which flourished four millennia ago in what is now Pakistan and western India—remain tantalizingly mysterious.

Unable to decipher the Indus script, archaeologists have pored over beads, slivers of pottery, and other artifacts for insights into one of the world's first city-building cultures.

Now scientists are turning to long-silent witnesses: human bones. In two new studies of skeletons from Indus cemeteries, researchers have found intriguing clues to the makeup of one city's population—and hints that the society was not as peaceful as it has been portrayed.

Peaceful or not, the Indus civilization accomplished great things. At its peak, its settlements spanned an area greater than that of ancient Egypt, a contemporary culture. Indus jewelry was so coveted that examples have been found as far as Mesopotamia, some 1,500 miles (2,500 kilometers) away. Indus cities boasted blocks of houses built on a grid pattern and drains that funneled sewage from homes to dumping grounds outside the city walls.

166. Which of the following adjectives best describe the people of Indus Civilization?

(*a*) Hardworking and intelligent

(*b*) Sophisticated and progressive

(*c*) Peaceful and result oriented

(*d*) Materialistic and boastful

167. Which of the following could be a reason that prompted the author to say "the society was not as peaceful as it has been portrayed"?

(*a*) Archeologists have excavated weapons of destruction from the site of the civilization.

(*b*) Scientists have found pottery depicting scenes from war.

(*c*) Scientists feel that the Indus script points to the fact that the people were aggressive in nature.

(*d*) Discoveries, which suggested the same, were made from the skeletal remains of the people of the Indus civilization.

168. The author calls the people of Indus Civilization as tantalizingly mysterious because

(*a*) archeologists have been unable to decipher the Indus script.

(*b*) archeologists have invested a lot of time in knowing about the civilisation, but to no avail.

(*c*) all that is known about the civilization is not true.

(*d*) they knew little about the people of Indus civilization.

169. What is the author's purpose in writing the last paragraph?

(*a*) To show that the civilization covered a greater area than the Egyptian civilisation.

(*b*) To show how great a culture Indus Civilization was.

(*c*) To show how it was better than its contemporary.

(*d*) To show that there is a dispute whether the civilization was peace loving or not.

170. Why does the author refer to the human bones as 'the long silent witnesses'?

(*a*) The scientists could not get substantial clues from the skeletal remains for a long time.

(*b*) The scientists did not try to dig clues from the skeletal remains earlier.

(*c*) It is now that they have found the skeletons of the people of the civilization.

(*d*) None of the above

Directions for question 171 and 172 : In the following question, a word is followed by four options. Select the option that best expresses the meaning of the given word.

171. Inferno

(*a*) Blaze (*b*) Mark

(*c*) Poison (*d*) Anger

172. Preposterous

(*a*) Realistic (*b*) Absurd

(*c*) Wise (*d*) Impulsive

Directions for questions 173 and 174 : In each of the following questions, a word is followed by four options. Select the option that is opposite in meaning to the given word.

173. Dovish

(*a*) Hawkish (*b*) Owlish

(*c*) Vulturous (*d*) Ravenous

174. Apposite

(*a*) Suitable (*b*) Harmonious

(*c*) Immaterial (*d*) Disagreeable

Directions: In question nos. **175** to **177**, Out of the four alternatives, choose the one which can be substituted for the given words/sentences.

175. I <u>was under the wrong impression</u> that the course was for complete beginners.

(*a*) fathomed (*b*) comprehended

(*c*) misapprehended (*d*) apprehended

176. They are celebrating their <u>fifth</u> wedding anniversary this year.

(*a*) triennial (*b*) quinquennial

(*c*) septennial (*d*) vicennial

177. <u>A person who writes historical records</u>

(*a*) Archaeologist (*b*) Annalist

(*c*) Radiologist (*d*) Genealogist

Direction for questions 178 to 180: In the following question four alternatives are given for the idioms/phrases given below. Choose the alternative which best expresses the meaning of the given idiom/phrase.

178. Miss Jasmine <u>kicked the bucket</u> last week.

(*a*) had a fit (*b*) died

(*c*) retired (*d*) fired someone

179. Mr. Hyde was killed by a gangster because he was thought to be a <u>stool pigeon</u>.

(*a*) lazy person. (*b*) spy for the police.

(*c*) forgetful person. (*d*) burden.

180. <u>To blow a fuse</u>

(*a*) To be nervous

(*b*) To lose one's temper

(*c*) To act in boastful way

(*d*) To beat someone to death

OMR SHEET

1. Ⓐ Ⓑ Ⓒ Ⓓ	26. Ⓐ Ⓑ Ⓒ Ⓓ	51. Ⓐ Ⓑ Ⓒ Ⓓ	76. Ⓐ Ⓑ Ⓒ Ⓓ
2. Ⓐ Ⓑ Ⓒ Ⓓ	27. Ⓐ Ⓑ Ⓒ Ⓓ	52. Ⓐ Ⓑ Ⓒ Ⓓ	77. Ⓐ Ⓑ Ⓒ Ⓓ
3. Ⓐ Ⓑ Ⓒ Ⓓ	28. Ⓐ Ⓑ Ⓒ Ⓓ	53. Ⓐ Ⓑ Ⓒ Ⓓ	78. Ⓐ Ⓑ Ⓒ Ⓓ
4. Ⓐ Ⓑ Ⓒ Ⓓ	29. Ⓐ Ⓑ Ⓒ Ⓓ	54. Ⓐ Ⓑ Ⓒ Ⓓ	79. Ⓐ Ⓑ Ⓒ Ⓓ
5. Ⓐ Ⓑ Ⓒ Ⓓ	30. Ⓐ Ⓑ Ⓒ Ⓓ	55. Ⓐ Ⓑ Ⓒ Ⓓ	80. Ⓐ Ⓑ Ⓒ Ⓓ
6. Ⓐ Ⓑ Ⓒ Ⓓ	31. Ⓐ Ⓑ Ⓒ Ⓓ	56. Ⓐ Ⓑ Ⓒ Ⓓ	81. Ⓐ Ⓑ Ⓒ Ⓓ
7. Ⓐ Ⓑ Ⓒ Ⓓ	32. Ⓐ Ⓑ Ⓒ Ⓓ	57. Ⓐ Ⓑ Ⓒ Ⓓ	82. Ⓐ Ⓑ Ⓒ Ⓓ
8. Ⓐ Ⓑ Ⓒ Ⓓ	33. Ⓐ Ⓑ Ⓒ Ⓓ	58. Ⓐ Ⓑ Ⓒ Ⓓ	83. Ⓐ Ⓑ Ⓒ Ⓓ
9. Ⓐ Ⓑ Ⓒ Ⓓ	34. Ⓐ Ⓑ Ⓒ Ⓓ	59. Ⓐ Ⓑ Ⓒ Ⓓ	84. Ⓐ Ⓑ Ⓒ Ⓓ
10. Ⓐ Ⓑ Ⓒ Ⓓ	35. Ⓐ Ⓑ Ⓒ Ⓓ	60. Ⓐ Ⓑ Ⓒ Ⓓ	85. Ⓐ Ⓑ Ⓒ Ⓓ
11. Ⓐ Ⓑ Ⓒ Ⓓ	36. Ⓐ Ⓑ Ⓒ Ⓓ	61. Ⓐ Ⓑ Ⓒ Ⓓ	86. Ⓐ Ⓑ Ⓒ Ⓓ
12. Ⓐ Ⓑ Ⓒ Ⓓ	37. Ⓐ Ⓑ Ⓒ Ⓓ	62. Ⓐ Ⓑ Ⓒ Ⓓ	87. Ⓐ Ⓑ Ⓒ Ⓓ
13. Ⓐ Ⓑ Ⓒ Ⓓ	38. Ⓐ Ⓑ Ⓒ Ⓓ	63. Ⓐ Ⓑ Ⓒ Ⓓ	88. Ⓐ Ⓑ Ⓒ Ⓓ
14. Ⓐ Ⓑ Ⓒ Ⓓ	39. Ⓐ Ⓑ Ⓒ Ⓓ	64. Ⓐ Ⓑ Ⓒ Ⓓ	89. Ⓐ Ⓑ Ⓒ Ⓓ
15. Ⓐ Ⓑ Ⓒ Ⓓ	40. Ⓐ Ⓑ Ⓒ Ⓓ	65. Ⓐ Ⓑ Ⓒ Ⓓ	90. Ⓐ Ⓑ Ⓒ Ⓓ
16. Ⓐ Ⓑ Ⓒ Ⓓ	41. Ⓐ Ⓑ Ⓒ Ⓓ	66. Ⓐ Ⓑ Ⓒ Ⓓ	91. Ⓐ Ⓑ Ⓒ Ⓓ
17. Ⓐ Ⓑ Ⓒ Ⓓ	42. Ⓐ Ⓑ Ⓒ Ⓓ	67. Ⓐ Ⓑ Ⓒ Ⓓ	92. Ⓐ Ⓑ Ⓒ Ⓓ
18. Ⓐ Ⓑ Ⓒ Ⓓ	43. Ⓐ Ⓑ Ⓒ Ⓓ	68. Ⓐ Ⓑ Ⓒ Ⓓ	93. Ⓐ Ⓑ Ⓒ Ⓓ
19. Ⓐ Ⓑ Ⓒ Ⓓ	44. Ⓐ Ⓑ Ⓒ Ⓓ	69. Ⓐ Ⓑ Ⓒ Ⓓ	94. Ⓐ Ⓑ Ⓒ Ⓓ
20. Ⓐ Ⓑ Ⓒ Ⓓ	45. Ⓐ Ⓑ Ⓒ Ⓓ	70. Ⓐ Ⓑ Ⓒ Ⓓ	95. Ⓐ Ⓑ Ⓒ Ⓓ
21. Ⓐ Ⓑ Ⓒ Ⓓ	46. Ⓐ Ⓑ Ⓒ Ⓓ	71. Ⓐ Ⓑ Ⓒ Ⓓ	96. Ⓐ Ⓑ Ⓒ Ⓓ
22. Ⓐ Ⓑ Ⓒ Ⓓ	47. Ⓐ Ⓑ Ⓒ Ⓓ	72. Ⓐ Ⓑ Ⓒ Ⓓ	97. Ⓐ Ⓑ Ⓒ Ⓓ
23. Ⓐ Ⓑ Ⓒ Ⓓ	48. Ⓐ Ⓑ Ⓒ Ⓓ	73. Ⓐ Ⓑ Ⓒ Ⓓ	98. Ⓐ Ⓑ Ⓒ Ⓓ
24. Ⓐ Ⓑ Ⓒ Ⓓ	49. Ⓐ Ⓑ Ⓒ Ⓓ	74. Ⓐ Ⓑ Ⓒ Ⓓ	99. Ⓐ Ⓑ Ⓒ Ⓓ
25. Ⓐ Ⓑ Ⓒ Ⓓ	50. Ⓐ Ⓑ Ⓒ Ⓓ	75. Ⓐ Ⓑ Ⓒ Ⓓ	100. Ⓐ Ⓑ Ⓒ Ⓓ

OMR SHEET

101. (A) (B) (C) (D)	121. (A) (B) (C) (D)	141. (A) (B) (C) (D)	161. (A) (B) (C) (D)
102. (A) (B) (C) (D)	122. (A) (B) (C) (D)	142. (A) (B) (C) (D)	162. (A) (B) (C) (D)
103. (A) (B) (C) (D)	123. (A) (B) (C) (D)	143. (A) (B) (C) (D)	163. (A) (B) (C) (D)
104. (A) (B) (C) (D)	124. (A) (B) (C) (D)	144. (A) (B) (C) (D)	164. (A) (B) (C) (D)
105. (A) (B) (C) (D)	125. (A) (B) (C) (D)	145. (A) (B) (C) (D)	165. (A) (B) (C) (D)
106. (A) (B) (C) (D)	126. (A) (B) (C) (D)	146. (A) (B) (C) (D)	166. (A) (B) (C) (D)
107. (A) (B) (C) (D)	127. (A) (B) (C) (D)	147. (A) (B) (C) (D)	167. (A) (B) (C) (D)
108. (A) (B) (C) (D)	128. (A) (B) (C) (D)	148. (A) (B) (C) (D)	168. (A) (B) (C) (D)
109. (A) (B) (C) (D)	129. (A) (B) (C) (D)	149. (A) (B) (C) (D)	169. (A) (B) (C) (D)
110. (A) (B) (C) (D)	130. (A) (B) (C) (D)	150. (A) (B) (C) (D)	170. (A) (B) (C) (D)
111. (A) (B) (C) (D)	131. (A) (B) (C) (D)	151. (A) (B) (C) (D)	171. (A) (B) (C) (D)
112. (A) (B) (C) (D)	132. (A) (B) (C) (D)	152. (A) (B) (C) (D)	172. (A) (B) (C) (D)
113. (A) (B) (C) (D)	133. (A) (B) (C) (D)	153. (A) (B) (C) (D)	173. (A) (B) (C) (D)
114. (A) (B) (C) (D)	134. (A) (B) (C) (D)	154. (A) (B) (C) (D)	174. (A) (B) (C) (D)
115. (A) (B) (C) (D)	135. (A) (B) (C) (D)	155. (A) (B) (C) (D)	175. (A) (B) (C) (D)
116. (A) (B) (C) (D)	136. (A) (B) (C) (D)	156. (A) (B) (C) (D)	176. (A) (B) (C) (D)
117. (A) (B) (C) (D)	137. (A) (B) (C) (D)	157. (A) (B) (C) (D)	177. (A) (B) (C) (D)
118. (A) (B) (C) (D)	138. (A) (B) (C) (D)	158. (A) (B) (C) (D)	178. (A) (B) (C) (D)
119. (A) (B) (C) (D)	139. (A) (B) (C) (D)	159. (A) (B) (C) (D)	179. (A) (B) (C) (D)
120. (A) (B) (C) (D)	140. (A) (B) (C) (D)	160. (A) (B) (C) (D)	180. (A) (B) (C) (D)

ANSWERS

1. (*c*)	2. (*d*)	3. (*a*)	4. (*c*)	5. (*a*)	6. (*d*)	7. (*a*)	8. (*a*)	9. (*c*)	10. (*d*)
11. (*b*)	12. (*d*)	13. (*b*)	14. (*b*)	15. (*a*)	16. (*a*)	17. (*b*)	18. (*a*)	19. (*c*)	20. (*b*)
21. (*b*)	22. (*c*)	23. (*b*)	24. (*b*)	25. (*c*)	26. (*c*)	27. (*b*)	28. (*c*)	29. (*d*)	30. (*a*)
31. (*b*)	32. (*a*)	33. (*c*)	34. (*c*)	35. (*c*)	36. (*b*)	37. (*c*)	38. (*d*)	39. (*d*)	40. (*c*)
41. (*c*)	42. (*c*)	43. (*d*)	44. (*a*)	45. (*c*)	46. (*d*)	47. (*a*)	48. (*b*)	49. (*c*)	50. (*c*)
51. (*b*)	52. (*a*)	53. (*c*)	54. (*c*)	55. (*b*)	56. (*c*)	57. (*a*)	58. (*b*)	59. (*b*)	60. (*c*)
61. (*c*)	62. (*a*)	63. (*c*)	64. (*d*)	65. (*d*)	66. (*b*)	67. (*c*)	68. (*d*)	69. (*c*)	70. (*d*)
71. (*c*)	72. (*a*)	73. (*a*)	74. (*d*)	75. (*b*)	76. (*a*)	77. (*b*)	78. (*a*)	79. (*c*)	80. (*b*)
81. (*c*)	82. (*d*)	83. (*c*)	84. (*b*)	85. (*a*)	86. (*b*)	87. (*c*)	88. (*a*)	89. (*b*)	90. (*b*)
91. (*c*)	92. (*b*)	93. (*a*)	94. (*a*)	95. (*b*)	96. (*a*)	97. (*b*)	98. (*d*)	99. (*b*)	100. (*d*)
101. (*a*)	102. (*a*)	103. (*c*)	104. (*a*)	105. (*b*)	106. (*b*)	107. (*a*)	108. (*b*)	109. (*a*)	110. (*a*)
111. (*c*)	112. (*c*)	113. (*a*)	114. (*a*)	115. (*d*)	116. (*c*)	117. (*d*)	118. (*c*)	119. (*c*)	120. (*c*)
121. (*b*)	122. (*d*)	123. (*b*)	124. (*c*)	125. (*c*)	126. (*d*)	127. (*b*)	128. (*c*)	129. (*a*)	130. (*a*)
131. (*d*)	132. (*b*)	133. (*b*)	134. (*c*)	135. (*c*)	136. (*a*)	137. (*c*)	138. (*c*)	139. (*d*)	140. (*c*)
141. (*c*)	142. (*b*)	143. (*b*)	144. (*d*)	145. (*c*)	146. (*d*)	147. (*a*)	148. (*b*)	149. (*b*)	150. (*b*)
151. (*a*)	152. (*b*)	153. (*a*)	154. (*d*)	155. (*b*)	156. (*b*)	157. (*d*)	158. (*c*)	159. (*a*)	160. (*c*)
161. (*a*)	162. (*c*)	163. (*b*)	164. (*a*)	165. (*d*)	166. (*b*)	167. (*d*)	168. (*d*)	169. (*b*)	170. (*b*)
171. (*a*)	172. (*b*)	173. (*a*)	174. (*c*)	175. (*c*)	176. (*b*)	177. (*b*)	178. (*b*)	179. (*b*)	180. (*b*)

EXPLANATIONS

1. R_e resistor works as a negative feedback which gives stability of the circuit. Therefore it stabilizes against the variation of both temperature and β.

2. The thevenin equivalent is shown below.

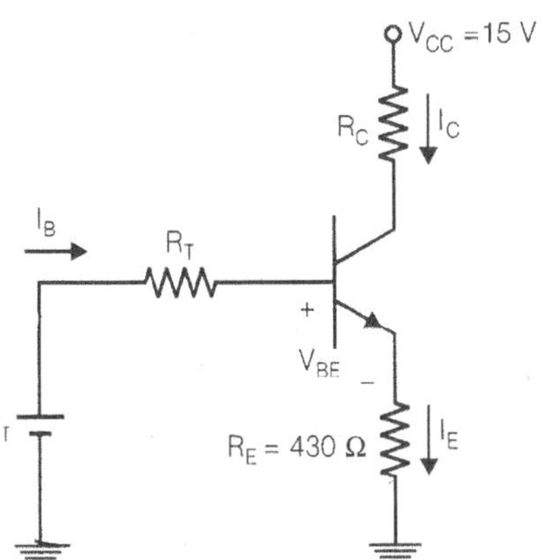

$$V_T = \frac{R_2}{R_1 + R_2} \times V_{CC}$$

$$= \frac{5}{10+5} \times 15 = 5V$$

Since β is large

$I_C = I_E$, $I_B \approx 0$ and

$$I_E = \frac{V_T - V_{BE}}{R_E} = \frac{5-0.7}{0.430k\Omega} = \frac{4.3}{0.430k\Omega} = 10mA$$

3. $V_{CC} - I_C R_C - V_{CE} = 0$

$$I_C = \frac{V_{CC} - V_{CE}}{R_C} = \frac{3-0.2}{1K} = 2.8mA$$

$$I_B = \frac{I_C}{\beta} = \frac{2.8mA}{50} = 56\mu A$$

4. $I_E = I_S\left[e^{V_{BE}/\eta VT} - 1\right]$

$$10^{-13}\left[e^{0.7/1\times 0.026}\right] = 49mA$$

5. Thermal run away will take place if $V_{CE} > \frac{1}{2}V_{CC}$

6. For the transistor to be in saturation

$$I_B > \frac{I_C}{\beta}$$

7. At high frequency

$$A_I = \frac{-g_m}{g_{b'e} + j\omega C}$$

$$A_I \propto \frac{1}{\text{capacitance}}$$

Thus due to the transistor capacitance current gain of a bipolar transistor drops.

8. $A_f = \frac{A}{1 + A\beta}$

Always negative feedback gain will be less than basic amplifier gain.

9. For two-wattmeter method, watt meter readings are

$w_1 = V_L I_L \cos(30 + \phi)$

$w_2 = V_L I_L \cos(30 - \phi)$

For $\phi = 30°$, or $(\pi/6)$

$w_1 = V_L I_L \cos 60°$

$w_1 = \frac{V_L I_L}{2}$ and $w_2 = V_L I_L \cos 0° = V_L I_L$

Therefore, $w_2 = 2w_1$

11. The average power consumed by the load

$P = V_1 I_1 \cos\phi,$

$$P = \frac{100}{\sqrt{2}} \cdot \frac{10}{\sqrt{2}} \cos 60°$$

$P = 250$ W

12. For an ideal operational amplifier, the voltage gain for the common mode signal is equal to zero.

13. For inverting operational amplifier

$$V_0 = -\left(\frac{R_f}{R_1}\right) V_i$$

For switch is open,

$$Y' = \frac{V_0}{V_i} = -\left(\frac{2R}{R}\right) = -2$$

For switch is closed,

$$Y' = \frac{V_0}{V_i} = -\left(\frac{R}{R}\right) = -1$$

$$Y' = \left(\frac{y}{2}\right)$$

14. $(\underline{1\,0\,0}\,\underline{1\,1\,0})_2$

$\Rightarrow (46)_8$

$\Rightarrow$100110 is equivalent to octal representation of 46.

15. Peak inverse voltage of bridge rectifier = V_m

Peak inverse voltage of full-wave rectifier = $2V_m$

16. By potential devider rule, the voltage across, the zener diode.

$$V_2' = \frac{20 \times 5}{5 + 5} \times 5$$

$V_2' = 10$ V

$V_2' = V_{0C} > V_2$

So zener diode will not be in brakdown mode, so open circuited.

Here $(V_2' < V_2)$ so, output voltage is equal to 10 volts.

17. The minimum value of 'R_L' for the zener diode as a voltage regulator.

$$R_L = \frac{15}{i_{max}}$$

$$i_{max} = \frac{24 - 15}{27}$$

$R_{L(min)} = 45\ \Omega$

18. $X = B + A \cdot \bar{B} + A \cdot B = B + A(\bar{B} + B) = A + B$

19. Output frequency a mod N counters is

$f = f_{in}/N$

$\therefore$ The output frequency of a decade counter (mod –10 counters) = 50 kHz 10 = 5 kHz

20. Collector modulation of transistor C amplifier has better dineasity than base modulation of a transistor.

21. For constant value of load current average and r.m.s. value of the current are equal.

22.

A \ BC	00	01	11	11
0	1	1		
1			1	1

From k – map

$y = \bar{A}\bar{B} + AB$ (EX-NOR Gate)

23. An operational amplifier is basically a high gain DC amplifier.

24. K-map of Boolean expression

AB \ CD	00	01	11	10
00				
01				
11			1	1
10			1	1

The equivalent of Boolean expression

$y = AC$

37. In saturation mode.

$h_{fe} I_b \geq I_{CS}$

35.

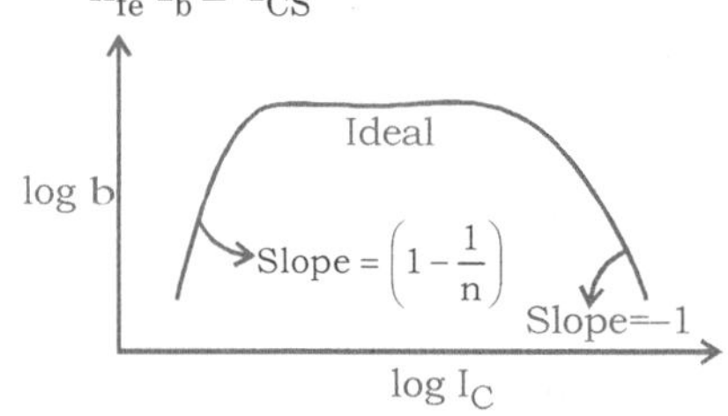

At low injection levels β is degeraded by poor emitter injection efficiency. At high currents, β decreases due to excess majority charge in the base.

36. $I_D = K(V_{GS} - V_T)^2$

38. In parallel connection, current through each resistor is 1/2.

$$W = \left(\frac{1}{2}\right)^2 R = \frac{I^2R}{4}$$

In series connection

$W' = I^2R = 4\ W$

42. The ripple frequency in the output of a full wave rectifier is double to the line frequency.

44. $A_V = \dfrac{V_0}{V_i} = \dfrac{-I_C R_L}{V_i}$

$A_V \propto I_C$

47. Number of Boolean function = 2^{2n}

where n = number of variables

49. Minimum possible SWR = 1

$SWR = Z_L/Z_0$

50. In parallel connection, $C = C_1 + C_2$

In series connection,

$$C = \frac{C_1 C_2}{C_1 + C_2}$$

51. Intrinsic semiconductor has negative temperature coefficient of resistance.

52. $i_c = C\dfrac{dv_c}{dt} = 1 \times 10^{-6} \times 2 = 2 \times 10^{-6}$ A

53. $V_L = L\dfrac{di_L}{dt} = 1 \times 10^{-3} \times 2 \times 10^{-3} = 2 \times 10^{-6}$ V

54. Reversing of the seeback effect is peltier effect.

64.

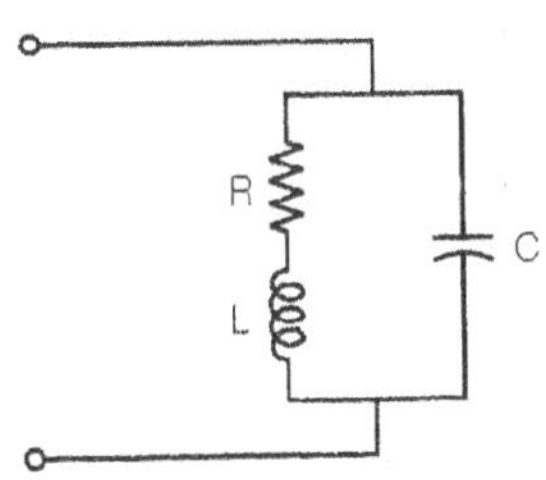

Resonant Frequency

$$\omega_0 = \frac{1}{\sqrt{LC}}\sqrt{1 - \frac{R^2C}{L}}$$

$$R < \sqrt{\frac{L}{C}}$$

Input Impedance

$$Z_{in} = \frac{R^2 + \omega_0^2 L^2}{R}$$

$$= R + \frac{1}{RLC}\left(1 - \frac{R^2C}{L}\right)L^2$$

$$= R + \frac{1}{RC}\left[L - R^2C\right] = R + \frac{L}{RC} - R$$

$$Z_{in} = \frac{L}{RC}$$

$$\text{as } R^2 < \frac{L}{C} \Rightarrow R < \frac{L}{RC}$$

Circuit behaves like a pure resistor of value much higher than R.

65. Minimum number of line current should be equal to the number of branches in the graph of a network.

66. In any linear, planer network number of independent KVL equations are equal to number of links and KCL equations are (n-1).

67. In a series resonance circuit.

$$Q = \frac{1}{R}\sqrt{\frac{L}{C}}$$

72. Tuned voltage amplifiers are not used in public address systems.

73. $\theta_{jc} = \dfrac{T_j - T_A}{P_D} = \dfrac{150° - 25°}{50W} = 2.5°C/W$

74. In crystal, frequency depends on

- Thickness of crystal.
- Angle of cut.
- Physical size of crystal.

75. In tuned oscillators, frequency of oscillations is determined by tank circuit.

76. The young one of a baboon is called an infant while the young one of a beaver is called a kitten. Hence, option (A) is the correct answer.

77.
B G L
↓+4 ↓+4 ↓+4
F K P
↓+4 ↓+4 ↓+4
J O T
↓+4 ↓+4 ↓+4
N S X

78.
B X
↓+3 ↓-3
E U
↓+3 ↓-3
H R
↓+3 ↓-3
K O
↓+3 ↓-3
N L
↓+3 ↓-3
Q I
↓+3 ↓-3
T F

79. By reversing the digits of each term, the series will be:

27, 64, 125, 216, 343.

The series is formed by the cubes of consecutive numbers starting from 3. Hence, the next term in the series will be $8^3 = 512$ and by reversing the digits of the same, we get 215 as the next term in the series.

80.

W	5	K
↓ –3	↓ +7	↓ +2
T	12	M
↓ –3	↓ +8	↓ +2
Q	20	O
↓ –3	↓ +9	↓ +2
N	29	Q
↓ –3	↓ +10	↓ +2
<u>K</u>	<u>39</u>	<u>S</u>

81.

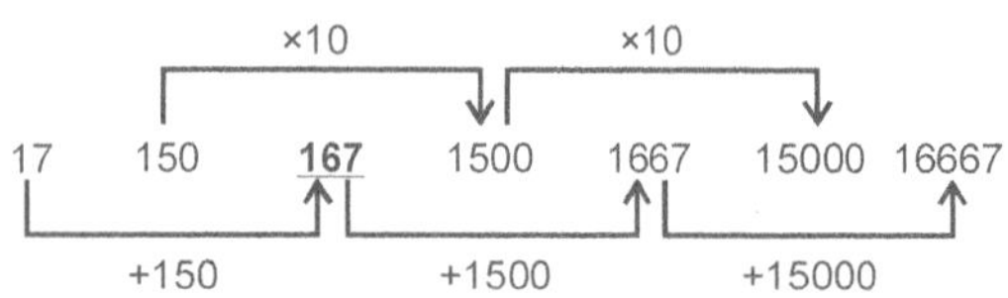

82. Gourd, Pumpkin and French Beans grow on vines while Jackfruit grows on trees. Hence, option (*d*) is the answer.

83. Fossil fuel is the answer since it is a non renewable source of energy, while energy resources in options (*a*), (*b*) and (*d*) are renewable energy sources.

84. As $(53)^2 < 2851 < (54)^2$

∴ The least number that must be added

$= (54)^2 - 2851 = 65.$

85. The order of the heights of five given persons will be:

Anil > Karan >Mukesh and Anil > Sunil > Rakesh

Hence, Anil is the tallest among the five.

86. Option (*b*) is the answer since the given word has a single 'N' while TOURNAMENT uses 'N' twice.

87. In the series D F J P X the number of letters skipped between adjacent letters is consecutive odd numbers starting from 1.

$$D \xrightarrow{+2} F \xrightarrow{+4} J \xrightarrow{+6} P \xrightarrow{+8} X$$

88. Total members = 2 + 3 + 4 × 2 + 4 × 5 = 33.

89. As (4 + 2) × (7 + 4) = 66 and (3 + 5) × (1 + 7) = 64

Similarly, (5 + 7) × (2 + 1) = **<u>36</u>**.

90. In column 1,

2 × 3 = 6, 2 × 5 = 10 ⇒ 2 + 3 + 2 + 5 = 12

In column 2,

3 × 5 = 15, 3 × 7 = 21 ⇒ 3 + 5 + 3 + 7 = 18

Similarly, in column 3,

5 × 7 = 35, 5 × 11 = 55 ⇒ 5 + 7 + 5 + 11 = **<u>28</u>**.

106. Let the amount be ₹x. Then,

$$\left(33000\left(1+\frac{20}{100}\right)-x\right)\left(1+\frac{20}{100}\right)=x$$

$$\Rightarrow \frac{x\times 5}{6}+x=33000\times\frac{6}{5}$$

$$\Rightarrow x\times\frac{11}{6}=39600$$

⇒ x = ₹21,600.

107. $2x^2 - 6x + 5 = 2x^2-6x+\frac{9}{2}+\frac{1}{2}=2\left(x-\frac{3}{2}\right)^2+2\left(\frac{1}{2}\right)^2$

$$\therefore q^2=\left(\frac{1}{2}\right)^2 \Rightarrow q=\pm\frac{1}{2}.$$

108. Difference between the two time limits = 15 hrs. = 15 × 60 × 60 = 54000 sec

⇒ Lag at the end of 54000 sec = 5400 sec = 1 hr 30 min

Hence, the time shown = 7:30 p.m.

109. $x=\cfrac{1}{3+\cfrac{3}{5+\cfrac{4}{5+\cfrac{7}{2}}}}$

$$\Rightarrow x=\cfrac{1}{3+\cfrac{3}{5+\cfrac{4}{\frac{17}{2}}}} \Rightarrow x=\cfrac{1}{3+\cfrac{3}{5+\cfrac{8}{17}}}$$

$$\Rightarrow x=\cfrac{1}{3+\cfrac{3}{\frac{93}{17}}} \Rightarrow x=\cfrac{1}{3+\cfrac{51}{93}}$$

$$\Rightarrow x=\frac{93}{330}=\frac{31}{110}.$$

110. $\frac{a}{b}=\frac{1}{4}$ and $\frac{b}{c}=\frac{2}{3} \Rightarrow a:b:c = 1:4:6$

$$\therefore \frac{12a^2-8c^2}{33a^2+c^2}=\frac{12\times1^2-8\times6^2}{33\times1^2+6^2}=\frac{12(1-24)}{3(11+12)}=-4.$$

111.

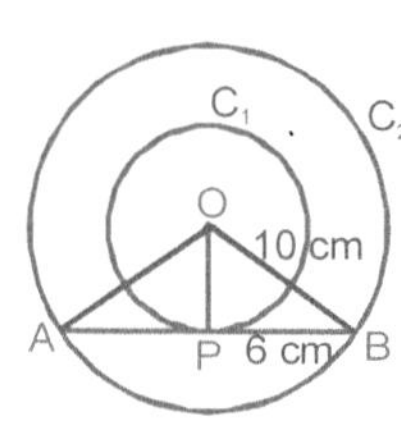

In ΔOPB, $PB=\sqrt{OB^2-OP^2}=\sqrt{10^2-6^2}$ = 8 cm

Area of triangle AOB = Area of ΔPOB + Area of ΔPOA

= 2 × Area of ΔPOB = $2\times\frac{OP\times PB}{2}$ = 6 × 8 = 48 cm².

112. 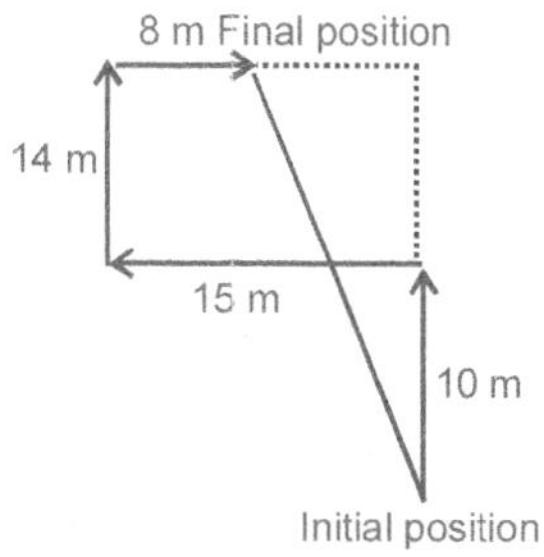

Hence, required distance

$$= \sqrt{(10+14)^2+(15-8)^2} = \sqrt{24^2+7^2} = 25 \text{ m}.$$

113. 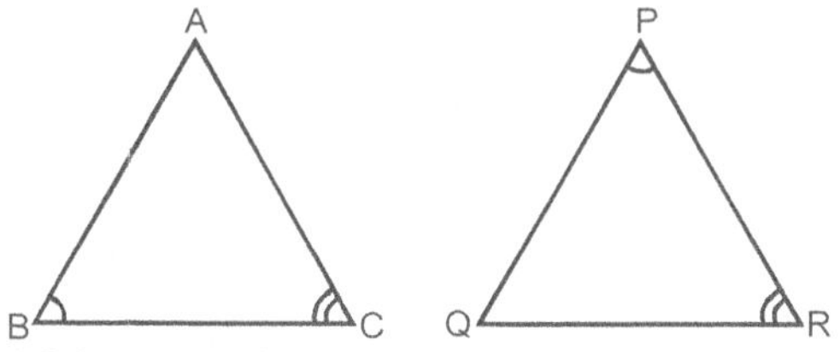

In ΔABC and ΔPQR,

$\angle ABC = \angle QPR$ and $\angle BCA = \angle PRQ$, then for $\Delta ABC \cong \Delta QPR$,

BC = PR (ASA property).

114. 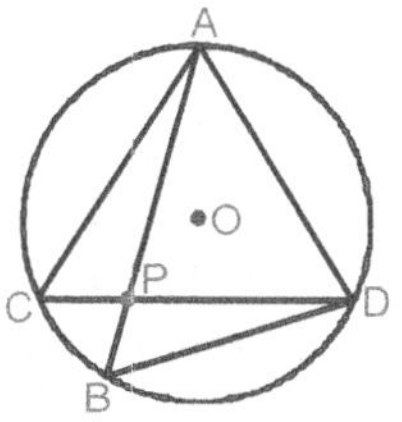

Let the point of intersection of CD and AB be P. Then,

$\angle DCA = \angle DBA = 28°$ (Angle in the same segment)

$\angle BPD = 180° - \angle APD = 180° - 62° = 118°$

In ΔBPD,

$\angle PDB = 180° - (\angle DBA + \angle BPD) = 180° - (28° + 118°) = 34°$.

115. Let the 3 digit number be 100x + 10y + z. Then, (100z + 10y + x) – (100x + 10y + z) = 99(z – x) = 9 × 11 × a number.

116. Let the capacity of the tank be (LCM of 12, 18 and 36) 36 units. Then,

Pipe A empties $\frac{36}{12}$ = 3 units, pipe B empties $\frac{36}{18}$ = 2 units while pipe C empties $\frac{36}{36}$ = 1 unit from the tank in 1 minute.

Hence, required time = $\frac{36}{3+2+1}$ = 6 minutes.

117. Let the work done by a man and a woman in 1 day be 'm' and 'w' units and the total work be T units. Then,

$$(4m + 10w) \times 4 = \frac{T}{3}$$

$$\Rightarrow 2m + 5w = \frac{T}{24} \quad \text{... (i)}$$

$$\text{Now, } (6m + 12w) \times 2 = \frac{2T}{9}$$

$$\Rightarrow 2m + 4w = \frac{T}{27} \quad \text{... (ii)}$$

From (i) and (ii), we have

$$w = \frac{T}{9\times 24} \text{ and } m = \frac{T}{9\times 12}$$

Remaining work = $T - \frac{T}{3} - \frac{2T}{9} = \frac{4T}{9}$ units

Let 'n' be the number of required woman. Then,

$$(6m + (12 + n)w) \times 4 = \frac{4T}{9}$$

$$\Rightarrow n = 0.$$

118. Let the level of the milk in the jar be 'h'. Then, volume of the conical vessel = volume of the cylinder jar

$$\Rightarrow \frac{1}{3}\pi(9)^2 \times 8 = \pi(3)^2 \times h$$

$$\Rightarrow h = 24 \text{ cm}.$$

119. Let the cost price be ₹x. Then,

$$\frac{(12.5+6)x}{100} = 51.80 \Rightarrow x = 280.$$

120. Total discount = $15 + 25 - \frac{15\times 25}{100}$ = 36.25%.

121. 'Exude' means to flow out slowly, so does seep. Hence, option (*b*) is the correct answer. 'Gush', which means to flow out qickly, is the antonym of 'exude'.

122. 'Imperious' refers to someone who is fond of ordering people around. Hence, its synonym will be option (*d*), 'authoritarian'.

123. 'Passive' is used to describe someone who allows things to happen or accepts what other people decide for him. 'Resistant' is used to describe someone who is opposed to something. Hence, option (*b*) is the correct answer. Something or someone is called 'yielding' when it is not hard or rigid. 'Forbearing' is someone who avoids saying or doing something.

124. 'Incorrigible' is someone who is not able to be corrected or changed. 'Reformable' is someone who favours improvement. Hence, option (*c*) is the correct answer. 'Unruly' is someone who cannot be easily ruled, disciplined or managed. 'Inveterate' is someone who is often doing something specified.

125. 'Agriculture' deals with the science of growing plants and breeding animals for farming. Hence, option (*c*) is the correct answer. 'Aviculture' is concerned with the raising and caring of birds, especially of wild birds in captivity. 'Horticulture' deals with the study and science of flowers, fruits and vegetables. 'Apiculture' refers to keeping of bees especially on large scale.

126. 'Glutton for work' is a person who enjoys doing difficult and unpleasant tasks. Hence, option (*d*) is the correct answer. A 'narcoleptic' is a person who falls in deep sleep when he is in relaxing surroundings. A 'pyromaniac' is someone who enjoys making or watching fires. A 'workaholic' is a person who works very hard and finds it difficult to stop and relax.

127. 'Evident' means clear to sight or mind. Hence, option (*b*) is the correct option. 'Witty' means funny or clever. 'Confirmed' means not likely to change. 'Chronic' is something that exists frequently.

128. 'Chasing one's tail' means to run around ineffectually. Hence, option (*c*) is the correct answer.

129. 'To run a tight ship' means to run an organisation in an orderly and disciplined manner. Hence, option (*a*) is the correct answer.

130. 'To catch some rays' means to tan in the sun. Hence, option (*a*) is the correct answer.

131. The correct spelling is 'Cassette'. Hence, the correct answer is option (*d*)

132. The word 'considering' is not followed by any preposition. The preposition 'to' in part (*b*) is redundant. Hence, part (*b*) is the answer.

133. Clauses containing 'if' are like hypothesis, suppositions or desires and in such cases, we use past tense forms to talk about the present and the future. Therefore, the correct sentence will be 'If I were working in Bandra, I would commute by the local train'. Hence, the correct answer is part (*b*).

134. Future perfect continuous tense is used for actions which will be in progress over a period of time and will end in the future. In the sentence, 'shall be married', which is in future continuous tense, should be replaced by 'shall have been married'. Hence, part (*c*) is the answer.

135. 'Under way' means happening now. Hence, option (*c*) is the correct answer. Options (*a*), (*b*) and (*d*) will make the sentence erroneous.

136. If a statement is positive, then the question tag (short questions that we put at the end of sentences) is negative and vice-versa. Also, the question tag takes the same verb as used in the preceding part of the statement. For example: He <u>did</u> his homework, <u>didn't</u> he? Similarly, in the given sentence, the verb is 'should', which is referring to 'we'. Therefore, the correct question tag will be 'shouldn't we'. Hence, option (*a*) is the correct answer.

137. The sentence can suggest only two things; either people have stopped going to Niagara falls because it has frozen or even more people have started going to the falls to see the spectacle. So, the only words that can fit in the first blank would be 'converged' and 'assembled'. 'Disunited' means separated and 'disband' means to end an organisation. Both will make the sentence logically incorrect. Option (*d*) can be negated because 'continued' is incorrect for the second blank. We need a word, which shows an action that is ongoing and not something that has happened in the past. Hence, 'continues' would be the appropriate word to fill in the second blank. Thus, option (*c*) is the correct answer.

138. The underlined word should technically describe her 'problems'. Hence, the correct word to be used is option (*c*) - 'arisen', which is a past participle. A participle is an English verbal that functions as an adjective. 'Arosen' and 'arised' are incorrect words that do not exist in the dictionary.

139. The given sentence is correct in itself and hence, option (*d*) is the answer. The statement refers to an action that is going on at some time in the past. In such cases, the past continues tense is used.

140. The given statement is incorrect because it is not clear whether the place from where they had been banished was a country or a car. The grammatically correct sentence is given in option (*c*), where the modifier 'where they had been banished' is correctly placed beside 'country' making the place, where they were banished from, clear.

141. Refer to the third sentence of the passage. It says that the main idea of a war has either been to expand power, independence, or joining a war to help allies. Hence, option (*c*) is the correct answer.

142. 'Infamy' refers to the condition of being known for having done an evil or terrible act. 'Infamous' means well known for evil act. Hence, option (*b*) is the correct answer.

143. Option (*a*) is incorrect because we cannot infer from the passage which war had more allies. Option (*b*) says that the degree of damage caused made World War I different from World War II, which can be inferred from the third last sentence of the passage. Hence, option (*b*) is the answer. Option (*c*) is out of the scope of the passage.

144. What really triggered the war was Germany invading Poland. Refer to the following lines of the passage: 'So they decided ... invaded Poland which triggered the start of the war'. Hence, option (*d*) is the correct answer.

145. The passage states that most of the war was fought in trenches. Flame throwers, tanks, etc. were used for the first time. Hence, option (*d*) is the correct answer. Nuclear bomb was first used in WWII. The passage makes no mention of the use of riffle in the wars.

146. 'Awash' means flooded in large amount of water or any other liquid. 'Soaked' means to make someone or something wet with a large amount of water or any other liquid. Hence, option (*d*) is the correct answer. 'Swamp' refers to the land that is always wet and often covered in water. 'Arid' means very dry. 'Dampish' refers to the slight wetness in the air.

147. 'Endue' means to provide. Hence, option (*a*) is the correct answer.

148. A person is called 'timid' when he shows lack of courage or confidence. 'Audacious' is used to describe someone who is very bold and confident. Hence, option (*b*) is the correct answer. A person is described as 'aggressive' when he shows aggression and uses forceful methods to succeed.

149. 'Precarious' refers to something that is not safe, strong or steady. Hence, its antonym will be option (*b*), stable.

150. An agent who negotiates contracts of purchase and sale is called a 'broker'. The contracts can be contracts of securities, commodities, real estate, etc. Hence, option (*b*) is the correct answer. A 'salesman' is a person whose job is to sell things. A 'bargainer' is a negotiator of the terms of a trade. A 'moderator' is someone who leads a discussion in a group and tells each person when to speak.

151. A person who believes that a particular type of behaviour is important is called a 'stickler'. Hence, option (*a*) is the correct answer. 'Perseverer' is a persistent person. Someone who is lazy and avoids work is called a 'slacker'. Person who has certain social values is said to be 'moral'.

152. 'Egotist' refers to a person who feels that he is too important. Hence, option (*b*) is the correct answer. 'Immodest' means having too high opinion of oneself. 'Presumptuous' means being too confident in a way that shows lack of respect. 'Inordinate' means going beyond what is usual or proper.

153. 'Hooked up' means to connect. 'Associate' means to make a connection between people or things in your mind. Hence, option (*a*) is the correct answer. 'Disperse' means to move apart in different directions. 'Snub' means to ignore someone deliberately.

154. 'To think on one's feet' means to be able to think and speak effectively without any preparation. Hence, option (*d*) is the correct answer.

155. If a remark is 'close to home', it means that it is accurate and it makes you feel uncomfortable and embarrassed. Hence, option (*b*) is the correct answer.

156. The correct spelling is 'Assassin'. Hence, option (*b*) is the correct answer.

157. The given sentence is correct. A preposition such as 'between' should be followed by an objective pronoun (such as me, him, her and us) rather than a subjective pronoun (such as I, he, she and we). In part (*c*), the correct expression is 'Somiya and me' since it is the object of the sentence. Hence, part (*d*) is the correct answer.

158. There is a parallelism error in the sentence. The correct sentence should be "... I'm having dinner with Dikshant". Hence, part (*c*) is the correct option.

159. The sentence is in present perfect continuous tense since the action (reading book) started in the past, i.e., two months back, and is continuing in the present. Therefore, 'I am' in part (*a*) should be 'I have been', making part (*a*) the answer.

160. 'To be a cut above the rest' means to be better than other things or people. Hence, option (*c*) is the correct answer. Options (*a*), (*b*) and (*d*) will make the sentence grammatically incorrect.

161. Since 'eating disorders' is plural, the blank will take a plural verb. Therefore, options (*c*) and (*d*) are incorrect. Moreover, eating disorders are not traits; they are life-threatening conditions. 'Trait' is a quality that makes one different from another. Hence, option (*a*) is the correct answer.

162. Option (*c*) is the correct answer. The blank needs to be filled in with an adjective that would describe the noun 'pressure'. The other forms of the word will make the sentence grammatically incorrect.

163. Simple present tense is used for official timetables and programmes. Hence, the correct answer is option (*b*). 'Will be arriving' is future continuous tense that talks about things that will be in progress at a time in the future. The sentence does not suggest this and is therefore, incorrect.

164. The given sentence is incorrect. Past perfect continuous is used to denote when an action began in the past and is still continuing. Hence, option (*a*) is the correct answer.

165. In order to express the unreal, the hypothetical, the speculative, or imagined, we move time one step backward. In such cases two verbs are involved: one in the clause stating the condition (the "if" clause) and one in the result clause. Here **present unreal** event is used, so we will put the verb in the condition clause one step back, i.e., into the past. Hence, the given sentence, option (*d*) is the correct answer.

166. The Indus Civilization can be best described as sophisticated and progressive. From the second sentence of the passage we can infer that the people of Indus civilization were sophisticated. The civilization can be said to be progressive since it was then adept at plumbing, it was one of the worlds first city building cultures and had a well planned sewage system. Hence, option (*b*) is the correct answer.

167. The passage says that the scientists found hints that suggested that the people of Indus civilization were not as peaceful as they are portrayed to be. Hence, option (*d*) is the correct answer.

168. Option (*d*) is the correct answer. It can be inferred from the passage that scientists are trying to learn more about the civilization. Hence, option (*d*) is the correct answer.

169. The last paragraph says that the civilization accomplished great things and then goes on to list its accomplishments. Hence, option (*b*) is the correct answer.

170. The passage says that the scientists are 'now' trying to unravel the mysteries of the civilization through human bones. This means that earlier attempts were not made to know about the civilization through bones. Hence, option (*b*) is the correct answer.

171. 'Inferno' refers to a very large and dangerous fire. 'Blaze' refers to an intensely burning fire. Hence, option (*a*) is the correct answer.

172. 'Preposterous' means very foolish or silly. Hence, option (*b*) is the correct answer.

173. 'Dovish' refers to a person who does not want war, but peace. 'Hawkish' is one who takes militant attitude. Hence, option (*a*) is the correct answer. 'Owlish' means resembling or suggesting an owl. 'Vulturous' means resembling a vulture especially in scavenging habits. 'Ravenous' means very hungry.

174. 'Apposite' means appropriate for a situation. 'Immaterial' means not important in a particular situation. Hence, option (*c*) is the correct answer.

175. 'Misapprehend' means to get a wrong idea about something. Hence, option (*c*) is the correct answer. 'Apprehend' means to notice and understand something. 'Fathom' means to understand the reason of something.

176. 'Quinquennial' refers to something that happens every five years. Hence, option (*b*) is the correct answer. 'Triennial' refers to something that happens every three years. 'Septennial' means happening every 7 years. 'Vicennial' means happening every 20 years.

177. An 'annalist' is a person who writes annals or historical records. Hence, option (*b*) is the correct answer. An 'archaeologist' is a person who deals with past human life. A 'radiologist' is a doctor trained in radiology. A 'genealogist' is someone who traces or studies the descent of persons or families.

178. 'Kick the bucket' means to die. Hence, option (*b*) is the correct answer.

179. A 'stool pigeon' is a criminal who helps the police to catch another criminal by spending time with them. Hence, option (*b*) is the correct answer.

180. 'To blow a fuse' means to lose one's temper. Hence, option (*b*) is the correct answer.

PRACTICE PAPER-7

ELECTRONICS ENGINEERING

Time Allowed : 2:15 Hours **Maximum Marks : 180**

Read the following instructions carefully before you begin to answer the questions. This Booklet contains questions in English.

INSTRUCTIONS

1. This Booklet contains 180 questions in total comprising of the following five parts:

Part – I:	Domain Knowledge	[1 – 75 Questions]
Part – II:	General Intelligence and Reasoning	[76 – 90 Questions]
Part – III:	General Awareness	[91 – 105 Questions]
Part – IV:	Quantitative Aptitude	[106 – 120 Questions]
Part – V:	General English	[121 – 180 Questions]

2. All questions are compulsory and carry 1 mark each.

3. Before you start to answer the questions you must check this Booklet to ensure that it contains all the pages and see that no page is missing or repeated.

4. For every wrong answer 1/3 marks will be deducted.

5. You are supplied with the OMR-Sheet separately.

6. Answers must be indicated by completely blackening one of the corresponding ovals on OMR-Sheet against the relevant question number, by black/blue ball pen only. Answers which are not indicated by black/blue ball pen will not be awarded any mark.

7. The OMR-Sheet must be handed over to the Invigilator before you leave the Examination Hall.

8. Failure to comply with any of the above instructions will render a candidate liable to such action/penalty as may be deemed fit.

PRACTICE PAPER – 7

DOMAIN KNOWLEDGE

1. If the maximum and minimum voltages on a transmission line are 12V and 4V respectively, for a typical load VSWR is-

(*a*) 1.0 (*b*) 0.5

(*c*) 3.0 (*d*) 8.0

2. Which resistive component is designed to be temperature sensitive?

(*a*) Thermistor (*b*) Rheostat

(*c*) Potentiometer (*d*) Photoconductive cell

3. VSB modulation is preferred in TV because

(*a*) it reduces the bandwidth requirement.

(*b*) it avoids phase distortion at low frequencies

(*c*) it results in better reception

(*d*) none of the above

4. In FM signal with a modulation index m_f is passed through a frequency tripler. The wave in the output of the tripler will have a modulation index of

(*a*) m_f (*b*) $3m_f$

(*c*) $m_f/3$ (*d*) $m_f/9$

5. A 400 W carrier is amplitude modulated with m = 0.75. The total power in AM is

(*a*) 400 W (*b*) 512 W

(*c*) 588 W (*d*) 650 W

6. In radar systems PRF stands for

(*a*) Power Return Factor

(*b*) Pulse Return Factor

(*c*) Pulse Repetition Frequency

(*d*) Pulse Response Factor

7. Which of the following is the indirect way of FM generation?

(*a*) Reactance bipolar transistor modulator

(*b*) Armstrong modulator

(*c*) Varactor diode modulator

(*d*) Reactance FM modulator

8. The velocity of sound waves in air

(*a*) is constant at all temperatures

(*b*) varies directly as temperature

(*c*) varies inversely as absolute temperature

(*d*) varies directly as square root of absolute temperature

9. The rate at which information can be carried through a communication channel depends on

(*a*) carrier frequency (*b*) bandwidth

(*c*) transmission loss (*d*) transmitted power

10. A carrier is simultaneously modulated by two sine waves having modulation indices of 0.4 and 0.3. The total modulation index will be

(*a*) 0.1 (*b*) 0.7

(*c*) 0.5 (*d*) 0.35

11. In a full-wave rectifier circuit with centre tap transformer, if voltage between one end of secondary winding and centre tap is 300 V peak, then PIV (peak inverse voltage) is

(*a*) 300 V (*b*) 150 V

(*c*) 600 V (*d*) 900 V

12. The effect of phase error of local oscillator on spectral components in coherent detection of SSB Amplitude modulated signal is

(*a*) The frequency of demodulated signals varies with fixed amplitude

(*b*) Amplitude of demodulated signals varies with fixed frequency

(*c*) Both amplitude and frequency vary

(*d*) The recovered signal has phase offset with fixed amplitude

13. The maximum transmission power efficiency of DSB-C Amplitude modulation is

(*a*) 25% (*b*) 33.33%

(*c*) 50% (*d*) 100%

14. For identical total transmitted power, the ratio of S/N for wide band FM to S/N of DSB-AM (with $m_a = 1$) is

(Given FM = 15 kHz and frequency deviation = ± 75 kHz)

(*a*) 112.5 (*b*) 300

(*c*) 75 (*d*) 450

15. A signal m(t)=3cos(6000πt) + 4cos (7000πt) + 5cos(10000πt) to be sampled. What is the minimum sampling rate for band pass sampling for truth full representation?

(*a*) 2000 Hz (*b*) 3000 Hz

(*c*) 5000 Hz (*d*) 10000 Hz

16. Minimum number of bits required to represent maximum value of given analogue signal with 0.1% accuracy is

(*a*) 8 bits (*b*) 9 bits
(*c*) 10 bits (*d*) 12 bits

17. Input voltage applied to a circuit is 1 V rms and the output is 1 mV rms. Net gain of the circuit is

(*a*) + 30 dB (*b*) – 30 dB
(*c*) – 60 dB (*d*) + 60 dB

18. The divergence of magnetic intensity is

(*a*) electric charge density
(*b*) electric field intensity
(*c*) zero
(*d*) conduction current density

19. A transmission line having characteristic impedance of 50 Ω has to deliver 10 kW power at 100 kHz to a load. Maximum permissible value of rms current anywhere along the line is 20 A. What is the maximum VSWR that can be tolerated on this line?

(*a*) 2 (*b*) 1
(*c*) 3 (*d*) 2.5

20. The current flowing through a capacitor in an AC circuit is

(*a*) non-existent
(*b*) conduction current
(*c*) displacement current
(*d*) none of the above

21. Conduction angle of a Class AB amplifier is

(*a*) <180°
(*b*) between 180° and 360°
(*c*) 360°
(*d*) 90°

22. A Zener diode, when used in voltage stabilization circuits, is biased in

(*a*) reverse bias region below the breakdown voltage
(*b*) reverse breakdown region
(*c*) forward bias region
(*d*) forward bias constant current mode

23. If for a silicon npn transistor, the base-to emitter voltage (V_{BE}) is 0.7 V and the collector-to-base voltage (V_{CB}) is 0.2 V, then the transistor is operating in the

(*a*) normal active mode
(*b*) saturation mode
(*c*) inverse active mode
(*d*) cut-off mode

24. The two numbers represented in signed 2's complement form are P = 11101101 and Q = 11100110. If Q is subtracted from P, the value obtained in signed 2's complement is

(*a*) 1000001111 (*b*) 00000111
(*c*) 11111001 (*d*) 111111001

25. Depletion type MOSFET operates in

(*a*) depletion mode only
(*b*) enhancement mode only
(*c*) both depletion and enhancement mode
(*d*) none of the above

26. A half-wave rectifier having a resistance load of 1 kΩ rectifiers an ac voltage of 325 V peak value and the diode has a forward resistance of 100Ω. What is the RMS value of the current?

(*a*) 100.4 mA (*b*) 150 mA
(*c*) 147.7 mA (*d*) 200 mA

27. In a centre tap full-wave rectifier, 200 V is the peak voltage between the centre tap and one end of the secondary. What is the maximum voltage across the reverse biased diode?

(*a*) 150 V (*b*) 200 V
(*c*) 350 V (*d*) 400 V

28. If the input to the circuit of figure is a sine wave. The output will be

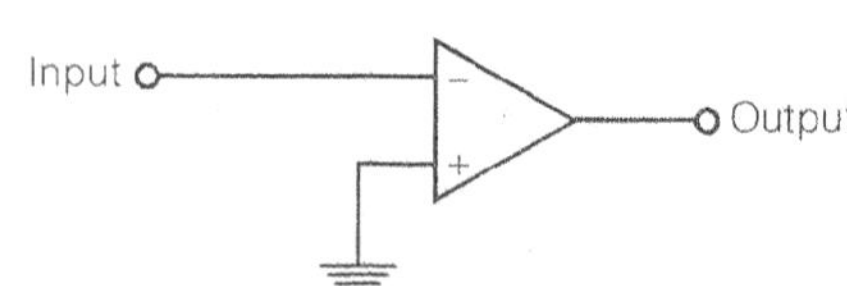

(*a*) a half wave rectified sine wave
(*b*) a full wave rectified sine wave
(*c*) triangular wave
(*d*) square wave

29. Frequency compensation is used in op-amp to increase its

(*a*) voltage gain (*b*) input resistance
(*c*) bandwidth (*d*) None

30. The ideal op-amp has the following characteristics

(*a*) $R_i = \infty$, $A = \infty$, $R_0 = 0$
(*b*) $R_i = 0$, $A = \infty$, $R_0 = 0$
(*c*) $R_i = \infty$, $A = \infty$, $R_0 = \infty$
(*d*) $R_i = 0$, $A = \infty$, $R_0 = \infty$

31. A first order instrument is characterized by

(*a*) Time constant only
(*b*) Static sensitivity and time constant
(*c*) Static sensitivity and damping coefficient
(*d*) Static sensitivity, damping coefficient and natural frequency of oscillations

32. The difference between the indicated value and the true value of a quantity is

(*a*) Gross error (*b*) Absolute error

(*c*) Dynamic error (*d*) Relative error

33. which of the following transducers is classified as an active transducer?

(*a*) Metallic strain gauge

(*b*) Capacitive microphone

(*c*) LVDT

(*d*) Piezoelectric transducer

34. Time division multiplexing is used when the data to be transmitted is

(*a*) slow changing

(*b*) of small bandwidth

(*c*) slow changing and has a small bandwidth

(*d*) fast changing and has a wide bandwidth

35. If an induction type energy meter runs fast, it can be slowed down by

(*a*) lag adjustment

(*b*) light load adjustment

(*c*) adjusting the position of braking magnet and moving it closer from the centre of the disc

(*d*) adjusting the position of braking magnet and moving it away from the centre of the disc

36. The Circuit generally used in digital instruments to convert sine waves into rectangular pulses is

(*a*) sawtooth generator

(*b*) differential amplifier

(*c*) sample and hold circuit

(*d*) Schmitt trigger

37. Electric field and magnetic field perpendicular to each other in

(*a*) Klystron (*b*) Magnetron

(*c*) TWTA (*d*) all of the above

38. The Maxwell's equation $\nabla \times \bar{E} = -\frac{\partial \bar{E}}{\partial t}$ is obtained from

(*a*) Ampere's Law (*b*) Faraday's Law

(*c*) Lenz's Law (*d*) Both (*b*) and (*c*)

39. A lossless line having characteristic impedance Z_0 is terminated with a load impedance of jZ_0. VSWR of the line will be

(*a*) 1 (*b*) 10

(*c*) infinite (*d*) none of the above

40. Which of the following modulation scheme is most bandwidth efficient?

(*a*) AM (b) FM

(*c*) PM (*d*) SSB-SC

41. A Colpitt's Oscillator is having tank capacitances of 1 nF and 10 nF, and inductance of 0.1 μH. The gain required by the circuit to start oscillating is

(*a*) 10 (*b*) 100

(*c*) 1 (*d*) 1000

42. Which modulation scheme would you suggest for high speed data transmission over a fading channel?

(*a*) MSK (*b*) BPSK

(*c*) FSK (*d*) OFDM

43. A semiconductor is uniformly doped with N_A acceptors and N_D donors. Let the free electron and hole concentrations be n and p respectively. Assume that the semiconductor is at thermal equilibrium and that 100% ionisation has taken place. Then which of the following is true?

(*a*) $N_A + N_D = p + n$ (*b*) $N_A - N_D = n - p$

(*c*) $N_A N_D = p\,n$ (*d*) $N_D - N_A = n - p$

44. In a uniformly doped abrupt p-n junction, the doping level of the p-side is ten times the doping level of the n-side. The ratio w_p/w_n of the duplication layer widths in the p- and n-regions respectively is

(*a*) 0.1 (*b*) 1

(*c*) 2 (*d*) 10

45. The breakdown in a Bipolar Transistor is characterised by two parameters; breakdown voltage with base open (BV_{CBO}) and breakdown voltage with emitter open (BV_{CEO}). Which one of the following is true?

(*a*) $BV_{CBO} > BV_{CEO}$

(*b*) $BV_{CBO} < BV_{CEO}$

(*c*) $BV_{CBO} = BV_{CEO}$

(*d*) Can be larger or smaller depending on other factors

46. Which one of the following is considered as an AM signal?

(*a*) Binary phase shift keying (BPSK).

(*b*) Differential phase shift keying (DPSK).

(*c*) Differential encoded PSK.

(*d*) Quadrature PSK.

47. A modulator is a device to

(*a*) separate two frequencies.

(*b*) impress the information on to a radio frequency carrier.

(*c*) extract information from the carrier.

(*d*) amplify the audio frequency Signal.

48. Which of the following modulated signals can be detected by an envelope detector?

(*a*) DSB suppressed carrier

(*b*) DSB full carrier

(*c*) Frequency modulated signal.

(*d*) SSB supported carrier

49. Which of the following desirable features are possessed by SSB in comparison to AM?

1. Less power is required
2. Bandwidth required is less
3. Antenna size can be reduced
4. Low frequency can be used

Select the correct answer using the codes given below codes.

(*a*) 3 and 4 (*b*) 2, 3 and 4

(*c*) 1 and 2 (*d*) 1, 2 and 4

50. The demodulation of a delta modulated signal is achieved by

(*a*) integration (*b*) differentiation

(*c*) sampling (*d*) band pass filtering

51. In frequency modulation

(*a*) the frequency of the carrier remains constant

(*b*) the amplitude of carrier remains constant

(*c*) the amplitude of the carrier wave is varied

(*d*) the frequency of the signal is made equal to the carrier frequency

52. In phase modulation, phase deviation is proportional to

(*a*) carrier amplitude

(*b*) carrier phase

(*c*) message signal

(*d*) message signal frequencies

53. Which one of the following addressing technique is not used in 8085 microprocessor?

(*a*) Register (*b*) Immediate

(*c*) Register indirect (*d*) Relative

54. A Direct Memory Access (DMA) transfer implies

(*a*) direct transfer of data between memory and accumulator

(*b*) direct transfer of data between memory and I/O devices without the use of microprocessor.

(*c*) transfer of data exclusively within microprocessor registers.

(*d*) A fast transfer of data between microprocessor and I/O devices.

55. A receiver connected to an antenna whose resistance is 50 ohm has an equivalent noise resistance of 30 ohm. Calculate the receivers noise figure in decibels and its equivalent noise temperature.

(*a*) 2.04 db, 174 K (*b*) 1.04 db, 164 K

(*c*) 4.04 db, 194 K (*d*) 3.04 db, 184 K

56. Which one of the following statements is correct? The wavelength of a wave propagating in a wave guide is

(*a*) smaller than the free space wavelength

(*b*) greater than the free space wavelength

(*c*) directly proportional to the group velocity

(*d*) inversely proportional to the phase velocity

57. Which one of the following statements is correct? A wave guide can be considered to be analogous to a

(*a*) Low pass filter (*b*) High pass filter

(*c*) Band pass filter (*d*) Band stop filter

58. What is the phase velocity of plane wave in a good conductor

(*a*) $\sqrt{\pi f \mu \sigma}$ (*b*) $\sqrt{\dfrac{\pi f \sigma}{(\mu \sigma)}}$

(*c*) $\sqrt{\dfrac{\pi f}{(\mu \sigma)}}$ (*d*) $2\sqrt{\dfrac{\pi f}{(\mu \sigma)}}$

59. What is the attenuation constant a for distortionless transmission line

(*a*) $\alpha = 0$ (*b*) $\alpha = R\sqrt{\dfrac{C}{L}}$

(*c*) $a = R\sqrt{\dfrac{L}{C}}$ (*d*) $\alpha = \sqrt{\dfrac{RL}{C}}$

60. A load impedance of $(75 - j50)\ \Omega$ is connected to a transmission line of characteristic impedance $Z_0 = 75\ \Omega$. The best method of matching comprises

(*a*) a short circuit stub at load

(*b*) a short circuit stub at some specific distance from load

(*c*) an open stub at load

(*d*) two short circuited stubs at specific distances from load

61. When a lossless transmission line is terminated by a resistance equal to surge impedance, then what is value of the reflection coefficient

(*a*) 1 (*b*) –1

(*c*) 0 (*d*) 0.5

62. Impedance inversion may be obtained with
(*a*) a short-circuited stub
(*b*) an open-circuited stub
(*c*) a quarter-wave line
(*d*) a half-wave line

63. What does the function $f(x\text{-}v_0t)$ represent?
(*a*) A stationary wave
(*b*) A wave motion in a reverse direction
(*c*) A wave motion in a forward direction
(*d*) Not a travelling wave

64. What does the standing wave ratio (SWR) of unity imply?
(*a*) Transmission line is open circuited
(*b*) Transmission line is short circuited
(*c*) Transmission line's characteristic impedance is equal to load impedance
(*d*) Transmission line's characteristic impedance is not equal to load impedance

65. What is the radiation resistance of a dipole antenna $\lambda/20$ long approximately equal to?
(*a*) 2 Ω (*b*) 40 Ω
(*c*) 0.6 Ω (*d*) 20 Ω

66. In a linear network, a 1Ω resistor consumes a power of 4 W when voltage source of 4 V is applied to the entire circuit, and 16 W when the voltage source is replaced by an 8 V source. The power consumed by the 1Ω resistor when 12 V is applied will be
(*a*) 0 W (*b*) 20 W
(*c*) 36 W (*d*) 144 W

67. The maximum power will be transferred from a voltage source to a load when
(*a*) the source impedance is half that of the load impedance.
(*b*) the source impedance is equal to that of the load impedance.
(*c*) the source impedance is twice that of the load impedance.
(*d*) both source and load impedance must be zero.

68. A 3-phase, 4-wire system supplies power to a balanced star-connected load. The current in each phase is 15 A The current in the neutral wire will be
(*a*) 15 A (*b*) 45 A
(*c*) 8.66 A (*d*) 0 A

69. The condition for reciprocity for a two-port transmission network is expressed by
(*a*) $\begin{vmatrix} A & B \\ C & D \end{vmatrix} = 0$ (*b*) $\begin{vmatrix} A & D \\ B & C \end{vmatrix} = 1$
(*c*) $\begin{vmatrix} A & C \\ B & D \end{vmatrix} = 0$ (*d*) $\begin{vmatrix} A & B \\ C & D \end{vmatrix} = 1$

70. The Norton equivalent between A and B for the circuit is

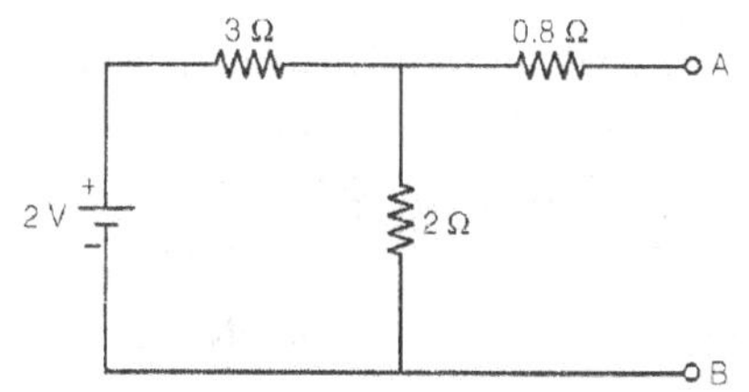

(*a*) 2 A and 2.5 Ω (*b*) 0.5 A and 1 Ω
(*c*) 1 A and 2 Ω (*d*) 0.4 A and 2 Ω

71. The voltage V_x across the 2Ω resistance in the circuit is

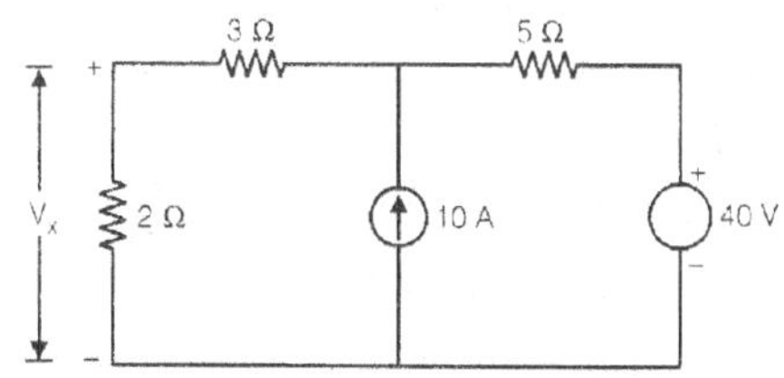

(*a*) 16V (*b*) 60V
(*c*) 18V (*d*) 10V

72. For a parallel RLC circuit, if R = 40Ω, L = 2 H and C = 0.5 F, the bandwidth and quality factor are respectively
(*a*) 20rad/s, 0.05 (*b*) 10rad/s, 20
(*c*) 20 rad/s, 10 (*d*) 0.05rad/s,20

73. A load is connected to an active network. At the terminals to which the load is connected, R_n = 10Ω and V_{th} = 60. Then the maximum power supplied to the load is
(*a*) 360 W (*b*) 90 W
(*c*) 60 W (*d*) 10 W

74. What is the voltage across the current source for the below shown circuit?

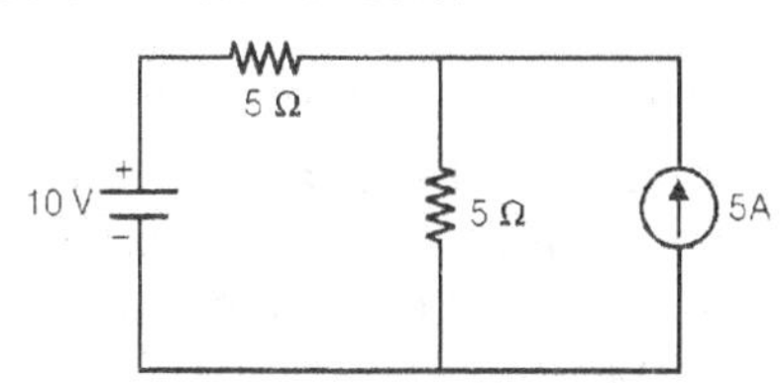

(*a*) 5.0 V (*b*) 7.5 V
(*c*) 12.5 V (*d*) 17.5 V

75. For a series RLC resonant circuit, what is the total reactance at the lower half power frequency?
(*a*) $\sqrt{2}R\angle 45^\circ$
(*b*) $\sqrt{2}R\angle -45^\circ$
(*c*) R
(*d*) -R

GENERAL INTELLIGENCE AND REASONING

Directions: In question nos. **76** and **77**, select the related letters /word /number from the given alternatives.

76. TALENT : VYNCPR :: NORMAL : ?
(*a*) PMYKBK (*b*) PMTKCJ
(*c*) PMTLBJ (*d*) PNTKDK

77. Sick : Dog :: Busy : ?
(*a*) Toad (*b*) Crocodile
(*c*) Cuckoo (*d*) Bee

78. Which one of the given responses would be a meaningful order of the following?
(1) Idea Generation and Screening
(2) Market Testing
(3) Concept Development/Testing
(4) Product Development
(5) Commercialization
(6) Analysis
(*a*) 1, 2, 4, 6, 5, 3 (*b*) 1, 2, 3, 4, 5, 6
(*c*) 1, 3, 6, 4, 2, 5 (*d*) 1, 4, 3, 2, 6, 5

Direction: In the following question which one set of letters when sequentially placed at the gaps in the given letter series shall complete it?

79. c_ba_bba_bbac_ba
(*a*) bbab (*b*) bccb
(*c*) abac (*d*) acab

Directions: In question nos. **80** to **82**, find the odd word/ number / number-pair from the given alternatives.

80. (*a*) Black gram (*b*) Groundnut
(*c*) Wheat (*d*) Fennel

81. (*a*) 253 – 351 (*b*) 478 – 692
(*c*) 123 – 22 (*d*) 971 – 816

82. (*a*) Primitive (*b*) Protest
(*c*) Problem (*d*) Project

Directions: In question nos. **83** and **84**, a series is given, with one term missing. Choose the correct alternative from the given ones that will complete the series.

83. ACE, GIK, MOQ, SUW, _?_, _?_.
(*a*) XZB, DFH (*b*) YZB, DFH
(*c*) XZA, CEG (*d*) YAC, EGI

84. 3 * 12 : 7 * 9 :: 6 * 8 : ?
(*a*) 6 * 9 (*b*) 5 * 7
(*c*) 7 * 12 (*d*) 9 * 11

85. Find the correct group of signs to solve the equation.
17 ? 13 ? 36 ? 6 ? 5
(*a*) = – × ÷ (*b*) – ÷ = +
(*c*) + = ÷ × (*d*) + – × ÷

86. Arun is older than Bhanu. Charu is older than Divya. Esha is older than Charu but younger than Bhanu. Who is the eldest among them?
(*a*) Arun (*b*) Charu
(*c*) Esha (*d*) Divya

Directions: In question nos. **87** and **88**, from the given alternatives, select the word which **cannot** be formed using the letters of the given word.

87. UNPARAGONED
(*a*) PRONE (*b*) DAGGER
(*c*) ROPED (*d*) ADORN

88. BIBLIOGRAPHIES
(*a*) GRAPES (*b*) BRIBE
(*c*) PROBABLE (*d*) SHARPER

89. If > stands for multiplication, × stands for addition, + stands for division, ÷ stands for subtraction, = stands for less than, – stands for greater than and < stands for equal to, state which of the following is true?
(*a*) 112 × 8 + 6 ÷ 10 < 5 (*b*) 15 ÷ 4 – 11 > 6 +2
(*c*) 11 > 2 < 51 + 3 × 4 (*d*) 34 × 52 + 4 – 27 ÷ 8

90. If each of the letters in the English alphabet is assigned an odd numerical value by giving A = 1, B = 3 and so on, what would be the total value of the letters for the word WATCH when similarly coded?
(*a*) 105 (*b*) 55
(*c*) 84 (*d*) 96

GENERAL AWARENESS

91. Who is named as the captain of the national team for the Rio Olympics by the Indian Hockey Federation?
(*a*) PR Ajees (*b*) PR Murugesh
(*c*) PR Rajesh (*d*) PR Sreejesh

92. Where is Kansai International Airport located?
(*a*) South Korea (*b*) North Korea
(*c*) Japan (*d*) China

93. Which of the following prize/award is also known as Asia's Nobel Prize?
(*a*) Ramon Magsaysay Award
(*b*) Simon Bolivar Prize
(*c*) Jnanpith Award
(*d*) Chameli Devi Award

94. Smog is a type of
(*a*) Air Pollution (*b*) Water Pollution
(*c*) Soil Pollution (*d*) Marine pollution

95. Which of the following is not true about Indian constitution?
(*a*) It is the longest written constitution of any sovereign nation.

(*b*) Jawaharlal Nehru is considered as the 'Father of Indian Constitution'.

(*c*) The constitution of India is federal in structure.

(*d*) Besides the English version, there is an official Hindi translation.

96. Barabati Stadium is situated at

(*a*) Kolkata (*b*) Cuttack

(*c*) Chennai (*d*) Karnataka

97. 74th amendment of Indian Constitution is associated with the

(*a*) Panchayati Raj Institutions

(*b*) Urban local Bodies

(*c*) Both (*a*) & (*b*)

(*d*) Reservation of women in Panchayat elections

98. Giffen goods are

(*a*) A product that people consume more often as the price rises

(*b*) A product that people consume more often as the price falls

(*c*) A product that people consume more or less in the same quantity irrespective of price changes

(*d*) None of the above

99. 'In the short run, especially during recessions, economic output is strongly influenced by aggregate demand.' This hypothesis was given by

(*a*) Paul Krugman (*b*) John Maynard Keynes

(*c*) Zoltan Acs (*d*) Franklin Allen

100. What is the name of the programme which is unveiled by the Union Urban Development Minister Venkaiah Naidu for 500 cities?

(*a*) Swachh Survekshan

(*b*) Swachh Surakshan

(*c*) Swachh Subikshan

(*d*) Swachh Sulokshan

101. Which of the following amendment of Indian constitution is known as 'Mini Constitution'?

(*a*) Ninth amendment, 1960

(*b*) Twenty first amendment, 1967

(*c*) Forty second amendment, 1976

(*d*) Forty fourth amendment, 1978

102. In which year 'Tana Bhagat Resistance Movement' started?

(*a*) 1910 (*b*) 1914

(*c*) 1918 (*d*) 1922

103. Which of the following year was not a part of 'Imperial Durbar' in India?

(*a*) 1877 (*b*) 1903

(*c*) 1911 (*d*) 1921

104. Which one of the following queen reigned Jammu & Kashmir?

(*a*) Rani Padmini (*b*) Didda

(*c*) Rani Durgavati (*d*) None of the above

105. Which one of the following is not true about the 'Golden Quadrilateral'?

(*a*) The largest highway project in India and the fifth longest in the world.

(*b*) It is the second phase of the National Highways Development Project (NHDP).

(*c*) The project was launched in 2001 by Atal Bihari Vajpayee under the NDA government and was planned to complete in January, 2012.

(*d*) The GQ project is managed by the National Highways Authority of India (NHAI) under the Ministry of Road, Transport and Highways.

QUANTITATIVE APTITUDE

106. The square root of $23 - 4\sqrt{15}$ is

(*a*) $\pm\left(\sqrt{5} - 2\sqrt{3}\right)$ (*b*) $\pm\left(2\sqrt{5} - \sqrt{3}\right)$

(*c*) $\pm\left(2\sqrt{5} + \sqrt{3}\right)$ (*d*) $\pm\left(\sqrt{5} + 2\sqrt{3}\right)$

107. What is the value of $\cot 3°\cot 6°\cot 9°...\cot 93°$?

(*a*) undefined (*b*) 0

(*c*) 1 (*d*) 93

108. The average age of 21 students and their teacher is 15 years. If the teacher's age is excluded, the average age reduces by 2. What is the teacher's age?

(*a*) 41 years (*b*) 51 years

(*c*) 47 years (*d*) 57 years

109. If the angle of elevation of the Sun changes from 30° to 60°, the length of the shadow of a tower decreases by 50 m. The height of the tower is

(*a*) 25 m (*b*) $25\sqrt{3}$ m

(*c*) $\frac{25}{\sqrt{3}}$ m (*d*) $\frac{50}{\sqrt{3}}$ m

110. If $x^2 - 4x + 1 = 0$, then the value of $x^3 + \frac{1}{x^3}$ will be

(*a*) 56 (*b*) 44

(*c*) 48 (*d*) 52

111. In the given figure, if MN is the diameter of the circle and $\angle MPQ = 120°$, then the value of $\angle QNM$ will be

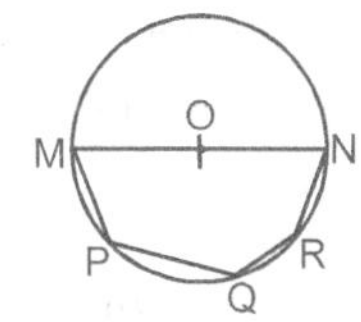

(*a*) 90° (*b*) 120°

(*c*) 30° (*d*) 60°

112. A square shaped field of edge 105 feet has to be fenced with wire. How many equal spaced supporting poles are required to fence it, if the poles should be fixed 7 feet apart?

(*a*) 60 (*b*) 64
(*c*) 62 (*d*) 56

113. If $a\cos\theta - b\sin\theta = c$, then the value of $a\sin\theta + b\cos\theta$ is

(*a*) $\pm\sqrt{a^2 - b^2 - c^2}$ (*b*) $\pm\sqrt{a^2 - b^2 + c^2}$
(*c*) $\pm\sqrt{-a^2 + b^2 + c^2}$ (*d*) $\pm\sqrt{a^2 + b^2 - c^2}$

114. If a\$b = $a - b + \frac{b^2}{a}$, then the value of 15\$3 is

(*a*) 87 (*b*) 12.6
(*c*) 11.6 (*d*) 12.4

115. If the length of the sides of a triangle are in the ratio of 3 : 4 : 5 and the circumradius of the triangle is 30 cm, then the length of the largest side is

(*a*) 60 cm (*b*) 55 cm
(*c*) 40 cm (*d*) 45 cm

116. A boy takes 2 minutes less time to cross a circular ground along its shortest path than to cover it once along the boundary. If his speed was 1.5 m/s, then the radius of the circular ground is

(Take $\pi = \frac{22}{7}$)

(*a*) 43.4 m (*b*) 157.5 m
(*c*) 78.5 m (*d*) 97.45 m

117. Ram gets 4 marks for each correct sum and loses 1 mark for each wrong sum. He attempts 40 sums and obtains 80 marks. The number of sums that was attempted wrong, is

(*a*) 22 (*b*) 20
(*c*) 18 (*d*) 16

118. A conveyor belt is attatched to two wheels of a machine of radius 7 cm and 2 cm. If the centres of the wheels are at a distance of 13 cm, then what is the total length (in cm) of the belt?

(*a*) $9\pi + 12$ (*b*) $9\pi + 24$
(*c*) $8\pi + 5$ (*d*) $8\pi + 3$

119. Two circles of equal radii 5 cm, intersect each other in such a way that a circle passes through the centre of the other circle. What is the area of the quadrilateral formed by joining the centers and the intersection points?

(*a*) $25\sqrt{2}$ cm^2 (*b*) $\frac{25\sqrt{3}}{2}$ cm^2
(*c*) $25\sqrt{3}$ cm^2 (*d*) $50\sqrt{3}$ cm^2

120. If $\frac{\operatorname{cosec}\theta - \cot\theta}{\operatorname{cosec}\theta + \cot\theta} = \frac{1}{9}$, then the value of $\sin\theta$ is equal to

(*a*) $\frac{2}{5}$ (*b*) $\frac{4}{5}$
(*c*) $\frac{2}{3}$ (*d*) $\frac{3}{5}$

GENERAL ENGLISH

Directions for questions 121 to 123 : Each of the following sentences contains a blank to be filled in with an appropriate word. Choose the correct alternative out of the four given options.

121. Night came ______. A full moon ______ high over the trees.

(*a*) by, risen (*b*) on, rose
(*c*) in, rise (*d*) by, raised

122. 'Feel free to contact me ______ my email address', said Flintstone.

(*a*) upon (*b*) on
(*c*) at (*d*) by

123. Even the president's own bipartisan commission can't agree ______ what to do.

(*a*) with (*b*) on
(*c*) to (*d*) at

Directions for questions 124 to 126: In each of the following questions, a part of the sentence is underlined. Beneath each sentence four different ways of phrasing the underlined part are given. Choose the grammatically correct option. In case no improvement is needed, your answer is option (*d*).

124. If your <u>child is thinking about a gap year, they</u> can get good advice from this book.

(*a*) child is thinking about a gap year, he
(*b*) children is thinking about a gap year, they
(*c*) child think about a gap year, he
(*d*) No improvement

125. By next December, we <u>will have lived</u> here for 6 months.

(*a*) shall have been living
(*b*) will be living
(*c*) shall be living
(*d*) No improvement

126. Tarak writes with just the right mix of <u>disdain</u> and empathy.

(*a*) admiration (*b*) favour
(*c*) affection (*d*) No improvement

Directions for questions 127 to 130 : You have given passage with 5 questions. Read the passage carefully and choose the best answer to each question out of the four alternatives given.

The murders of Andrew Borden and Abby Borden shocked the community. Borden was one of the wealthiest—and most unpopular—men in town. Frugal to a fault, he was a self-made man who had become the head of one of the town's largest banks and a substantial property owner. The dour businessman had also made many enemies on his rise to the top, and rumors swirled that Andrew and Abby had perhaps been killed as revenge for Andrew's shady business dealings. For the police, their daughter, Lizzie, was the prime suspect of the case. However, there was no physical evidence linking her to the murders. A hatchet had been discovered in the basement of the Borden home, but its blade was clean and the handle had been broken off—by Lizzie, according to police. The police's reluctance to use any sort of forensic testing hampered the investigation.

127. Find an option that is opposite in meaning to the word given in bold.

Dour

(*a*) Hostile (*b*) Cold
(*c*) Grim (*d*) Gentle

128. Andrew was not liked by many people in town because

(*a*) people were jealous that he was a self made man.
(*b*) he used his money carefully.
(*c*) he indulged in mendacious business practices.
(*d*) people were jealous of his wealth.

129. According to the passage, which of the following was not one of the reasons that held back the investigation?

(*a*) The police was not comfortable with forensic testing.
(*b*) Lizzie distorted the evidence that interfered with the investigation.
(*c*) The police was unable to find any physical evidence linking Lizzie to the case.
(*d*) Andrew Borden had many enemies and so it was difficult to determine who the murderer was.

130. Which of the following is not a trait describing Andrew Borden?

(*a*) Unpopular (*b*) Affluent
(*c*) Economical (*d*) Prudent

Directions for questions 131 and 132 : In each of the following questions, a word is followed by four options. Select the option that best expresses the meaning of the given word.

131. Pinnacle

(*a*) Bottom (*b*) Crest
(*c*) Minimum (*d*) Abyss

132. Sleuth

(*a*) Butcher (*b*) Executioner
(*c*) Invigilator (*d*) Investigator

Directions for questions 133 and 134 : In the following question, a word is followed by four options. Select the option that is opposite in meaning to the given word.

133. Detrimental

(*a*) Wicked (*b*) Damaging
(*c*) Benign (*d*) Indecisive

134. Nibble

(*a*) Devour (*b*) Pester
(*c*) Gnaw (*d*) Dull

Directions: In question nos. **135** to **137**, Out of the four alternatives, choose the one which can be substituted for the given words/sentences.

135. Mahatma Gandhi was <u>the one who advocated the removal of inequalities among people</u>.

(*a*) an egalitarian (*b*) a federalist
(*c*) an absolutist (*d*) a reformist

136. <u>A male servant who has charge of other employees</u>

(*a*) Assistant (*b*) Minion
(*c*) Retainer (*d*) Butler

137. The evidence suggests that he ran <u>an enterprise of embezzlement, bribery and blackmail</u>.

(*a*) a clatter (*b*) a rumpus
(*c*) a racket (*d*) a squabble

Directions for questions 138 to 140: Four alternatives are given for the idioms/phrases given below. Choose the alternative which best expresses the meaning of the given idiom/phrase.

138. Until now, Aladdin couldn't figure out how to <u>get his foot in the door</u>, but now he has the perfect contact.

(*a*) enter an organisation as an initial step towards success
(*b*) enter into a competition that offers huge prize money
(*c*) step out of a difficult situation which involves money
(*d*) use the available opportunity for one's own good

139. Dr Jekyll, the chief of critical care at the hospital, said that the medical procedure was <u>in the home stretch.</u>

(*a*) time consuming. (*b*) started.
(*c*) mid-way. (*d*) nearly completed.

140. Hearing the train whistle gives me <u>itchy feet</u>.

(*a*) a feeling to move to a different place.
(*b*) a feeling of irritation.
(*c*) a desire to run away.
(*d*) a desire to stay at one place.

Direction : In the following group of four different words, one word is correctly spelt. Find the correctly spelt word and mark your answer accordingly.

141. (*a*) Bearucrat (*b*) Beareaucrat
(*c*) Bureaucrat (*d*) Bureacrat

Directions for questions 142 to 144: Read each of the following sentences to find out whether there is any grammatical error in it. The error, if any, will be in one part of the sentence. The number of that part is the answer. If there is 'No error' the answer is (*d*).

142. 'Cheer up father, (*a*) / I'll go and will finish the (*b*)/ remaining work,' said Supandi. (*c*)/ No error (*d*)

143. Archie is one of (*a*)/ the cleverest boys that (*b*)/ has passed through the school. (*c*)/ No error (*d*)

144. Billu's taste in (*a*)/ comic books is (*b*)/ different than Pinky's (*c*)/ No error (*d*)

Directions for questions 145 to 147 : Each of the following sentences contains a blank to be filled in with an appropriate word. Choose the correct alternative out of the four given options.

145. Even _____ the price of gas is starting to come down now, many people _____ to take the bus or the train.
(*a*) if, had chosen (*b*) though, had chosen
(*c*) if, are choosing (*d*) though, are choosing

146. Tomorrow is Halloween, ____ everybody gets dressed up in a fancy costume. So we will go trick-or-treating _____ is a traditional practice for children in many countries.
(*a*) where, that (*b*) where, which
(*c*) when, that (*d*) when, which

147. I called _____ my professor when I last visited the town.
(*a*) to (*b*) forth
(*c*) on (*d*) after

Directions for questions 148 to 150: In each of the following questions, a part of the sentence is underlined. Beneath each sentence four different ways of phrasing the underlined part are given. Choose the grammatically correct option. In case no improvement is needed, your answer is option (*d*).

148. <u>Less than 20 children suffer from the disease each year</u>.
(*a*) Less than 20 children suffer from the diseases each year
(*b*) Fewer than 20 children suffer from the disease each year.
(*c*) Fewer than 20 children suffers from the disease each year.
(*d*) No improvement

149. He found a <u>blue small square photo-frame</u> in the store room.
(*a*) blue square and small photo-frame
(*b*) small square blue photo-frame
(*c*) damaged small blue photo-frame
(*d*) No improvement

150. Pluto is the <u>eldest</u> dog in the pack.
(*a*) oldest (*b*) older
(*c*) elder (*d*) No improvement

Directions: In question nos. **151** to **155**, you have given a passages with 5 questions. Read the passage carefully and choose the best answer to each question out of the four alternatives.

One of the mysteries of Stonehenge is why our ancestors chose to use bluestones that had to be hauled hundreds of kilometres from the Preseli Hills in Pembrokeshire to the site in Wiltshire. But new research from London's Royal College of Art suggests sound might have played a role. Researchers tested thousands of stones on CarnMenyn in the Preseli Hills, and found a large number of the rocks ring when they are struck. Usually, stones produce a disappointing clunk when hit, with microscopic cracks making it difficult for vibrations to travel within the rock. But certain bluestones have the right microscopic structure - and sound like a metallic gong.

They also found a few of the rocks remaining at Stonehenge rang as well. The challenge they now face is providing good evidence that the bluestones were used for their musical quality. Sound is ephemeral, and disappears as soon as it is made, so it is difficult to know for sure that our ancestors used the stones as percussion instruments.

151. Which of the following cannot be inferred from the passage?
(*a*) The reason for using bluestones to create Stonehenge is still unknown.
(*b*) It is wrong to say that all the stones at Stonehenge produce a clunk sound when stuck together.
(*c*) Time has not been able to preserve the Stonehenge the way it was built by our ancestors.
(*d*) Scientists are performing tests to know more about Stonehenge.

152. Which of the following was suggested by the London's Royal College of Art?
(*a*) Ocular properties of the stone led our ancestors to choose to use bluestones.
(*b*) Acoustic properties of the stone led our ancestors to choose to use bluestones.
(*c*) Our ancestors liked the sound of bluestones over the sound of the other stones.
(*d*) Our ancestors knew that microscopic cracks in bluestones make it easy for the vibrations to travel.

153. Which of the following best expresses the meaning of the word 'ephemeral'?

(*a*) Brief (*b*) Timeless
(*c*) Continuing (*d*) Ancient

154. Which of the following is opposite in meaning to the word 'haul'?

(*a*) Lift (*b*) Push
(*c*) Drag (*d*) Carry

155. Why is it difficult for us to know for sure whether or not the stones were used as percussion instruments?

(*a*) There are no written records to conform our doubts.
(*b*) Researchers have not been able to find evidence suggesting that the stones were used for their musical ability.
(*c*) Sound does not last for a long period of time and disappears as soon as it is made.
(*d*) Since Stonehenge is not preserved in its original form, it is difficult to research on it.

Directions (Q. 156 and 157) : *In each of the following questions, a word is followed by four options. Select the option that best expresses the meaning of the given word.*

156. Cherubic

(*a*) Angelic (*b*) Snobbish
(*c*) Restless (*d*) Honest

157. Copious

(*a*) Powerful (*b*) Delaying
(*c*) Plentiful (*d*) Foolish

Directions (Q. 158 and 159) : *In each of the following questions, a word is followed by four options. Select the option that is opposite in meaning to the given word.*

158. Choleric

(*a*) Calm
(*b*) Ill-tempered
(*c*) Hygienic
(*d*) Sober

159. Harried

(*a*) Relaxed (*b*) Harassed
(*c*) Instigated (*d*) Pitch forked

Directions (Q. 160 to 162) : *Out of the four alternatives, choose the one which can be substituted for the given words.*

160. A person who is difficult to please

(*a*) Rhapsodic (*b*) Fastidious
(*c*) Succulent (*d*) Rambunctious

161. A cinema show, which is held in the afternoon.

(*a*) Proscenium (*b*) Matinee
(*c*) Claire (*d*) Dirge

162. A person who knows everything

(*a*) Omniscient (*b*) Omnipresent
(*c*) Ignoramus (*d*) Omnipotent

Directions (Q. 163 to 165) : *Four alternatives are given for the Idiom/Phrase. Choose the alternative which best expresses the meaning of the Idiom/Phrase.*

163. To crease something up

(*a*) to iron something
(*b*) to put things in order
(*c*) to disfigure something
(*d*) to wrinkle one's clothing

164. Skimp on something

(*a*) to use too little of something
(*b*) to gorge on a feast
(*c*) to skip some duty or task
(*d*) to do something easily

165. To die in harness

(*a*) to die while in service
(*b*) to die in prison
(*c*) to die lonely
(*d*) None of these

166. *Group of four words are given. One word is correctly spelt. Find the correctly spelt word.*

(*a*) Committment (*b*) Comittment
(*c*) Commitment (*d*) Comitment

Directions (Q. 167 to 169) : *Some of the sentences have errors and some are correct. Find out the part of a sentence which has an error. If there is no error, mark your answer is (d).*

167. He took to drink (*a*)/ to lessen (*b*)/ his mental worries (*c*)/. No error(*d*)

168. Despite of his protests (*a*)/ I decided to buy (*b*)/ the jeans which he did not like. (*c*) / No error(*d*)

169. The Poet and Scholar (*a*)/ have been(*b*)/ honoured. (*c*)/ No error(*d*)

Directions (Q. 170 to 172) : *Sentences are given with blanks to be filled in with an appropriate word. Four alternatives are suggested for each question. Choose the correct alternative out of the four.*

170. A bullet ________ his cheek.

(*a*) glazed (*b*) grazed
(*c*) throttled (*d*) amputated

171. Tourists forget their ________ ideas as soon as they visit our country.

(*a*) preconceived (*b*) ordained
(*c*) gestalt (*d*) ruffled

172. He has a reputation for being a ________ critic.

(*a*) sacramental (*b*) cyclopean

(*c*) unleashed (*d*) forthright

Directions (Q. 173 to 175) : *In the following questions a part of the sentence is* ***bold****. Below are given alternatives to the bold part at (a), (b) and (c) which may improve the sentence. Choose the correct alternative. In case no Improvement is needed your answer is (d).*

173. You are asked to copy this letter **word by word.**

(*a*) word for word (*b*) word with word

(*c*) word to word (*d*) No improvement

174. The weak man is a slave to his **sensuous** pleasures.

(*a*) sensory (*b*) sensual

(*c*) secondary (*d*) No improvement

175. As soon as she noticed the workmen, she asked them **what they have been doing**.

(*a*) have done (*b*) had been

(*c*) are doing (*d*) No improvement

Directions (Q. 176 to 180) : *Read the following passage carefully and choose the best answer to each question out of four alternatives and mark it the correct answer.*

The problem of water pollution by pesticides can be understood only in context, as part of the whole to which it belongs-the pollution of the total environment of mankind. The pollution entering our waterways comes from many sources, radioactive wastes from reactors, laboratories, and hospitals; fallout from nuclear explosions; domestic wastes from cities and towns; chemical wastes from factories. To these is added a new kind of fallout-the chemical sprays applied to crop lands and gardens, forests and fields. Many of the chemical agents in this alarming melange initiate and augment the harmful effects of radiation, and within the groups of chemicals themselves there are sinister and little-understood interactions, transformations, and summations of effect.

Ever since the chemists began to manufacture substances that nature never invented, the problem of water purification have become complex and the danger to users of water has increased. As we have seen, the production of these synthetic chemicals in large volume began in the 1940'ss. It has now reached such proportion that an appalling deluge of chemical pollution is daily poured into the nation's waterways. When inextricably mixed with domestic and other wastes discharged into the same water, these chemicals sometimes defy detection by the methods in ordinary use by purification plants. Most of them are so complex that they cannot be identified. In rivers, a really incredible variety of pollutants combine to produce deposits that sanitary engineers can only despairingly refer to as "gunk".

176. All the following words mean 'chemicals' except :

(*a*) sands (*b*) substances

(*c*) pesticides (*d*) deposits

177. The main argument of paragraph 1 is :

(*a*) That there are sinister interaction in the use of chemicals

(*b*) that there are numerous reasons for contamination of water supplies

(*c*) that there are many dangers from nuclear fallout

(*d*) that pesticides are dangerous

178. The word 'gunk' in the last line refers :

(*a*) to the waste products deposited by sanitary engineers

(*b*) to the debris found in rivers

(*c*) to unidentifiable chemicals found in water

(*d*) to the domestic water supplies

179. Water pollution can only be understood :

(*a*) in relation to world contamination

(*b*) by the whole human race

(*c*) in context

(*d*) in relation to the number of pesticides that exist

180. Water contamination has become serious :

(*a*) since water pollution was difficult to assess

(*b*) since nature has taken a hand in pollution

(*c*) since chemists began to use new substances

(*d*) since businessmen authorised the use of chemicals

OMR SHEET

1. Ⓐ Ⓑ Ⓒ Ⓓ	26. Ⓐ Ⓑ Ⓒ Ⓓ	51. Ⓐ Ⓑ Ⓒ Ⓓ	76. Ⓐ Ⓑ Ⓒ Ⓓ
2. Ⓐ Ⓑ Ⓒ Ⓓ	27. Ⓐ Ⓑ Ⓒ Ⓓ	52. Ⓐ Ⓑ Ⓒ Ⓓ	77. Ⓐ Ⓑ Ⓒ Ⓓ
3. Ⓐ Ⓑ Ⓒ Ⓓ	28. Ⓐ Ⓑ Ⓒ Ⓓ	53. Ⓐ Ⓑ Ⓒ Ⓓ	78. Ⓐ Ⓑ Ⓒ Ⓓ
4. Ⓐ Ⓑ Ⓒ Ⓓ	29. Ⓐ Ⓑ Ⓒ Ⓓ	54. Ⓐ Ⓑ Ⓒ Ⓓ	79. Ⓐ Ⓑ Ⓒ Ⓓ
5. Ⓐ Ⓑ Ⓒ Ⓓ	30. Ⓐ Ⓑ Ⓒ Ⓓ	55. Ⓐ Ⓑ Ⓒ Ⓓ	80. Ⓐ Ⓑ Ⓒ Ⓓ
6. Ⓐ Ⓑ Ⓒ Ⓓ	31. Ⓐ Ⓑ Ⓒ Ⓓ	56. Ⓐ Ⓑ Ⓒ Ⓓ	81. Ⓐ Ⓑ Ⓒ Ⓓ
7. Ⓐ Ⓑ Ⓒ Ⓓ	32. Ⓐ Ⓑ Ⓒ Ⓓ	57. Ⓐ Ⓑ Ⓒ Ⓓ	82. Ⓐ Ⓑ Ⓒ Ⓓ
8. Ⓐ Ⓑ Ⓒ Ⓓ	33. Ⓐ Ⓑ Ⓒ Ⓓ	58. Ⓐ Ⓑ Ⓒ Ⓓ	83. Ⓐ Ⓑ Ⓒ Ⓓ
9. Ⓐ Ⓑ Ⓒ Ⓓ	34. Ⓐ Ⓑ Ⓒ Ⓓ	59. Ⓐ Ⓑ Ⓒ Ⓓ	84. Ⓐ Ⓑ Ⓒ Ⓓ
10. Ⓐ Ⓑ Ⓒ Ⓓ	35. Ⓐ Ⓑ Ⓒ Ⓓ	60. Ⓐ Ⓑ Ⓒ Ⓓ	85. Ⓐ Ⓑ Ⓒ Ⓓ
11. Ⓐ Ⓑ Ⓒ Ⓓ	36. Ⓐ Ⓑ Ⓒ Ⓓ	61. Ⓐ Ⓑ Ⓒ Ⓓ	86. Ⓐ Ⓑ Ⓒ Ⓓ
12. Ⓐ Ⓑ Ⓒ Ⓓ	37. Ⓐ Ⓑ Ⓒ Ⓓ	62. Ⓐ Ⓑ Ⓒ Ⓓ	87. Ⓐ Ⓑ Ⓒ Ⓓ
13. Ⓐ Ⓑ Ⓒ Ⓓ	38. Ⓐ Ⓑ Ⓒ Ⓓ	63. Ⓐ Ⓑ Ⓒ Ⓓ	88. Ⓐ Ⓑ Ⓒ Ⓓ
14. Ⓐ Ⓑ Ⓒ Ⓓ	39. Ⓐ Ⓑ Ⓒ Ⓓ	64. Ⓐ Ⓑ Ⓒ Ⓓ	89. Ⓐ Ⓑ Ⓒ Ⓓ
15. Ⓐ Ⓑ Ⓒ Ⓓ	40. Ⓐ Ⓑ Ⓒ Ⓓ	65. Ⓐ Ⓑ Ⓒ Ⓓ	90. Ⓐ Ⓑ Ⓒ Ⓓ
16. Ⓐ Ⓑ Ⓒ Ⓓ	41. Ⓐ Ⓑ Ⓒ Ⓓ	66. Ⓐ Ⓑ Ⓒ Ⓓ	91. Ⓐ Ⓑ Ⓒ Ⓓ
17. Ⓐ Ⓑ Ⓒ Ⓓ	42. Ⓐ Ⓑ Ⓒ Ⓓ	67. Ⓐ Ⓑ Ⓒ Ⓓ	92. Ⓐ Ⓑ Ⓒ Ⓓ
18. Ⓐ Ⓑ Ⓒ Ⓓ	43. Ⓐ Ⓑ Ⓒ Ⓓ	68. Ⓐ Ⓑ Ⓒ Ⓓ	93. Ⓐ Ⓑ Ⓒ Ⓓ
19. Ⓐ Ⓑ Ⓒ Ⓓ	44. Ⓐ Ⓑ Ⓒ Ⓓ	69. Ⓐ Ⓑ Ⓒ Ⓓ	94. Ⓐ Ⓑ Ⓒ Ⓓ
20. Ⓐ Ⓑ Ⓒ Ⓓ	45. Ⓐ Ⓑ Ⓒ Ⓓ	70. Ⓐ Ⓑ Ⓒ Ⓓ	95. Ⓐ Ⓑ Ⓒ Ⓓ
21. Ⓐ Ⓑ Ⓒ Ⓓ	46. Ⓐ Ⓑ Ⓒ Ⓓ	71. Ⓐ Ⓑ Ⓒ Ⓓ	96. Ⓐ Ⓑ Ⓒ Ⓓ
22. Ⓐ Ⓑ Ⓒ Ⓓ	47. Ⓐ Ⓑ Ⓒ Ⓓ	72. Ⓐ Ⓑ Ⓒ Ⓓ	97. Ⓐ Ⓑ Ⓒ Ⓓ
23. Ⓐ Ⓑ Ⓒ Ⓓ	48. Ⓐ Ⓑ Ⓒ Ⓓ	73. Ⓐ Ⓑ Ⓒ Ⓓ	98. Ⓐ Ⓑ Ⓒ Ⓓ
24. Ⓐ Ⓑ Ⓒ Ⓓ	49. Ⓐ Ⓑ Ⓒ Ⓓ	74. Ⓐ Ⓑ Ⓒ Ⓓ	99. Ⓐ Ⓑ Ⓒ Ⓓ
25. Ⓐ Ⓑ Ⓒ Ⓓ	50. Ⓐ Ⓑ Ⓒ Ⓓ	75. Ⓐ Ⓑ Ⓒ Ⓓ	100. Ⓐ Ⓑ Ⓒ Ⓓ

OMR SHEET

101. Ⓐ Ⓑ Ⓒ Ⓓ	121. Ⓐ Ⓑ Ⓒ Ⓓ	141. Ⓐ Ⓑ Ⓒ Ⓓ	161. Ⓐ Ⓑ Ⓒ Ⓓ
102. Ⓐ Ⓑ Ⓒ Ⓓ	122. Ⓐ Ⓑ Ⓒ Ⓓ	142. Ⓐ Ⓑ Ⓒ Ⓓ	162. Ⓐ Ⓑ Ⓒ Ⓓ
103. Ⓐ Ⓑ Ⓒ Ⓓ	123. Ⓐ Ⓑ Ⓒ Ⓓ	143. Ⓐ Ⓑ Ⓒ Ⓓ	163. Ⓐ Ⓑ Ⓒ Ⓓ
104. Ⓐ Ⓑ Ⓒ Ⓓ	124. Ⓐ Ⓑ Ⓒ Ⓓ	144. Ⓐ Ⓑ Ⓒ Ⓓ	164. Ⓐ Ⓑ Ⓒ Ⓓ
105. Ⓐ Ⓑ Ⓒ Ⓓ	125. Ⓐ Ⓑ Ⓒ Ⓓ	145. Ⓐ Ⓑ Ⓒ Ⓓ	165. Ⓐ Ⓑ Ⓒ Ⓓ
106. Ⓐ Ⓑ Ⓒ Ⓓ	126. Ⓐ Ⓑ Ⓒ Ⓓ	146. Ⓐ Ⓑ Ⓒ Ⓓ	166. Ⓐ Ⓑ Ⓒ Ⓓ
107. Ⓐ Ⓑ Ⓒ Ⓓ	127. Ⓐ Ⓑ Ⓒ Ⓓ	147. Ⓐ Ⓑ Ⓒ Ⓓ	167. Ⓐ Ⓑ Ⓒ Ⓓ
108. Ⓐ Ⓑ Ⓒ Ⓓ	128. Ⓐ Ⓑ Ⓒ Ⓓ	148. Ⓐ Ⓑ Ⓒ Ⓓ	168. Ⓐ Ⓑ Ⓒ Ⓓ
109. Ⓐ Ⓑ Ⓒ Ⓓ	129. Ⓐ Ⓑ Ⓒ Ⓓ	149. Ⓐ Ⓑ Ⓒ Ⓓ	169. Ⓐ Ⓑ Ⓒ Ⓓ
110. Ⓐ Ⓑ Ⓒ Ⓓ	130. Ⓐ Ⓑ Ⓒ Ⓓ	150. Ⓐ Ⓑ Ⓒ Ⓓ	170. Ⓐ Ⓑ Ⓒ Ⓓ
111. Ⓐ Ⓑ Ⓒ Ⓓ	131. Ⓐ Ⓑ Ⓒ Ⓓ	151. Ⓐ Ⓑ Ⓒ Ⓓ	171. Ⓐ Ⓑ Ⓒ Ⓓ
112. Ⓐ Ⓑ Ⓒ Ⓓ	132. Ⓐ Ⓑ Ⓒ Ⓓ	152. Ⓐ Ⓑ Ⓒ Ⓓ	172. Ⓐ Ⓑ Ⓒ Ⓓ
113. Ⓐ Ⓑ Ⓒ Ⓓ	133. Ⓐ Ⓑ Ⓒ Ⓓ	153. Ⓐ Ⓑ Ⓒ Ⓓ	173. Ⓐ Ⓑ Ⓒ Ⓓ
114. Ⓐ Ⓑ Ⓒ Ⓓ	134. Ⓐ Ⓑ Ⓒ Ⓓ	154. Ⓐ Ⓑ Ⓒ Ⓓ	174. Ⓐ Ⓑ Ⓒ Ⓓ
115. Ⓐ Ⓑ Ⓒ Ⓓ	135. Ⓐ Ⓑ Ⓒ Ⓓ	155. Ⓐ Ⓑ Ⓒ Ⓓ	175. Ⓐ Ⓑ Ⓒ Ⓓ
116. Ⓐ Ⓑ Ⓒ Ⓓ	136. Ⓐ Ⓑ Ⓒ Ⓓ	156. Ⓐ Ⓑ Ⓒ Ⓓ	176. Ⓐ Ⓑ Ⓒ Ⓓ
117. Ⓐ Ⓑ Ⓒ Ⓓ	137. Ⓐ Ⓑ Ⓒ Ⓓ	157. Ⓐ Ⓑ Ⓒ Ⓓ	177. Ⓐ Ⓑ Ⓒ Ⓓ
118. Ⓐ Ⓑ Ⓒ Ⓓ	138. Ⓐ Ⓑ Ⓒ Ⓓ	158. Ⓐ Ⓑ Ⓒ Ⓓ	178. Ⓐ Ⓑ Ⓒ Ⓓ
119. Ⓐ Ⓑ Ⓒ Ⓓ	139. Ⓐ Ⓑ Ⓒ Ⓓ	159. Ⓐ Ⓑ Ⓒ Ⓓ	179. Ⓐ Ⓑ Ⓒ Ⓓ
120. Ⓐ Ⓑ Ⓒ Ⓓ	140. Ⓐ Ⓑ Ⓒ Ⓓ	160. Ⓐ Ⓑ Ⓒ Ⓓ	180. Ⓐ Ⓑ Ⓒ Ⓓ

ANSWERS

1. (c)	2. (a)	3. (a)	4. (b)	5. (b)	6. (c)	7. (b)	8. (d)	9. (b)	10. (c)
11. (c)	12. (a)	13. (d)	14. (a)	15. (d)	16. (c)	17. (c)	18. (c)	19. (a)	20. (c)
21. (b)	22. (b)	23. (a)	24. (b)	25. (c)	26. (c)	27. (d)	28. (d)	29. (c)	30. (a)
31. (b)	32. (b)	33. (d)	34. (c)	35. (d)	36. (d)	37. (a)	38. (d)	39. (c)	40. (d)
41. (a)	42. (d)	43. (d)	44. (a)	45. (a)	46. (d)	47. (b)	48. (b)	49. (c)	50. (a)
51. (b)	52. (c)	53. (d)	54. (b)	55. (a)	56. (b)	57. (b)	58. (d)	59. (b)	60. (d)
61. (c)	62. (c)	63. (c)	64. (c)	65. (a)	66. (c)	67. (b)	68. (d)	69. (d)	70. (c)
71. (c)	72. (d)	73. (b)	74. (d)	75. (b)	76. (b)	77. (d)	78. (c)	79. (b)	80. (c)
81. (a)	82. (a)	83. (d)	84. (c)	85. (c)	86. (a)	87. (b)	88. (d)	89. (d)	90. (a)
91. (d)	92. (c)	93. (a)	94. (a)	95. (b)	96. (b)	97. (b)	98. (a)	99. (b)	100. (a)
101. (c)	102. (b)	103. (d)	104. (b)	105. (b)	106. (b)	107. (b)	108. (d)	109. (b)	110. (d)
111. (d)	112. (a)	113. (d)	114. (b)	115. (a)	116. (b)	117. (d)	118. (b)	119. (b)	120. (d)
121. (b)	122. (c)	123. (b)	124. (a)	125. (a)	126. (d)	127. (d)	128. (c)	129. (d)	130. (d)
131. (b)	132. (d)	133. (c)	134. (a)	135. (a)	136. (d)	137. (c)	138. (a)	139. (d)	140. (a)
141. (c)	142. (d)	143. (c)	144. (c)	145. (d)	146. (c)	147. (c)	148. (b)	149. (b)	150. (a)
151. (d)	152. (b)	153. (a)	154. (b)	155. (b)	156. (a)	157. (c)	158. (a)	159. (a)	160. (b)
161. (b)	162. (a)	163. (d)	164. (a)	165. (a)	166. (c)	167. (a)	168. (a)	169. (b)	170. (b)
171. (a)	172. (d)	173. (a)	174. (b)	175. (d)	176. (a)	177. (b)	178. (c)	179. (c)	180. (c)

EXPLANATIONS

1. $V_{max} = 12$

$V_{min} = 4$

$$V_{SWR} = \frac{V_{max}}{V_{min}} = \frac{12}{4} = 3$$

3. VSB (vestigial side band) transmission transmits one side band fully and the other side band partially thus, reducing the bandwidth requirement.

5. $400\left(1+\frac{0.75^2}{2}\right) = 512\text{W}$

6. PRF in radar systems means pulse repetition frequency.

7. It generates FM through phase modulation.

8. $V \propto \sqrt{T}$

10. $m_1 = 0.4,\ m_2 = 0.3$

$$m_t = \sqrt{m_1^2 + m_2^2}$$

11. $PIV = 2\ V_m$

(for centre-tap full wave rectifier)

$= 2 \times 300 = 600$ V

13. DSC-SC is an amplitude modulation but without carrier wave,

$$\eta_{\text{power efficiency}} = \frac{\text{Carrier power}}{\text{Total transmitted power}} = 0$$

$$\eta_{\text{transmission efficiency}} + \eta_{\text{power efficiency}} = 1$$

$$\eta_{\text{transmission efficiency}} = 1 = 100\%$$

14. $(FOM)_{\text{wide bandFM}} = \frac{3}{2}\beta^2 = \frac{(S/N)_o}{(S/N)_i}$

$(FOM)_{\text{DSB-AM}} = \frac{1}{3} = \frac{(S/N)_o}{(S/N)_i}$

Since the total transmitted power is same for both FM and DSB-AM, then $\left(\frac{S}{N}\right)_{\text{inpu}}$ for both will also be same.

$\beta = \frac{\Delta f}{fm} = \frac{75}{15} = 5 \quad \therefore \quad \frac{(S/N)_{\text{wide band FM}}}{(S/N)_{\text{DSB-AM}}} = \frac{3/2\beta^2}{1/3}$

$= \frac{9}{2} \times 25 = 112.5$

15. $m(t) = 3\cos(6000\pi t) + 4\cos(7000\pi t) + 5\cos(10000\pi t)$

Minimum sampling rate

$= 2 \text{ max. } (fm_1, fm_3, fm_3)$

$= 2 \times (5000) = 10000$ Hz

16. Accuracy $= 0.1\% = 0.001 = \frac{1}{2^n}$

$\frac{1}{10^3} = \frac{1}{2^n}$

So, $\boxed{n \approx 10 \text{ bits}}$

17. Gain (dB) $= 20\log_{10}\frac{V_o}{V_i}$

$= 20\log_{10}\frac{1}{10^{-3}} = 20\log_{10} 10^3$

$= 20 \times 3$ dB

$= 60$ dB

18. Using Gausss law in manetostatics

$\nabla \cdot \vec{B} = 0$

or $\boxed{\iint \vec{B} \cdot \vec{ds} = 0}$

Thus, divergence of magnetic field is zero, as no isolated poles exist.

19. $\Gamma = \frac{Z_L - Z_0}{Z_L + Z_0}$

Also, $P = I^2 Z_L$

$Z_L = \frac{10 \times 10^3}{20 \times 20} = 25$

$\Gamma = \frac{Z_L - Z_0}{Z_L + Z_0} = \frac{25 - 50}{25 + 50} = -\frac{1}{3}$

20. Displacement current is the result of permittivity and exists wherever the electric field E change with time

$\boxed{\Delta \times \vec{H} = J_C + J_D}$

$J_D \rightarrow$ Displacement current density

Thus, in capacitor no direct connection exists between the two plates but an insulator is placed of definite permittivity, so when an AC field is applied, capacitor gets charged and a path for current is estabilished which is displacement current.

21. In class AB amplifier, the collector current flows for more than 180 but less than 360

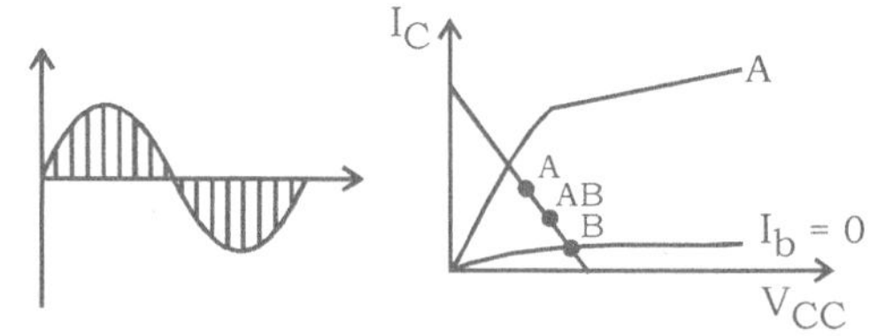

The Q point is located in the active region but very close to cut-off region.

22. Zener diode characteristics are shown as

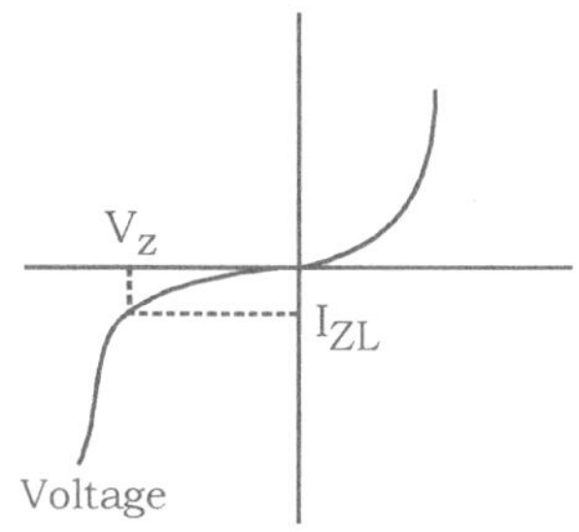

The diode in the reverse biased region above breadown voltage will now regulate the load voltage against variations in load current and against the variations in supply voltage.

23. $V_{BE} = 0.7$

(base to emitter is forward biased)

$V_{CB} = 0.2$ V

Since in common emitter configuration, emitter is genrally grounded,

So $V_E = 0$

also $V_B = 0.7$

$V_{CB} = V_C - V_B = 0.2$

$\Rightarrow V_C = 0.2 + 0.7$

$V_C = 0.9$ V

$V_{CE} = 0.9$ V

Thus, the transistor is in normal active-mode.

24. Substraction using 2's complement

(i) Equate the digital of minuend and subtrahend.

(ii) Take the 2's complement of subtrahend.

(iii) Add it to the minuend.

(iv) If there is carry, it means that answer is positive and neglect the carry.

(v) If there is no carry, then answer is negative so taken the 2's complement of the result to get the answer

$$P - Q$$

$$P = 111101101 \rightarrow \text{minuend}$$

$$Q = 11100110 \rightarrow \text{subtrahend}$$

2's complement of Q = 00011010

$$\text{Add it to P} = \frac{11101101}{\textcircled{1}00000111}$$

Since, there is carry, so answer is positive and neglect the carry.

$\therefore$ P – Q = 00000111.

25. Depletion type MOSFET operates in both deplection as well as enhancement mode by changing the gate voltage applied.

27. The PIV rating of full-wave rectifier with centre tap is $2V_m = 2 \times 100 = 200$ V

28. Output of op-amp will be square wave.

29. Frequency compensation is a method use in op-amp to increase the bandwidth.

38. Faraday discovered that the induced emf, V_{emf} (in volts) is every colsed circuit is equal to the time rate of change of the magnetic flux linkage by the circuit

$$V_{emf} = -\frac{d\lambda}{dt} = -N\frac{d\Psi}{dt}$$

The negative sign shows that the induced voltage acts in such a way to oppose the flux producing it as given by Lenz's law.

$$\therefore V_{emf} = -\frac{\partial \Psi}{\partial t}$$

$$V_{emf} = \oint_L \vec{E}\cdot dl = -\frac{\partial}{\partial t}\int_s \vec{B}\cdot dt$$

$$\therefore \boxed{\nabla \times \vec{E} = -\frac{\partial \vec{E}}{\partial t}}$$

39. $Z_L = jZ_0$

Reflection coefficient,

$$\Gamma = \frac{Z_L - Z_0}{Z_L + Z_0} = \frac{jZ_0 - Z_0}{jZ_0 + Z_0}$$

$$\Gamma = \frac{j-1}{j+1}$$

$$|\Gamma| = 1$$

$$\text{Now, VSWR} = \frac{1+|\Gamma|}{1-|\Gamma|} = \frac{1+1}{1-1} = \infty$$

40. Bandwidth requirement for various schemes

$BW_{AM} = 2f_m$, $BW_{PM} = 2(\beta_P + 1)f_m$

$BW_{FM} = 2(\beta_F + 1)F_m$, $BW_{SSB} = f_m$

Thus, SSB requires minimum bandwidth.

41. For colpitts oscillator,

$$A \geq \frac{C_2}{C_1}$$

$\therefore A \geq 10$

(where $C_2 = 10$ nF, $C_1 = 1$ nF)

So gain required for oscillation is 10.

42. In OFDM, data stream is divided into several parallel data-streams. Due to all parallel data streams being orthogonal, it is robust to multipath fading.

43. Sicne the semiconductor is electrically neutral, therefore, at thermal equilibrium,

$$N_D + p = N_A + n$$

$$N_D - N_A = n - p$$

44. Since the net charge must be zero, therefore

$$N_A W_P = N_D W_n$$

$$\frac{W_p}{W_n} = \frac{N_D}{N_A} = \frac{1}{10} = 0.1$$

45. The breakdown voltage in the common base case BV_{CBO} is significantly greater than BV_{CEO}.

46. The formula is valid for all the four configurations.

$$\boxed{R_0 = R_C = 1\,k\Omega}$$

55.

$$F = 1 + \frac{R_{eq}}{R_a} = 1 + \frac{30}{50}$$

$$= 1 + 0.6$$

$$= 1.6$$

$$= 10 \log 1.6 = 10 \times 0.24$$

$$= 2.04 \text{ dB}$$

$$T_{eq} = T_0(F - 1) = 290\,(1.6 - 1)$$

$$= 290 \times 0.6$$

$$\boxed{T_{eq} \Rightarrow 174 \text{ k}}$$

56. Guide wave length of a wave propagating in a wave guide is given by

$$\lambda_g = \frac{\lambda}{\sqrt{1-\left(\frac{f_c}{f}\right)^2}}$$

where $\lambda \rightarrow$ free space wave length

$f_c \rightarrow$ cutoff frequency for a particular mode

$f \rightarrow$ frequency of operation

$$\because \sqrt{1-\left(\frac{f_c}{f}\right)^2} < 1$$

$$\therefore \lambda_g > \lambda$$

57. A waveguide allows the waves (EM waves) to travel only if its frequency than a cutoff frequency defined for that mode of propagation so this is analogues to high pass filter where frequencies having value greater than certain cutoff value f_e will present at the output only.

58. For a good conducting medium

$$\alpha = \beta = \sqrt{\frac{\omega\mu\sigma}{2}} = \sqrt{\pi f\mu\sigma}$$

∴ Phase velocity,

$$V_p = \left(\frac{\omega}{\beta}\right) = \frac{\omega}{\sqrt{\frac{\omega\mu\sigma}{2}}}$$

$$= \sqrt{\frac{2\omega}{\mu\sigma}} = \sqrt{\frac{4\pi f}{\mu\sigma}} = 2\sqrt{\frac{\pi f}{\mu\sigma}}$$

59. For distortion less transmission line

RC = LG

and $\alpha = R\sqrt{\frac{C}{L}} = R\sqrt{\frac{G}{R}} = \sqrt{RG}$

60. Best matching will be short circuit stub (because it has better matching option and there is no radiation losses as in open circuit stub).

Best method will be two short circuit stub at specific distances from load because matching options will be two one from each stub so easily match to characteristic impedance of transmission line.

61. Since load impedance is equal to surge impedance

i.e. $Z_L = Z_0$

∴ Reflection coefficient

$$\Gamma = \frac{Z_L - Z_0}{Z_L + Z_0}$$

$$\Gamma = \frac{Z_0 - Z_0}{Z_0 + Z_0} = 0$$

62. For quarter wave line

where impedance $Z_{in} = \frac{Z_0^2}{Z_L}$

where Z_L is laod impedance

so if $Z_L = \infty$ (OC)

$Z_{in} = 0$ (SC)

and if $Z_L = 0$ (SC)

$Z_{in} \to \infty$ (OC)

so impedance inversion takes place.

63. $f(x - v_0 t) \Rightarrow \boxed{\frac{x}{t} = V_0}$

$$\frac{dx}{dt} = v_0$$

So, it represents at forward travelling wave

$$x = v_0 t + K$$

64. SWR = 1

∴ |Γ| (reflection coefficient)

$$= \frac{VSWR - 1}{VSWR + 1}$$

i.e. there is no reflection from load that is local is perfectly matched with characteristic impedance of line i.e. $Z_L = Z_0$.

65. $dl = \frac{\lambda}{20}$

Since length is very small compare to wave length so we can treat is as hertizian dipole therefore radiation resistance for hertizian diople.

$$R_r = 80\pi^2\left(\frac{dl}{\lambda}\right)^2$$

$$R_r = 80\pi^2 \times \frac{1}{400}$$

$$R_r = 1.973\Omega \simeq 2\Omega$$

66.

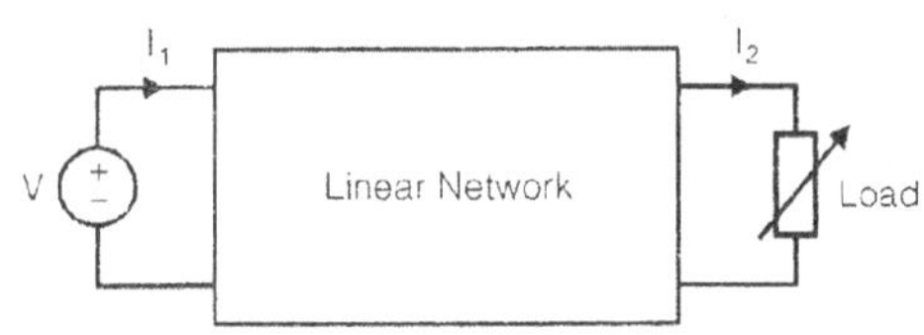

$V_1 = AV_2 + BI_2$ ...(i)

$I_1 = CV_2 + DI_2$ ...(ii)

When V = 4 V and P = 4 W

$$4 = A\sqrt{PR} + B\sqrt{\frac{P}{R}}$$

as R = 1Ω

$\therefore 4 = \sqrt{P}A + \sqrt{P}B$

4 = 2A + 2B ...(iii)

When V = 8 V and P = 16 W

8 = 4A + 4B ...(iv)

Adding equation (iii) and (iv)

12 = 6 A+ 6 B

V2 = 6 V and I_2 = 6 A

Therefore consumed by 1Ω resistor

$$= I_2^2 R = 36W$$

67. According to maximum power transfer theorem, the power transferred from voltage source to a load will be maximum when source impedance is equal to that of the load impedance.

68.

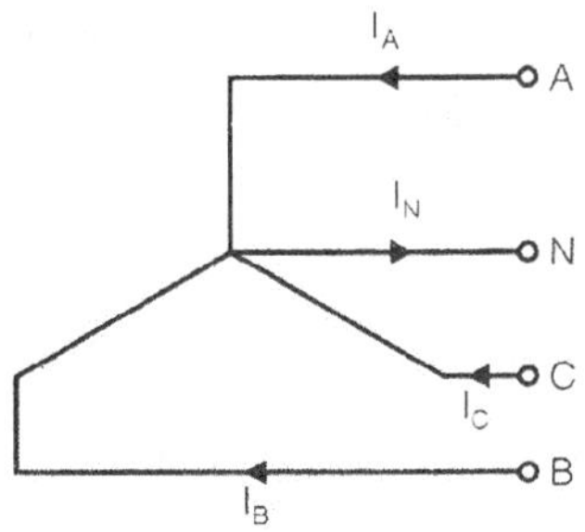

Current in neutral wire

$I_N = I_A + I_B + I_C$

$15 \angle 0^\circ + 15 \angle 120^\circ + 15 \angle 24^\circ = 0$

69. Condition for a network to be reciprocal.

$$AD - BC = 1 \text{ or } \begin{vmatrix} A & B \\ C & D \end{vmatrix} = 1$$

70.

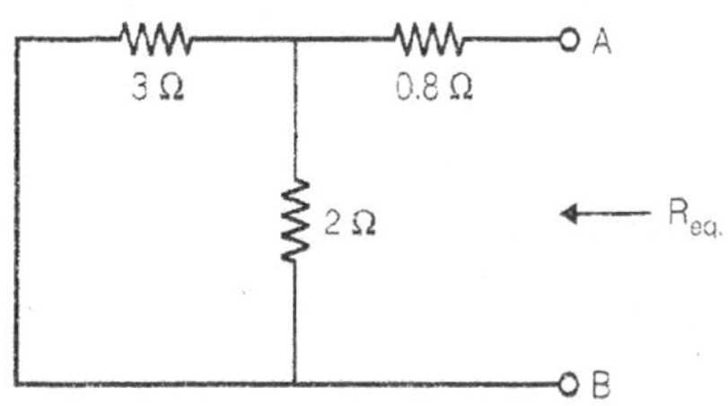

$$R_{eq} = 3 \,||\, 2 + 0.8 = 2\Omega$$

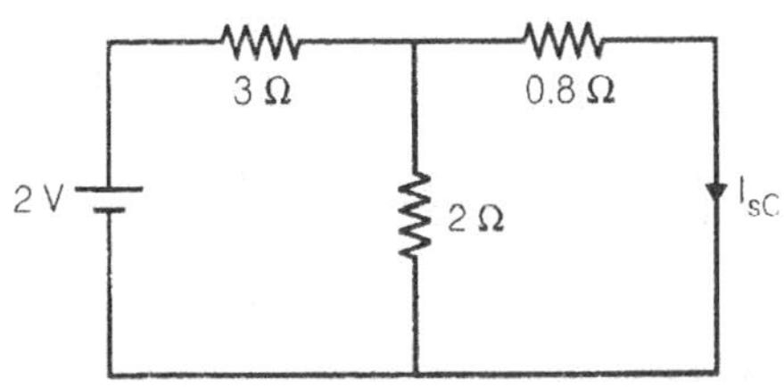

$$I_{SL} = \left(\frac{2}{3 + 2 \,||\, 0.8}\right) \times \left(\frac{2}{2 + 0.8}\right) = 0.4 \text{ A}$$

71.

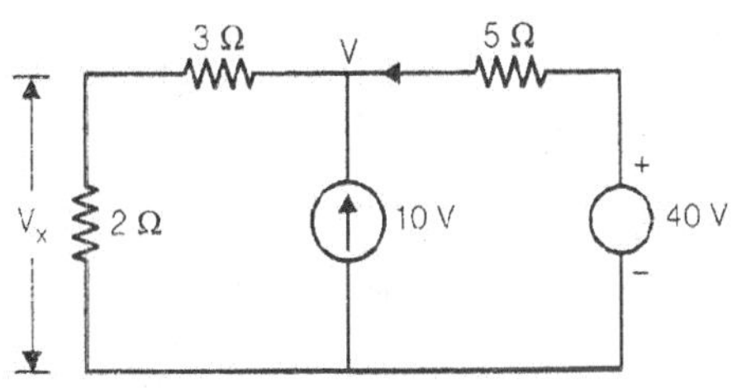

Applying KCL

$$\frac{V}{5} + \frac{V - 40}{5} - 10 = 0$$

V = 45

$$V_X = \frac{V}{5} \times 2 = \frac{45}{5} \times 2$$

= 18V

72. Characteistic equation for parallel resonance

$$S^2 + \frac{1}{RC} S + \frac{1}{LC} = 0$$

$$\omega_0^2 = \frac{1}{LC}, \quad \text{B.W.} \frac{1}{RC}$$

$$\omega_0 = \frac{1}{\sqrt{LC}} = 1 \text{ rad / sec}$$

$$\text{B.W.} = \frac{1}{20} \text{ rad / sec}$$

$$Q - \text{factor} = \frac{\omega_0}{\text{B.W.}} = \frac{1}{1/20} = 20$$

73. $R_n = 10\Omega$

$V_{th} = 60V$

$$P_{max} = \frac{V_{th}^2}{4R^{th}} = \frac{60 \times 60}{4 \times 10} = 90\Omega$$

74.

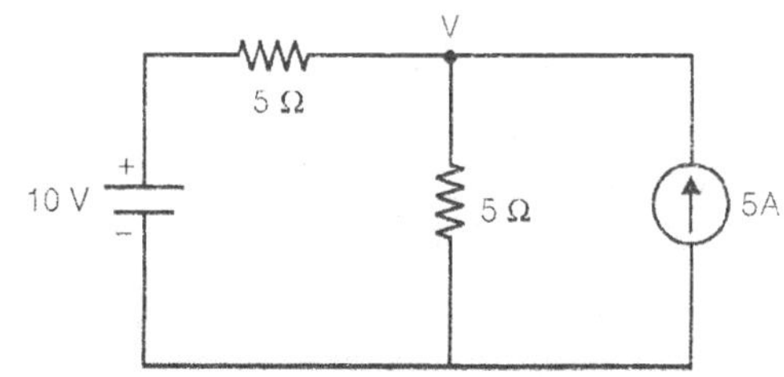

Applying Nodal Analysis

$$\frac{10 - V}{5} + 5 = \frac{V}{5}$$

10 – V + 25 = V

V = 17.5 V

76.

(20)	(1)	(12)	(5)	(14)	(20)
T	A	L	E	N	T
+2↓	–2↓	+2↓	–2↓	+2↓	–2↓
V	Y	N	C	P	R
(22)	(25)	(14)	(3)	(16)	(18)

Similarly,

(14)	(15)	(18)	(13)	(1)	(12)
N	O	R	M	A	L
+2↓	–2↓	+2↓	–2↓	+2↓	–2↓
P	M	T	K	C	J
(16)	(13)	(20)	(11)	(3)	(10)

77. This analogy is about the idioms which are used to draw comparisons. The idioms are as sick as a dog and as busy as a bee. Hence, option (*d*) is the correct answer.

78. The sequence shows the steps involved in a new product development. The correct answer is option (*c*), i.e., 1, 3, 6, 4, 2, 5.

79. The series is cbba, cbba, cbba,....

80. Wheat is a rabi crop while black gram, groundnut and fennel are kharif crops. Hence, option (*c*) is the correct answer.

81. 253 – 351 ⇒ 2 + 5 + 3 – (3 + 5 + 1) = 1

478 – 692 ⇒ 4 + 7 + 8 – (6 + 9 + 2) = 2

123 – 22 ⇒ 1 + 2 + 3 – (2 + 2) = 2

971 – 816 ⇒ 9 + 7 + 1 – (8 + 1 + 6) = 2

Hence, 253 – 351 is an odd pair.

82. Options (*b*), (*c*) and (*d*) have two vowels viz. 'o' and 'e'. Option (*a*) has 4 vowels, three 'i's' and an 'e'. Hence, option (*a*) is the correct answer.

83. The series is A(1) C(3) E(5), G(7) I(9) K(11), and so on. So, the required terms will be Y(25) A(1) C(3), E(5) G(7) I(9).

84. 3 * 12 ⇒ 3 × 12 = 36

7 * 9 ⇒ 7 × 9 = 63,

which is reverse of 36.

Similarly, 6 * 8 ⇒ 6 × 8 = 48

Then, 84 = 7 × 12 ⇒ 7 * 12.

85. From option (*c*), we get

17 + 13 = 36 ÷ 6 × 5, which is true.

Hence, answer is option (*c*).

86. The sequence will be

Arun > Bhanu > Esha > Charu > Divya

Hence, Arun is the eldest.

87. Option (*b*) is the answer since the given word has a single 'G' while 'DAGGER' has two Gs.

88. Option (*d*) is the correct answer since 'SHARPER' has two 'Rs' while the given word has a single 'R'.

89. The given relation is true for option (*d*). Equation can be written as:

34 + 52 ÷ 4 > 27 – 8 ⇒ 47 > 19, which is true.

90. WATCH = 45 + 1 + 39 + 5 + 15 = 105.

106. $23 - 4\sqrt{15} = (2\sqrt{5})^2 + (\sqrt{3})^2 - 2.2\sqrt{5}.\sqrt{3}$

$$= (2\sqrt{5} - \sqrt{3})^2$$

Hence, square root of $23 - 4\sqrt{15} = \pm(2\sqrt{5} - \sqrt{3})$.

107. One of the terms in the given expression will be cot90° which is equal to zero. Hence, the value of the given expression is 0.

108. Sum of ages of the students and the teacher = 22 × 15 = 330 years

Sum of ages of students only = 21 × 13 = 273 years

∴ Age of the teacher = 330 – 273 = 57 years.

109.

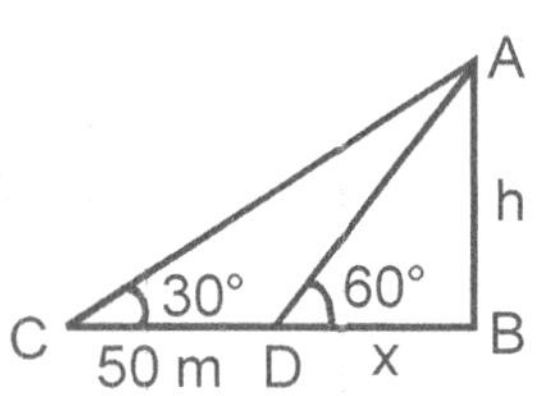

Let h be the height of the tower AB.

Then, in ΔABD,

$$\tan 60° = \frac{h}{x}$$

$$\Rightarrow \sqrt{3} = \frac{h}{x}$$

$$\Rightarrow x = \frac{h}{\sqrt{3}} \qquad \dots (i)$$

In ΔABC,

$$\tan 30° = \frac{h}{50 + x}$$

$$\Rightarrow \frac{1}{\sqrt{3}} = \frac{h}{50 + x}$$

$$\Rightarrow \sqrt{3}h = 50 + x \qquad \dots (ii)$$

From (i) and (ii),

$$h\sqrt{3} - \frac{h}{\sqrt{3}} = 50 \Rightarrow h = 25\sqrt{3} \text{ m.}$$

110. We have, $x^2 - 4x + 1 = 0$

$$\Rightarrow x + \frac{1}{x} = 4$$

Cubing both sides, we get

$$x^3 + \frac{1}{x^3} + 3.x.\frac{1}{x}\left(x + \frac{1}{x}\right) = (4)^3$$

$$\Rightarrow x^3 + \frac{1}{x^3} + 3(4) = 64$$

$$\Rightarrow x^3 + \frac{1}{x^3} = 52.$$

111. We have, $\angle MPQ = 120°$, $\angle MPN = 90°$ and $\angle MQN = 90°$

(As angle in a semi circle is a right angle.)

Hence, $\angle NPQ = 120° - 90° = 30°$

Also, $\angle NPQ = \angle NMQ$ (As angles in the same segment are equal.)

In ΔMQN,

$$\angle QNM = 180° - (\angle MQN + \angle NMQ)$$

Hence, $\angle QNM = 180° - 90° - 30° = 60°$.

112. Total poles required on each edge other than corner poles

$$= \frac{105}{7} - 1 = 14$$

Hence, total number of poles including 4 corner poles = 14 × 4 + 4 = 60.

113. We have, $a\cos\theta - b\sin\theta = c$

By squaring both sides,

$$(a\cos\theta - b\sin\theta)^2 = c^2$$

$$\Rightarrow a^2\cos^2\theta + b^2\sin^2\theta - 2ab\sin\theta\cos\theta = c^2$$

$\Rightarrow a^2(1-\sin^2\theta)+b^2(1-\cos^2\theta)-2ab\sin\theta\cos\theta = c^2$

$\Rightarrow a^2\sin^2\theta + b^2\cos^2\theta + 2ab\sin\theta\cos\theta = a^2+b^2-c^2$

$\Rightarrow (a\sin\theta + b\cos\theta)^2 = a^2+b^2-c^2$

Hence, $a\sin\theta + b\cos\theta = \pm\sqrt{a^2+b^2-c^2}$.

114. We have, a$b $= a - b + \frac{b^2}{a}$

$\Rightarrow 15\$3 = 15 - 3 + \frac{3^2}{15} = 12.6.$

115. Let the length of the sides of the triangle be 3x, 4x and 5x.

Semi perimeter(s) = $\frac{3x+4x+5x}{2} = 6x$

Area of Δ = $\sqrt{6x(3x)(2x)(x)} = 6x^2$

Circumradius $= \frac{abc}{4\times(\text{Area of }\Delta)}$

$\Rightarrow 30 = \frac{3x\times4x\times5x}{4\times6x^2} \Rightarrow x = 12$

Hence, the length of the largest side = 5x = 60 cm.

Alernate Method:

Given sides form a right angle triangle.

As, circumradius = $\frac{\text{hypotenuse}}{2}$

$\therefore$ Hypotenuse = 2 × 30 = 60 cm.

116. Let the radius of the circular ground be r m.
Time taken to cover the distance along the shortest path i.e. diameter = $\frac{2r}{1.5}$

And along the boundary = $\frac{\pi r}{1.5}$

$\therefore \frac{\pi r}{1.5} - \frac{2r}{1.5} = 2\times60 \Rightarrow r\left(\frac{22}{7}-2\right) = 2\times60\times1.5$

$\Rightarrow r = \frac{180\times7}{8} = 157.5$ m.

117. Let the total number of wrong sums be x.

$\Rightarrow$ Correct sum = 40 – x.

Then, $4(40-x) - x = 80 \Rightarrow x = 16.$

118. Length of the conveyor belt = Sum of perimeters of semicircle + 2(length of direct common tangent)

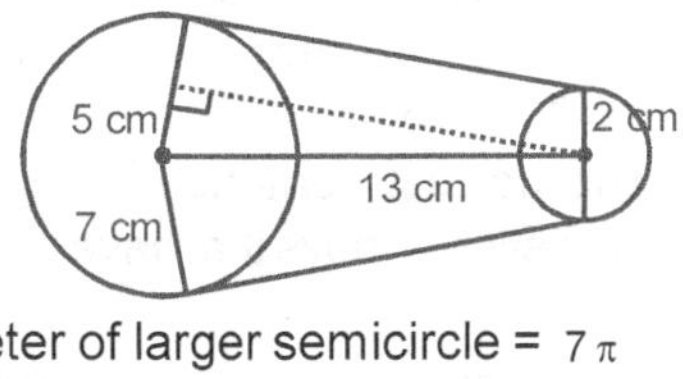

Perimeter of larger semicircle = 7π

Perimeter of smaller semicircle = 2π

Length of direct common tangent = $\sqrt{d^2-(r_1-r_2)^2}$

$= \sqrt{13^2-(7-2)^2} = 12\text{cm}$

Hence, length of the conveyor belt (in cm) = $9\pi + 24$.

119.

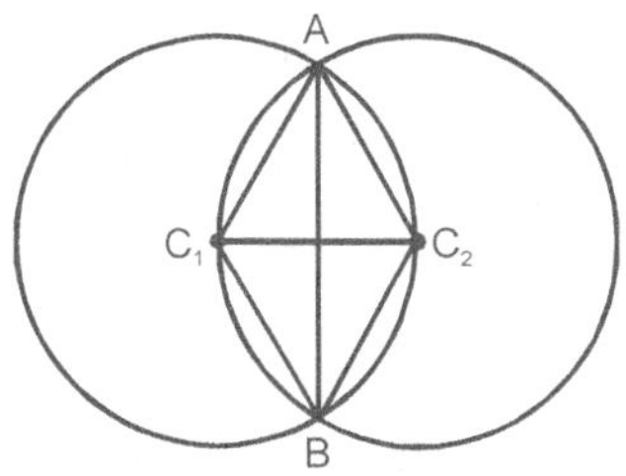

Let C_1 and C_2 be the centers of the two circles.
Here, AC_1C_2 and BC_1C_2 are equilateral triangles.
Hence, area of the quadrilateral AC_1BC_2

$= 2\times\frac{\sqrt{3}}{4}\times5^2 = \frac{25\sqrt{3}}{2}\text{ cm}^2.$

120. $\frac{\operatorname{cosec}\theta - \cot\theta}{\operatorname{cosec}\theta + \cot\theta} = \frac{1}{9}$

$\Rightarrow \frac{\frac{1}{\sin\theta} - \frac{\cos\theta}{\sin\theta}}{\frac{1}{\sin\theta} + \frac{\cos\theta}{\sin\theta}} = \frac{1}{9} \Rightarrow \frac{1-\cos\theta}{1+\cos\theta} = \frac{1}{9}$

$\Rightarrow \cos\theta = \frac{4}{5}$

Hence, $\sin\theta = \frac{3}{5}$.

121. The first sentence is in past tense. Therefore, the second blank will take a word that is in past tense. Thus, option (*c*) is incorrect. Option (*d*) is incorrect since 'raise' refers to lifting something manually. Option (*a*) is incorrect since 'risen' is always followed by 'had/have'. 'Rose' is simple past tense and is consistent with the structure of the sentence. 'Come on' means start to arrive or happen. Hence, option (*b*) is the correct answer.

122. Option (*c*), 'at', is the correct preposition to be filled in the blank. The preposition 'at' is used to indicate presence or occurrence, intended goal, the means, cause or manner, age or position in time, etc.

123. The correct preposition is 'on'. 'On' is used with 'agree' when we talk about an issue or a point of debate. Hence, option (*b*) is the correct answer. 'Agree with' is used when we accept the point of someone or something. 'Agree to' is used when we accept the demands or queries of a person. 'Agree at' is grammatically incorrect.

124. The given sentence is incorrect since 'child' is singular while 'they' refers to more than one. In option (*c*), 'child' is singular. Therefore, 'think', which is plural, should be replaced by 'thinks'. Option (*b*) is incorrect since 'children', which is plural, should be followed by 'are' and not 'is'. Only option (*a*) is correct since it takes one of the two prepositions (he or she) that go with 'child'.

125. Future perfect continuous tense is used for actions that will be in progress over a period of time and will end in future. Hence, option (*a*) is the correct answer. The given sentence is in future perfect tense, which is used to talk about actions that will be completed by a certain time in the future. Options (*b*) and (*c*) are in future continuous tense, which is used to talk about actions that will be in progress at a time in the future.

126. The given statement is correct. With 'empathy' we will use a word that contrasts with it since the sentence talks about the right mix of two things. 'Disdain' refers to a feeling of strong dislike or disapproval of someone or something you think does not deserve respect. 'Empathy', on the other hand, refers to the feeling that you understand and share another person's experiences and emotions. Hence, option (*d*) is the correct answer.

127. 'Dour' refers to someone who is serious and unfriendly. A person is called 'gentle' if he has a silent and kind nature. Hence, the correct answer is option (*d*) – gentle. 'Hostile' is someone who is not friendly. 'Grim' is someone who has serious appearance or manner. A person is described as 'cold' if he lacks normal human emotions, friendliness, compassion, etc.

128. Andrew was a self made man and was careful about spending money. But the reason why he was not liked by many people was that he engaged in dishonest business practices. Hence, option (*c*) is the correct answer. Option (*d*) is incorrect since people were not jealous of his wealth; rather he made some enemies while climbing the ladder to success.

129. Options (*a*), (*b*) and (*c*) are mentioned in the passage as a cause that hampered investigation. Although Andrew Borden had many enemies, it is nowhere stated that this posed a problem during the investigation. Hence, option (*d*) is the answer.

130. The passage does not suggest that Andrew Borden was careful while making judgments and decisions. Hence, option (*d*), prudent, is the correct answer. The passage says that he owned many properties. From this we can infer that he had a lot of money, i.e., he was affluent. The passage clearly says that he was 'frugal to a fault', which means that he spent his money carefully. This means that he was economical. The second sentence of the passage clearly states that he was unpopular.

131. 'Pinnacle' means the highest point of something and so does 'crest'. Hence, option (*b*) is the correct answer. Options (*a*), (*c*) and (*d*) are the antonyms of 'pinnacle'.

132. 'Sleuth' is a person who looks for information to solve crimes. An 'investigator' does the same. Hence, option (*d*) is the correct answer.

133. 'Detrimental' means causing damage or injury. 'Benign' refers to something that is harmless. Hence, option (*c*) is the correct answer.

134. 'Nibble' means to eat slowly or gently. Hence, option (*a*), devour, is the correct answer. 'Devour' refers to eating of food hungrily or quickly.

135. 'Egalitarian' is a person who aims for equal wealth, status, etc., for all people. Hence, option (*a*) is the correct answer. 'Federalist' is a supporter of the federal system of government. 'Absolutist' is a person who believes that a ruler or government should have total power at all times. 'Reformist' is someone who wants to change political or social situations.

136. A 'butler' is the main male servant in the house of a wealthy person and has charge of other employees. Hence, option (*d*) is the correct answer.

137. 'Racket' refers to an illegitimate business or enterprise. Hence, it is the correct answer. Racket also means confused clattering noise. 'Clatter' means to make quick series of short loud sounds. 'Rumpus' refers to a noisy argument or fight. 'Squabble' is a noisy altercation or quarrel usually over petty matters.

138. 'To get one's foot in the door' means to manage to enter an organisation that could bring one success. Hence, option (*a*) is the correct answer.

139. 'In the home stretch' refers to the last part of an activity, etc. Hence, option (*d*) is the correct answer.

140. 'Itchy feet' means to want to travel or move to a different place; to want to do something different. Hence, option (*a*) is the correct answer.

141. The correct spelling is 'Bureaucrat'. Hence, option (*c*) is the correct answer.

142. The sentence does not have any grammatical error. Hence, part (*d*) is the correct answer.

143. When the subject of the verb is a relative pronoun ('that', in this case), then the verb agrees in number and person with the antecedent of the relative (boys). Therefore, 'has' in part (*c*) is grammatically incorrect, rendering part (*c*) the answer. The correct sentence should be 'Archie is one of the cleverest boys that have passed through the school.'

144. The preposition 'than' is used with 'different' when a noun is compared with a clause. For example: Her appearance was very different than I'd expected. When two nouns are compared, like in this case, then the preposition 'from' is used with 'different'. Hence, part (*c*) is the correct answer.

145. 'Even if' means whether or not and has to do with the conditions that may apply. 'Even though' means despite the fact. The sentence means the latter and therefore, options (*a*) and (*c*) are incorrect. The sentence says that the price of gas is starting to come down 'now'. The action is continuing to happen at present. Therefore, the second blank should logically take present continuous tense, i.e., 'are choosing'. Hence, option (*d*) is the correct answer.

146. The first blank will take 'when' because we are talking about a time. 'Where' is used when we talk of a place. The second blank will take 'that' and not 'which'. 'Which' is used if the sentence does not need the clause that the word in question is connecting. If it does, 'that' is used instead. Hence, option (*c*) is the correct answer.

147. 'Call on' means to visit someone for a short time. Hence, option (*c*) is the correct answer. 'Called to my professor' is grammatically incorrect. 'Call forth' means to produce a particular reaction. Since the sentence does not suggest that the professor is taken somewhere, option (*b*) can be negated. 'Call after' means to give a person the same name as someone else. For example: He was called after his great-grandfather.

148. 'Less' is used for uncountable nouns. For example: Less money, less time, etc. 'Fewer' is used for countable nouns. Since 'children' is a countable noun, we will use 'fewer' instead of 'less'. Option (*c*) is incorrect since 'suffers' is singular, while 'children' is plural. Hence, option (*b*) is the correct answer.

149. The sentence deals with the placement of adjectives. Generally, a noun is preceded by no more than three adjectives. These adjectives have to follow a certain order, which is: opinion, dimension/size/weight, age, shape, colour, country of origin, material and purpose. Only option (*b*) follows this order, i.e., dimension, shape, colour. Hence, option (*b*) is the correct answer.

150. 'Elder' and 'eldest' are used only for persons, not for animals or things. 'Older' and 'oldest' are used for both persons and things. 'Older' is used when we compare two things. For example: older than, older of the two, etc. 'Oldest' is used when more than two things are compared. Hence, option (*a*) is the correct answer.

151. Option (*d*) cannot be inferred from the passage since the option specifically states that it is the scientists who are performing the tests. The passage, however, does not suggest that the tests are performed by 'scientists'. Moreover, 'researchers', in the passage, refers to the researchers of London's Royal College of Art. Hence, option (*d*) is the correct answer. Option (*a*) can be inferred from the last line of the passage which says that it is difficult to say for sure that our ancestors used the stones as percussion instruments. This means that the reason for using bluestones is unknown. Option (*b*) can be directly inferred from the first line of the second paragraph - They also found a few of the rocks remaining at Stonehenge rang as well. The first sentence of the second paragraph states that a few rocks 'remaining' at Stonehenge cannot ring. From this we can infer that not all the stones that were originally put there could be preserved. Thus, option (*c*) can also be inferred.

152. 'Acoustic' is something that is related to sound. First paragraph clearly states that sound might have played the role. Hence, option (*b*) is the correct answer. 'Ocular' is something that is related to the eye and so, option (*a*) cannot be the answer.

153. 'Ephemeral' refers to something that lasts a very short time. Hence, option (*a*) is the correct answer.

154. 'Haul' means to move (something) with effort. Hence, its opposite will be option (*b*), push.

155. The last paragraph says that the challenge is providing good evidence that the bluestones were used for their musical ability. Hence, option (*b*) is the correct answer.

156. "Cherubic" means a winged celestial being. "Angelic" is the word most similar in meaning to cherubic. "Snobbish" means pretentious. "Restless" means marked by a lack of quiet, repose, or rest. "Honest" which means marked by or displaying integrity is inappropriate in the given context and thus negated.

157. "Copious" means yielding or containing plenty. "Plentiful" is the word most similar in meaning to the given word. "Powerful" means having or capable of exerting power. "Delaying" means to postpone until a later time. "Foolish" means lacking or exhibiting a lack of good sense or judgment.

158. "Choleric" means easily moved to often unreasonable or excessive anger. "Calm" is the word opposite in meaning to the given word. "Ill tempered" is negated as it is a synonym. "Hygienic" means having or showing good hygiene. "Sober" means not addicted to intoxicating drink.

159. "Harried" means beset by problems or harassed. "Relaxed" is the word most opposite in meaning to the given word. "Instigated" means provoked. "Pitchforked" means to thrust (someone) unwillingly into a position. "Harassed" is negated as it is a synonym.

160. "Rhapsodic" means extravagantly emotional and is thus negated. "Fastidious" which means having high and often capricious standards or difficult to please can be substituted for the given words. "Succulent" means full of juice. "Rambunctious" means marked by uncontrollable exuberance.

161. "Proscenium" which means the stage of an ancient Greek or Roman theater is negated. "Matinee" which means a musical or dramatic performance or social or public event held in the daytime and especially in the afternoon is the correct answer. "Dirge" means a funeral hymn or lament. "Claire" means a small enclosed pond for growing or observing the growth of oysters.

162. "Omniscient" which means having infinite awareness, understanding, and insight is the correct answer. "Omnipresent" means present everywhere simultaneously. "Ignoramus" means an ignorant person. "Omnipotent" means one having unlimited power or authority.

163. "To crease something up" means to wrinkle one's clothing.

164. "Skimp on something" means to use too little of something.

165. "To die in harness" means to die while actively engaged in work or duty.

166. "Commitment" is the correct spelling.

167. The correct sentence is "he took to drinking to lessen his mental worries". The gerund "drinking" is the object of the preposition "to".

168. The correct sentence is "despite his protests I decided to buy the jeans which he did not like." "Despite of" is idiomatically incorrect.

169. The correct sentence is "the poet and scholar has been honoured." Here the poet and the scholar refer to the same person; thus singular verb is used.

170. A bullet bruised his cheek. "Glazed" which means a substance coated with a layer of a vitreous substance is negated. "Grazed" which means an area of skin roughened or worn away by harsh rubbing against another surface is appropriate for the blank. "Throttled" means to keep (someone) from breathing by exerting pressure on the windpipe. "Amputated" means to cut off (a projecting body part), especially by surgery.

171. Tourist forget their biased ideas as soon as they visit our country. "Preconceived" which means to form (an opinion, for example) before possessing full or adequate knowledge is appropriate for the blank. "Ordained" means to establish or order by appointment, decree, or law. "Gestalt" is a German word for form or shape. "Ruffled" means troubled.

172. "Sacramental" means of, relating to, or having the character of a sacrament. "Cyclopean" means massive. "Unleashed" means to throw, shoot, or set in motion forcefully. "Forthright" which means straightforward is appropriate for the blank.

173. **word for word** means : In exactly the same words or when translated exactly equivalent words.

174. The word **sensual (Adjective)** means : connected with your physical feelings; giving pleasure to your physical senses; espe cially to sexual pleasures.

CBSE ACCOUNTANCY

[Class XII]